Equine-Assisted Services: Studies Illustrating the Potential Benefits for Children, Adolescents and Adults

Equine-Assisted Services: Studies Illustrating the Potential Benefits for Children, Adolescents and Adults

Edited by

Lori R. Kogan

CABI

CABI is a trading name of CAB International

CABI
Nosworthy Way
Wallingford
Oxfordshire OX10 8DE
UK

Tel: +44 (0)1491 832111
E-mail: info@cabi.org
Website: www.cabi.org

CABI
200 Portland Street
Boston
MA 02114
USA

T: +1 (617)682-9015
E-mail: cabi-nao@cabi.org

A catalogue record for this book is available from the British Library, London, UK.

ISBN-13: 9781836991588 (hardback)
9781836991595 (ePDF)
9781836991601 (ePub)

DOI: 10.1079/9781836991601.0000

Commissioning Editor: Alexandra Lainsbury
Editorial Assistant: Theresa Regueira
Production Editor: Helen Elliott

Typeset by Exeter Premedia Services Pvt Ltd, Chennai, India
Printed and bound in the UK by CPI Group (UK) Ltd, Croydon, CR0 4YY

Contents

Supplementary Content

https://www.cabidigitallibrary.org/doi/book/10.1079/9781836991601.0000
Chapter 4 includes supplementary material:
Figure S4.1 Scaled Illustration of the Architectural Spaces within the Equine Environment
Table S4.1 Coding Scheme
Table S4.2 Average Proportion of Total and Segment Session Time Spent in Architectural Spaces
Table S4.3 Average Proportion of Total Session and Segment Time Devoted to Occupational Opportunities
Chapter 16 includes supplementary material:
Table S16.3 Definitions of qualitative codes and sub codes

Original Publication Details

1 **Exploring Psychotherapy Incorporating Equine with Sexually Exploited Children**
Jenna M. Hershberger, Erika L. Berg, Jodi L. Tangen, Laura Feldmann and Les Barnhart
Previously published in Human-Animal Interactions 2024
https://doi.org/10.1079/hai.2024.0031

2 **Experimental Trial Demonstrates Positive Effects of Equine Facilitated Learning on Child Social Competence**
Patricia Pendry and Stephanie Roeter
Previously published in Human-Animal Interaction Bulletin 2013
https://doi/10.1079/hai.2013.0004

3 **Randomized Trial Examines Effects of Equine Facilitated Learning on Adolescents' Basal Cortisol Levels**
Patricia Pendry, Annelise N. Smith and Stephanie M. Roeter
Previously published in Human-Animal Interaction Bulletin 2014
https://doi.org/10.1079/hai.2014.0003

4 **A Transactional Analysis of Occupational Therapy in an Equine Environment for Youth with Autism Spectrum Disorder**
B. Caitlin Peters, Christof Bentele and Wendy Wood
Previously published in Human-Animal Interaction Bulletin 2021
https://doi.org/10.1079/hai.2021.0033

5 **The Impact of an Equine Facilitated Learning Program on Youth with Autism Spectrum Disorder**
Phyllis Erdman, Darcy Miller and Sue Jacobson
Previously published in Human-Animal Interaction Bulletin 2015
https://doi.org/10.1079/hai.2015.0011

6 **Impact of Adaptive Devices on Horses, Assisting Persons, and Riders with Cerebral Palsy in a Therapeutic Riding Program**
Lana Kaiser, Mathew J. Edick, Katherine Smith Buckler, Camie Heleski, Sherman Gorbis, Bonnie DePue, Liz Lewis, and LeeAnn J. Kaiser
Previously published in Human-Animal Interaction Bulletin 2019
https://doi.org/10.1079/hai.2019.0001

15 Behavioral and Physiological Effects of Equine-Assisted Early Intervention for Mother-Child Dyads with Insecure Attachment
Denise V. Hebesberger, Andrea Beetz and Kurt Kotrschal
Previously published in Human-Animal Interaction Bulletin 2019
https://doi.org/10.1079/hai.2019.0003

16 A Logic Model of a Dementia-Specific Program of Equine-Assisted Activities
Rebecca Lassell, Beth Fields, Sarah Busselman, Tauren Hempel and Wendy Wood
Previously published in Human-Animal Interaction Bulletin 2019
https://doi.org/10.1079/hai.2019.0001

Contributors

Cindy Adams: University of Calgary, Faculty of Veterinary Medicine, Canada

Nichole C. Anderson: School of Veterinary Medicine, Texas Tech University, 7671 Evans Drive, Amarillo, TX, 79106, USA

Cheryl Arratoon: Canadian Centre on Substance Abuse, Ottawa, ON K1P 5E7, Canada

Amber R. Bach-Gorman: Counseling Center, North Dakota State University, Fargo, North Dakota

Les Barnhart: Home on the Range, USA

Andrea Beetz: ISER, Department for Special Education, University of Rostock, Rostock, Germany

Christof Bentele: Department of Occupational Therapy, Colorado State University, Fort Collins, Colorado

Erika L. Berg: Department of Animal Sciences, North Dakota State University, Dept 2650, PO Box 6050, Fargo, ND 58108, USA

Thomas D. Berry: Department of Psychology, Christopher Newport University, Virginia

Janice Boucher: Cartier Equine Learning Centre, Prince Albert, Saskatchewan, Canada

Bjarne O. Braastad: Norwegian University of Life Sciences, As, Norway

Sarah Busselman: Fort Collins, Colorado, USA

Catharina Carlsson: Department of Social Work, Linnaeus University, Kalmar, Sweden

Gail Cartier: Cartier Equine Learning Centre

Darlene Chalmers: University of Regina, Faculty of Social Work, Canada

Catherine A. Cunningham: La Ronda Equestrian Club Cold Spring, MN 56320 USA

Colleen Anne Dell: University of Saskatchewan, Department of Sociology & School of Public Health, Canada

Debra Dell: Youth Solvent Addiction Committee, Saskatoon, SK S7K 1H8, Canada

Bonnie DePue: CHUM Therapeutic Riding

Dominique Dryka: Cartier Equine Learning Centre, Prince Albert, Saskatchewan, Canada

Randy Duncan: University of Saskatchewan, Department of Sociology, Canada

Kathryn Dunn: University of Saskatchewan, School of Public Health, Canada

Lee Dudley: The Equine Partnership Program, 40495 County Road 13 Elizabeth, CO, 80107, USA

Mathew J. Edick: CHUM Therapeutic Riding

Phyllis Erdman: College of Education, Washington State University, Pullman, Washington

Laura Feldmann: Home on the Range, USA

Beth Fields: Center for Caregiving Research, Education, and Policy, University of Pittsburgh and the Center for Health Equity Research and Promotion, VA Pittsburgh Healthcare System, Virginia

Angela K. Fournier: Department of Psychology, Bemidji State University, Minnesota

Judith L. Gibbons: Saint Louis University, Missouri

Sherman Gorbis: Michigan State University College of Osteopathic Medicine, East Lansing, Michigan

Nathan Hall: Department of Animal and Food Sciences, Texas Tech University, 2500 Broadway, Lubbock, TX, 79409, USA

Carolyn J. Hammer: Department of Animal Sciences, North Dakota State University, Fargo, North Dakota

Hilde Hauge: Norwegian University of Life Sciences, As, Norway

Camie Heleski: Michigan State University Department of Animal Science, College of Natural Science, East Lansing, Michigan

Tauren Hempel: Department of Equine Sciences, Colorado State University, Fort Collins, Colorado

Denise V. Hebesberger: Department of Biology, Anglia Ruskin University, Cambridge, United Kingdom

Jenna M. Hershberger: North Dakota State University, Dept 2650, PO Box 6050, Fargo, ND 58108, USA

Carol Hopkins: National Native Addictions Partnership Foundation, Bothwell, ON N0P 1C0, Canada

Sue Jacobson: Washington State University, Pullman, Washington

Lana Kaiser: Michigan State University Department of Medicine, College of Human Medicine, East Lansing, Michigan

LeeAnn J. Kaiser: Michigan State University College of Veterinary Medicine, East Lansing, Michigan

Kurt Kotrschal: Department of Behavioural Biology, University of Vienna, Vienna, Austria; and Core Facility Konrad Lorenz Research Station for Behaviour and Cognition, University of Vienna, Gruenau im Almtal, Austria

Ingela L. Kvalem: University of Oslo, Norway

Rebecca Lassell: Department of Occupational Therapy, Colorado State University, Fort Collins, Colorado

Elizabeth A. Letson: Eagle Vista Ranch & Wellness Center, Bemidji, MN 56601, USA

Liz Lewis: Michigan State University Department of Animal Science, College of Natural Science, East Lansing, Michigan

Loni Longclaws: White Buffalo Youth Inhalant Treatment Centre P.O. Box 2500 Prince Albert, SK S6V 7G3, Canada

Tamara MacKinnon: Cartier Equine Learning Centre, Prince Albert, Saskatchewan, Canada

Darcy Miller: Washington State University, Pullman, Washington

Marco Antonio Montufar Cardenas: Fundación Equinos Sanos Para El Pueblo, Guatemala

Daniel Nilsson Ranta: Department of Social Work, Mid Sweden University, Östersund, Sweden

Emma Leigh Pasiuk: Department of Animal and Food Sciences, Texas Tech University, 2500 Broadway, Lubbock, TX, 79409, USA

Leslie Paiz: La Ronda Equestrian Club, Cold Spring, MN 56320 USA

Ingeborg Pedersen: Norwegian University of Life Sciences, As, Norway

Patricia Pendry: Department of Human Development, Washington State University, Pullman, Washington

B. Caitlin Peters: Department of Animal Sciences, Colorado State University, Fort Collins, Colorado

Katelyn E. Poelker: Saint Louis University, Missouri

Stephanie M. Roeter: Department of Human Development, Washington State University, Pullman, Washington

Christine Rudd: Department of Animal and Food Sciences, Texas Tech University, 2500 Broadway, Lubbock, TX, 79409, USA

Ernie Sauve: White Buffalo Youth Inhalant Treatment Centre

Katy Schroeder: Department of Counselor Education, University of Iowa, N350 Lindquist Center North, Iowa City, IA 52242, USA

Mary Slouka: The Equine Partnership Program, 40495 County Road 13 Elizabeth, CO, 80107, USA

Annelise N. Smith: Department of Human Development, Washington State University, Pullman, Washington

Katherine Smith Buckler: Michigan State University Department of Medicine, College of Human Medicine, East Lansing, Michigan

Serene Spence: University of Saskatchewan, Department of Sociology, Canada

Lorann Stallones: Department of Psychology, 1876 Campus Delivery, Colorado State University, Fort Collins, CO, 80523-1876, USA; One Health Institute, Colorado State University, Fort Collins, CO, 80523, USA

Anastasia C. Stellato: Department of Animal and Food Sciences, Texas Tech University, 2500 Broadway, Lubbock, TX, 79409, USA

Jodi L. Tangen: North Dakota State University, Dept 2650, PO Box 6050, Fargo, ND 58108, USA

Bente Traeen: Department of Psychology, University of Oslo, Oslo, Norway

Emily Webberson: Department of Animal and Food Sciences, Texas Tech University, 2500 Broadway, Lubbock, TX, 79409, USA

Wendy Wood: Departments of Animal Sciences and Occupational Therapy and Director of Research at the Temple Grandin Equine Center, Colorado State University, Fort Collins, Colorado

Mallory Wuttunee: University of Saskatchewan, Department of Sociology, Canada

Ashleigh Zaker: Department of Psychology, 1876 Campus Delivery, Colorado State University, Fort Collins, CO, 80523-1876, USA

Introduction

Lori R. Kogan
Colorado State University, USA

The field of animal assisted services (AAS), and equine-assisted services (EAS) specifically, has grown exponentially over the last couple of decades, emerging as a popular intervention to help support human health and well-being (Fine and Andersen, 2021). Multitudes of studies have validated what we have intuitively known for years: interactions with animals can have a profound positive impact on human health and functioning.

In recent years, the breadth of animal-involved interventions has expanded in numerous areas, offering a wide array of options for human-animal teams. Although most of the AAS studies have focused on dog-assisted interventions (Bert *et al.*, 2016), EAS represent an emerging field within AAS. As the field of EAS has grown, it has become apparent that standardized terminology is needed to have a common language for horse-related services to facilitate collaboration across professional disciplines, reduce terminological confusion, and create the framework needed to implement and study EAS (Wood *et al.*, 2021). To this end, a consensus document was published in 2020 that provides guidance regarding optimal terminology for the diverse services that incorporate horses to benefit people (Wood *et al.*, 2021). This optimal terminology began with suggesting the use of the term *Equine-assisted services (EAS)* to reference the multiple services in which professionals incorporate equines to benefit people. Twelve types of EAS were subsequently categorized under three large umbrellas: therapy, learning and horsemanship.

Under the umbrella of therapy, five types were identified including counseling, occupational therapy, physical therapy, psychotherapy, and speech-language pathology. Counseling is performed by licensed therapy professionals and refers to the incorporation of horses into mental health counseling to address psychological and emotional challenges. Occupational therapy refers to the inclusion of horses to support functional skill development and improving daily living and work-related capabilities. Physical therapy uses equine movement to support rehabilitation, physical functioning, and overall health and wellness. Psychotherapy refers to the integration of horses into psychological treatment. Speech-language pathology utilizes horse-related activities to support the development of communication skills by providing sensory and interactive experiences that target speech and language development (Wood *et al.*, 2021). Regardless of the type of therapy, each one involves licensed professionals who work within the scope of practice of their particular discipline. In addition, it is strongly recommended that these professionals obtain specialized training focused on incorporating horses into their treatments (Wood *et al.*, 2021).

The umbrella of learning includes three types of services including equine-assisted learning in education, organizations, and personal development. Equine-assisted learning in education is tailored for each specific type of educational setting, and focuses on the development of academic skills, character building, problem-solving and critical thinking. Equine-assisted learning in organizations is designed to support team building and leadership development by providing experiential learning opportunities to enhance organizational dynamics and performance. Lastly, equine-assisted learning in personal development helps participants in addressing personal life challenges and providing growth opportunities. In addition to training within their particular field, all professionals engaged in equine-assisted learning should obtain horse behavior and handling training and experience, as well as training related to how to successfully implement horses into experiential learning activities (Wood *et al.*, 2021).

The last umbrella of equine-assisted services is horsemanship, and includes four types of services adapted from traditional equine disciplines: adaptive equestrian sport, adaptive riding or therapeutic riding, driving, and interactive vaulting. Adaptive equestrian sport supports athletic development and social inclusion for participants with diverse needs who want to engage in competitive events. Adaptive/therapeutic riding is used to help improve the physical, cognitive, and emotional functioning of individuals with diverse needs by adapting traditional riding. Driving is an activity that allows individuals who may not be able to ride a horse to still engage with horses and experience the therapeutic benefits of equine interaction through the use of horse-drawn vehicles. Lastly, interactive vaulting involves gymnastic movements around horses to help develop physical coordination and teamwork. Equine professionals within these types of horsemanship typically have specialized training to ensure they have the competencies needed to work with participants with diverse needs. In addition, these professionals should have training in the use of adaptive equipment and the teaching techniques needed based on each individual participant's personal abilities and goals (Wood *et al.*, 2021).

Regardless of the specific type, there is growing evidence of EAS benefits for a wide array of populations, including, but not limited to, individuals with autism (Gabriels *et al.*, 2015; Nieforth *et al.*, 2023; O'Haire, 2017; Trzmiel *et al.*, 2019), neurological disorders (Berardi *et al.*, 2022; Pálsdóttir *et al.*, 2020), cerebral palsy (Menor-Rodríguez *et al.*, 2021; Qin *et al.*, 2024), and psychiatric disorders (Bator *et al.*, 2020; Earles *et al.*, 2015; Nurenberg and Schleifer, 2011). To highlight this growing body of research, this book includes exemplary EAS research involving children, adolescents, and adults.

Children

The book begins with six chapters describing EAS with children. These chapters include sexually exploited children, children's social competence, children's cortisol levels, children with autism, and children with cerebral palsy.

The first chapter 'Exploring Psychotherapy Incorporating Equine with Sexually Exploited Children' by Hershberger *et al.*, outlines the application of Psycho-Equine Therapy (P-EQ) for survivors of Commercial Sexual Exploitation of Children (CSEC). Using interpretative phenomenological analysis, the researchers documented the experiences of adolescent CSEC survivors who participated in a seven-week Rhythmic Riding program at a therapeutic ranch in the Midwestern United States. The study collected data through semi-structured individual interviews, reflective journals, observational field notes, and a concluding focus group. Participants engaged in weekly 90-120 minute P-EQ sessions where they were assigned a specific horse with which they did activities that included catching, grooming, bridling, and riding bareback. At the same time, they were given therapeutic prompts designed to help them connect with their thoughts, feelings, and bodily sensations.

Results revealed that CSEC survivors experienced significant positive outcomes through P-EQ, including improved emotional regulation, increased distress tolerance, enhanced self-confidence, and restored ability to trust others. Participants specifically noted the healing connection with horses as the best part of their experience, and described how the animals provided understanding and patience in ways they had not encountered in traditional therapy formats. The researchers found that participants demonstrated growth in equine competence, developed a sense of family and belonging with the horses, and reported reduced symptoms associated with trauma, including those related to PTSD, anxiety, depression, and substance use. The study advocates for P-EQ as a valuable therapeutic approach for CSEC survivors, and highlights the unique benefits of human-horse partnerships in facilitating trauma recovery and personal growth.

The next chapter 'Experimental Trial Demonstrates Positive Effects of Equine Facilitated Learning on Child Social Competence,' by Pendry and Roeter, describes a randomized controlled trial investigating the efficacy of an 11-week equine facilitated learning program in improving social competence among 5th-8th grade children. The study involved 64 children who were randomly assigned to either an experimental group or a waitlisted control condition. Participants engaged in weekly 90-minute sessions that combined mounted and unmounted activities that included observation of equine behavior, grooming, in-hand horsemanship, riding, and both personal and group reflection.

Results demonstrated a positive effect on children's social competence following program participation, including improvements in self-awareness, self-management, personal responsibility, decision making, goal-directed behavior, and relationship skills. The positive effects remained significant even after controlling for pretest levels of social competence, referral status, gender, and age. Additionally, children in the waitlisted control group showed similar gains after they completed the program, providing additional evidence for the intervention's effectiveness. The researchers suggested several potential mechanisms underlie the positive outcomes, including the use of horse-human interactions as analogies for human social contexts, stress reduction activities, and the benefits of a structured small-group setting that allows for positive peer modeling.

The third chapter 'Randomized Trial Examines Effects of Equine Facilitated Learning on Adolescents' Basal Cortisol Levels' by Pendry, Smith and Roeter, explores another dimension of an equine facilitated learning (EFL) program. This chapter presents findings from a randomized controlled trial examining the effects of an 11-week equine facilitated learning program on basal hypothalamic-pituitary-adrenal (HPA) activity in adolescents, as measured through salivary cortisol. The study involved 113 children (mean age 11.35 years) from ten different schools in the rural Pacific Northwest. Participants were randomly assigned to either an experimental group or waitlisted control condition. Those in the experimental group engaged in weekly 90-minute sessions featuring both mounted and unmounted activities. Each participant worked with the same equine team (consisting of one horse, two child participants, an equine specialist, and a facilitator) throughout the program. In-home salivary cortisol samples were collected before the program assignment and 12 weeks post-test to assess changes in HPA activity.

Results demonstrated moderate treatment effects on participants' average daily cortisol levels, with those in the EFL program showing significantly lower levels compared to waitlisted controls. These effects remained significant even after controlling for baseline cortisol levels, gender, age, and referral status. The researchers suggested that lower basal cortisol levels may represent a protective influence against the development of psychopathology and health problems. This study was the first randomized controlled trial to demonstrate treatment effects of a prevention program on basal HPA activity in predominantly typical adolescents.

The next chapter is 'A Transactional Analysis of Occupational Therapy in an Equine Environment for Youth with Autism Spectrum Disorder' by Peters, Bentele and Wood. In this study, a detailed analysis of videotaped sessions from their pilot study was reported. As part of their phased approach to intervention development, this study focused on protocol development and manualization. Seven children with autism spectrum disorder (ASD) ages 6-13 participated in the intervention, which consisted of 10 weekly sessions lasting 45-60 minutes. The intervention was delivered by occupational therapists with activities individualized to address each participant's goals related to social participation and self-regulation.

The researchers utilized a transactional perspective of occupation to analyze the complex nature of the intervention, identifying several key components including architectural spaces utilized, progression of occupational opportunities with and without horses, and the actions of participants and practitioners. Their analysis revealed that participants were significantly less likely to disengage during mounted activities, supporting the hypothesis that horses capture participant attention while also supporting physiological regulation. The study highlights the importance of creating an intervention manual that standardizes intervention while also allowing for individualization.

'The Impact of an Equine Facilitated Learning Program on Youth with Autism Spectrum Disorder' chapter by Erdman, Miller, and Jacobson also focused on youth with autism spectrum disorder (ASD). This exploratory, mixed-methods case study examined the impact of an equine facilitated, peer-assisted learning program on the social competence of youth with ASD. Three male participants with moderate ASD were paired with three typically developing female peers. Together, with a horse handler and a therapeutic riding horse, they participated in a 10-week program of weekly 75-minute sessions. The program emphasized groundwork activities, with weekly themes centered on establishing trust, safety, respect, nonverbal communication, assertiveness, teamwork, boundaries, and confidence. Data collection included behavioral observations (video-recorded at weeks 2, 6, and 9) coded for social awareness, cognition, communication, motivation, restricted interests/repetitive behavior, humor, and confidence, alongside parent-completed Social Responsiveness Scale assessments and qualitative interviews before and after the program.

Results from both the qualitative and quantitative measures indicated improvements in multiple areas of social competence including social awareness, cognition, communication, motivation, humor, and confidence. The researchers noted that participants demonstrated increased comfort with humorous comments and exhibited greater confidence in handling the horses as the program progressed. The study makes a unique contribution to the field by focusing on groundwork rather than riding and intentionally pairing youth with ASD with unfamiliar typically developing peers.

The last chapter in this section is 'Impact of Adaptive Devices on Horses, Assisting Persons, and Riders with Cerebral Palsy in a Therapeutic Riding Program' by Kaiser *et al.* This study evaluated the clinical effects of different adaptive riding devices on individuals with cerebral palsy (CP) and assessed horse stress behaviors and side walker exertion during therapeutic riding sessions. Ten participants with CP who were currently participating in weekly therapeutic riding sessions were included in the study. Three adaptive devices were examined: English saddle (ES), bareback pad (BP), and Independence Saddle (IS). Clinical outcomes were evaluated through osteopathic musculoskeletal examination (OME) performed before and after riding by a single osteopathic physician. Six horses participated in the study, with the PATH certified instructor making all rider-horse-adaptive device pairing decisions based on safety guidelines and horsemanship knowledge.

Results revealed clinical improvement in 63% of participants after a riding lesson, with no single adaptive device showing a higher likelihood of improving clinical outcomes than the others. The study found no significant differences in the average number of horse stress behaviors across the different adaptive devices, supporting previous research indicating that therapeutic riding is not mentally stressful for properly matched horses. Pressure measurements beneath each device showed the highest pressures when using the BP, while the ES and IS devices exhibited similar pressure magnitudes, none of which were concerning when compared to pressure from able-bodied riders. Importantly, side walker exertion was found to be significantly less with the IS than with other adaptive devices, suggesting that the Independence Saddle could be a viable option to reduce workload for assisting people while maintaining similar clinical outcomes for riders and without increasing stress for horses.

Adolescents

The next section of the book explores EAS with adolescents and includes chapters on adolescents' life skills, persistence, authentic relationships, and anxiety, as well as youth in treatment for substance misuse.

The first chapter in this section is 'Integration of Equine Assisted Services and Social-Emotional Learning: A Preliminary Evaluation of a Life Skills Training Program for At-Risk Youth' by Zaker *et al.*, and evaluates the effectiveness of an equine-assisted life skills training program on the social-emotional development of at-risk youth. Participants were students aged 18-21 years, selected by school administrators, from a transition program in the Western United States. The 9-week intervention followed the Equine Assisted Growth and Learning Association model, incorporating horses as part of a therapeutic team that included a mental health professional and an equine specialist. The Depression, Anxiety, Pain, and Stress (DAPS) visual analogue self-assessment tool was used to measure outcomes before and after the intervention.

Results revealed significant improvements in depression and anxiety scores among participants, but no significant changes in emotional pain and stress scores. The hands-on, experiential nature of the program allowed participants to engage in structured problem-solving sessions and skill-building activities that promoted social-emotional learning competencies such as teamwork, communication, and conflict resolution. These findings suggest that equine-assisted services can be beneficial for at-risk youth with mental health or behavioral diagnoses who struggle with emotional regulation and social interactions.

The next chapter is 'Equine-Assisted Activities for Adolescents: Ethogram-Based Behavioral Analysis of Persistence during Horse-Related Tasks and Communication Patterns with the Horse' by Hauge *et al.* It investigated the activity patterns of Norwegian adolescents aged 12-15 years during a four-month equine intervention, focusing on persistence during challenging tasks and communication patterns between adolescents and horses. Twenty-nine participants were video-recorded both early and late in their intervention period and analyzed using ethological methods. The intervention consisted of weekly 2-hour sessions over approximately 16 weeks, where participants engaged in four categories of activities: horse-related tasks, riding or handling horses from the ground, grooming, and stable work.

Results showed that while participants maintained the same active attitude and time budget of behavior throughout the intervention, they demonstrated increased persistence in challenging tasks during later sessions, choosing to try again more frequently when initially unsuccessful. Physical contact, particularly petting, was the main type of interaction initiated by participants toward the horses, and horses predominantly responded positively or neutrally to this contact. These findings suggest that equine interventions may enhance task-specific self-efficacy among adolescents during activities with horses, potentially contributing to the development of positive self-beliefs that could transfer to other areas. The farm setting, which emphasized experience rather than skill acquisition, and the supportive role of the instructor who allowed participants to attempt tasks independently, were identified as important environmental factors that may have influenced the positive outcomes.

The third chapter in this section is 'Equine Assisted Social Work as a Means for Authentic Relations between Clients and Staff' by Carlsson, Ranta and Traeen. It is a qualitative study exploring the meaning of Equine-Assisted Social Work (EASW) for young women with self-harm problems, focusing on the role of horses in fostering authentic relationships. Nine female clients aged 15-21 years participated, alongside eight staff members with diverse professional backgrounds who all had previous horse experience. The EASW sessions were offered one hour weekly as a voluntary complement to traditional treatment approaches, designed to increase self-esteem, modify behavior, reduce fears, and provide relaxation. The study employed multiple data collection methods including in-depth interviews with both clients and staff, and video-recorded observations of client-staff-horse interactions conducted three times for each pair working together.

Results revealed that the horses functioned as 'emotional mirrors,' reflecting participants' emotions and necessitating emotional self-regulation by both clients and staff to ensure safe interactions. Clients perceived the horses as more honest, trustworthy, and non-judgmental than humans, which allowed them to express their true emotions and authentic self without fear of rejection. The stable environment transformed the traditional therapeutic framework, creating a more informal, dynamic, and power-balanced relationship between clients and staff, where direct feedback became more socially acceptable. This setting facilitated emotional awareness and regulation, reducing anxiety and aggression while promoting calmness. The findings suggest that EASW provides a unique social context that supports learning and wellbeing for clients with self-harm problems by fostering authentic relationships characterized by empathy, honesty, respect, and responsiveness.

'The Helping Horse: How Equine Assisted Learning Contributes to the Wellbeing of First Nations Youth in Treatment for Volatile Substance Misuse' by Adams *et al.* is the next chapter. This study examines how Equine-Assisted Learning (EAL) contributes to the wellbeing of First Nations female youth in treatment for volatile substance misuse. Using a community-based research approach, the study employed a case study design to explore the Cartier EAL program's impact on 66 First Nations girls (average age 15) receiving treatment at a treatment center. The collaborative research team included frontline workers, community leaders, Elders, researchers, and multiple organizations, working within a holistic bio-psycho-social-spiritual framework that aligns with First Nations medicine wheel teachings.

The findings revealed that EAL participation positively influenced multiple dimensions of wellbeing. Biologically, participants experienced healthy physical touch, improved coordination, and connection with nature. Psychological benefits included enhanced self-esteem, improved mood, and better problem-solving skills. Socially, involved youth developed stronger interpersonal relationships and community connections. Most notably, the program fostered spiritual wellbeing through the human-equine relationship, characterized by mutual respect and trust—particularly significant for First Nations youth. This exploratory study offers valuable insights into a culturally aligned EAL program for Indigenous youth struggling with substance misuse.

The last chapter in this section is 'Psychotherapy Incorporating Horses to Address Anxiety in Women College Students: A Pilot Study' by Berg, Bach-Gorman, and Hammer. This chapter presents a pilot study examining the effectiveness of equine-assisted learning for female undergraduate students experiencing anxiety. Using a one-group, pre-and-post-test design, the research followed seven women (ages 18-24) who were clients at a US university counseling center. Participants attended seven weekly 90-minute group psychotherapy sessions incorporating the EAGALA model, facilitated by a certified mental health professional and equine specialist. The study measured changes in self-compassion and anxiety using the Self-Compassion Scale and Depression Anxiety Stress Scale.

The findings revealed a statistically significant improvement in the common humanity subscale of self-compassion, suggesting that equine-assisted therapy may help participants perceive their experiences as part of the larger human experience. While other measures did not show statistically significant changes, facilitators observed higher attendance rates compared to office-based therapy groups, with only a 14% attrition rate. Participants reported that the connection with horses motivated their continued attendance and increased their mindfulness of intrapersonal experiences. This study demonstrates the promising potential of equine-assisted services to address the growing mental health needs of college students.

Adults

The last section of this book includes chapters focused on adults and includes psychotherapy clients, adults' mood changes, mother-child dyads, the reduction of community and family violence, and adults with dementia.

The first chapter in this section 'Human-Animal Interaction and Metaphor in Equine-Assisted Psychotherapy: Empirical Support for the EAGALA Model' by Fournier *et al.*, is a mixed-methods study that investigates the impact of an intervention, based on the Equine-Assisted Growth and Learning Association (EAGALA) model of psychotherapy, on 112 adult clients. The study examined how interactions with horses, the emergence of metaphors, and psychosocial learning are interconnected. The study tested three key hypotheses: that human-horse interactions would predict decreased subjective distress, that these interactions would be associated with increased metaphor use, and that metaphors would mediate the relationship between horse interactions and psychological learning.

The findings provided empirical support for several aspects of the EAGALA model. Researchers discovered that the quantity of behavioral interactions between humans and horses correlated with reduced subjective distress and increased psychosocial learning. Notably, clients developed metaphors during the sessions that represented various aspects of their lives, including feelings, relationships, and personal challenges. The researchers found that the degree to which clients interpreted horses as metaphorical representations was strongly related to their learning, suggesting that the therapeutic value lies not in specific interactions, but in the client's personal interpretation of the experience.

The next chapter is 'Investigating Human Experiences and Perceptions of Unmounted Activities Commonly Practiced in Equine-Assisted Services' by Pasiuk *et al.* This study investigates human-horse interactions (HHI) through three groundwork activities: grooming, leading, and navigating a maze. The activities were designed to progressively build participants' comfort and skills with horses, starting with grooming, then leading, and finally completing a maze. Researchers documented participants' mood states, behaviors, and experiences through the use of wrist sensors and mood surveys completed before and after each activity.

The results revealed that participants generally experienced positive mood states during the interactions, with the grooming activity being particularly notable for its calming effects and its potential to build emotional connections. While the activity type did not significantly impact mood arousal, participants reported enjoying the maze activity the most, possibly due to the confidence they gained from previous activities. The study also highlighted how individual horse characteristics, such as size and behavior, can influence the human interaction experience. The researchers concluded that different activities and horse characteristics might be strategically used to support specific therapeutic or educational goals.

"Before, he fought every day with the horse and with me': Reducing Violence in a Guatemalan Community through a Horse-Handling Program' by Gibbons *et al.*, is the next chapter. This study was conducted in the rural Guatemalan community of Las Cumbres, where researchers implemented a Join-Up®-inspired horse-handling curriculum to explore whether nonviolent equine training could reduce aggressive attitudes and behaviors. The study involved 18 participants (16 males) who learned to interact with horses using gentle, non-coercive methods. By measuring participants' attitudes toward violence, interviewing their female relatives, and observing the horses' reactions, the researchers sought to explore if the program could transform interpersonal interactions and reduce violent tendencies.

The study found that participants reported less endorsement of aggression, demonstrated improved horse care, and had calmer family interactions after the program. Female relatives reported the participants exhibited positive behavioral changes, and the horses demonstrated more relaxed responses to handling. The researchers concluded that this type of intervention can be of benefit, especially for youth who are at a critical stage for developing interpersonal relationships, to avoid future violent pathways.

The next chapter is 'Behavioral and Physiological Effects of Equine-Assisted Early Intervention for Mother-Child Dyads with Insecure Attachment' by Hebesberger, Beetz, and Kotrschal. In this study, conducted in Augsburg, Germany, researchers investigated the effects of an equine-assisted intervention (EAI) on mother-child dyads from high-risk backgrounds. The study involved 20 mother-child pairs, randomly assigned to either an equine-assisted intervention or a play-based intervention group. Participants included mothers aged 19-46 and children aged 11-24 months, many of whom had backgrounds involving psychological disorders, abuse, or other challenging circumstances. The intervention lasted eight weeks, with weekly sessions designed to promote positive interactions and attachment between mothers and children.

Researchers found that mothers in the EAI group demonstrated more physical contact and verbal interaction with their children, particularly when engaging with horses. While the study did not find significant differences in cortisol levels or heart rate variability, it did show promising results in improving maternal sensitivity and child attachment. Notably, 65% of mothers preferred the equine-assisted intervention, leading researchers to suggest a combined approach of play-based and equine-assisted interventions might be most effective for supporting insecurely attached mother-child dyads.

The last chapter in this section, 'A Logic Model of a Dementia-Specific Program of Equine-Assisted Activities' by Lassell *et al.*, is a secondary analysis of a mixed-methods study, whereby researchers developed a comprehensive logic model to describe an equine-assisted activities program for older adults with dementia. The study aimed to bridge the gap between research and community-based animal-assisted interventions by explicitly mapping the program's purpose, assumptions, activities, and outcomes. Conducted using a pragmatic research approach, the study sought to provide a framework for understanding how and why the program worked, with providers emphasizing the importance of connections with horses, nature, and social participation as key mechanisms for improving participants' quality of life.

The resulting logic model revealed several insights into the program's design and implementation. Providers offered dynamic, person-centered interactions with horses, tailoring experiences to individual participants' preferences and abilities across different cognitive and physical levels. The study highlighted the holistic nature of the intervention, including careful screening, staff training, and varied occupational opportunities such as grooming, petting, and observing horses. The researchers proposed several directions for future research, including investigating the program's theory of change, examining implementation fidelity, exploring optimal intervention dosages, and documenting both short-term and long-term outcomes.

The chapters in this book offer a broad array of scientific studies that explore EAS in various settings for diverse populations. With a variety of methodologies, they offer insights into the potential of EAS and provide a solid foundation for its future growth.

References

Bator, D., Jańczyk, M., Samek, I., Wójcik, M., Dąbrowska, J. and Łopuszańska, U. (2020). Effectiveness of Equine Assisted Activities or Therapies in relation to mental illness and mental disorders—A review. *Journal of Education, Health and Sport*, *10*(6), Article 6. DOI: 10.12775/JEHS.2020.10.06.033.

Berardi, A., Di Napoli, G., Ernesto, M., Fabbrini, G., Conte, A., Ferrazzano, G., Viselli, F. and Galeoto, G. (2022). The Effectiveness of Equine Therapy Intervention on Activities of Daily Living, Quality of Life, Mood, Balance and Gait in Individuals with Parkinson's Disease. *Healthcare*, *10*(3), Article 3. DOI: 10.3390/healthcare10030561.

Bert, F., Gualano, M.R., Camussi, E., Pieve, G., Voglino, G. and Siliquini, R. (2016). Animal assisted intervention: A systematic review of benefits and risks. *European Journal of Integrative Medicine*, *8*(5), 695–706. DOI: 10.1016/j.eujim.2016.05.005.

Earles, J.L., Vernon, L.L. and Yetz, J.P. (2015). Equine-Assisted Therapy for Anxiety and Posttraumatic Stress Symptoms. *Journal of Traumatic Stress*, *28*(2), 149–152. DOI: 10.1002/jts.21990.

Gabriels, R.L., Pan, Z., Dechant, B., Agnew, J.A., Brim, N. and Mesibov, G. (2015). Randomized Controlled Trial of Therapeutic Horseback Riding in Children and Adolescents With Autism Spectrum Disorder. *Journal of the American Academy of Child and Adolescent Psychiatry*, *54*(7), 541–549. DOI: 10.1016/j.jaac.2015.04.007.

Menor-Rodríguez, M.J., Sevilla Martín, M., Sánchez-García, J.C., Montiel-Troya, M., Cortés-Martín, J. and Rodríguez-Blanque, R. (2021). Role and Effects of Hippotherapy in the Treatment of Children with Cerebral Palsy: A Systematic Review of the Literature. *Journal of Clinical Medicine*, *10*(12), Article 12. DOI: 10.3390/jcm10122589.

Nieforth, L.O., Schwichtenberg, A.J. and O'Haire, M.E. (2023). Animal-Assisted Interventions for Autism Spectrum Disorder: A Systematic Review of the Literature from 2016 to 2020. *Review Journal of Autism and Developmental Disorders*, *10*(2), 255–280. DOI: 10.1007/s40489-021-00291-6.

Nurenberg, J.R. and Schleifer, S.J. (2011). P03-156—Equine assisted therapy for patients with severe chronic psychiatric disorders. *European Psychiatry*, *26*, 1325. DOI: 10.1016/S0924-9338(11)73030-9.

O'Haire, M. (2017). Research on animal-assisted intervention and autism spectrum disorder, 2012-2015. *Applied Developmental Science*, *21*(3), 200–216. DOI: 10.1080/10888691.2016.1243988.

Pálsdóttir, A.M., Gudmundsson, M. and Grahn, P. (2020). Equine-Assisted Intervention to Improve Perceived Value of Everyday Occupations and Quality of Life in People with Lifelong Neurological Disorders: A Prospective Controlled Study. *International Journal of Environmental Research and Public Health*, *17*(7), Article 7. DOI: 10.3390/ijerph17072431.

Qin, Z., Guo, Zizhao and Wang, J. (2024). Multiple Equine Therapies for the Treatment of Gross Motor Function in Children with Cerebral Palsy: A Systematic Review. *Journal of Community Health Nursing*, *41*(2), 82–95. DOI: 10.1080/07370016.2024.2304825.

Trzmiel, T., Purandare, B., Michalak, M., Zasadzka, E. and Pawlaczyk, M. (2019). Equine assisted activities and therapies in children with autism spectrum disorder: A systematic review and a meta-analysis. *Complementary Therapies in Medicine*, *42*, 104–113. DOI: 10.1016/j.ctim.2018.11.004.

Wood, W., Alm, K., Benjamin, J., Thomas, L., Anderson, D., Pohl, L. and Kane, M. (2021). Optimal Terminology for Services in the United States That Incorporate Horses to Benefit People: A Consensus Document. *The Journal of Alternative and Complementary Medicine*, *27*(1), 88–95. DOI: 10.1089/acm.2020.0415.

Part 1: Children

1 Exploring Psychotherapy Incorporating Equine with Sexually Exploited Children

Jenna M. Hershberger[1], Erika L. Berg[1], Jodi L. Tangen[1*], Laura Feldmann[2] and Les Barnhart[2]
[1]North Dakota State University, Dept 2650, PO Box 6050, Fargo, ND 58108, USA; [2]Home on the Range, USA

Abstract

The International Labor Organization estimates roughly 1.7 million children are trafficked for sex globally at any given time, yet there is a dearth of research on therapeutic interventions. In this qualitative study, we utilized Psychotherapy Incorporating Equine to explore the experiences of commercially sexually exploited youth. We analyzed the participants' experiences using interpretative phenomenological analysis and identified the following themes: shifting relationships with self and others, equine therapeutic factors, and learning new skills. We provide implications for future research and practice.

Keywords: trafficking, psychotherapy, equine, interpretative phenomenological analysis

Introduction

Human sex trafficking is a grave human rights violation that impacts the dignity of millions. Sex trafficking is defined as commercial sex acts induced by force, fraud, or coercion, or in which the person induced to perform the sex act has not attained 18 years of age (TVPA, 2000). Although the exact prevalence is difficult to determine due to the illegal and clandestine nature of trafficking, the International Labor Organization (ILO,2022) estimates roughly 1.7 million children are trafficked for sex globally at any given time. Within the United States, cases involving the sex trafficking of minors accounted for approximately two in five human trafficking offenses reported to federally funded human trafficking task forces from 2008 to 2010 (Banks and Kyckelhahn, 2011).

The ILO (2022) defines Commercial Sexual Exploitation of Children (CSEC) as 'the use, procuring, or offering of children for the production of child sexual abuse materials', which may include sexual exploitation, prostitution, survival sex, pornography, and/or sex tourism. These youth are exploited by traffickers who are legally defined as individuals who engage in 'the recruitment, harboring, transportation, provision, or obtaining of a person for labor or services, through the use of force, fraud, or coercion for the purpose of subjection to involuntary servitude, peonage, debt bondage, or slavery' (TVPA, 2000, p. 8). Traffickers may include 'boyfriends,' romantic partners, family, 'pimps,' and gang

*Corresponding author: jodi.tangen@ndsu.edu

members (Jordan *et al.*, 2013). Individuals who are trafficked may suffer significant physical, sexual, emotional, mental, and spiritual traumas (Rafferty, 2018). Sex trafficking risk factors have been highlighted in existing literature, due in part to survivor testimony and service provider clinical reports (Roe-Sepowitz *et al.*, 2014; Cole and Sprang, 2015). Although these research contributions are valuable, there is a dearth of literature regarding effective therapeutic interventions for the healing of CSEC, and no research has evaluated the use of animal-assisted therapy.

According to Trotter and Baggerly (2019), horses are social herd creatures who communicate as prey animals, which makes them uniquely attuned to therapeutic work with humans. Horses are vigilant for any threat to safety, communicating non-verbally through the limbic region of their brains (Trotter and Baggerly, 2019). From subtle cues, including the adjustment of their ears or widening of pupils, to more overt indicators, such as flaring of nostrils, whinnying, or kicking, horses respond to their environment and may signal danger to those around them (Waring, 1983). This attunement also means horses have highly adapted means of non-verbal communication that can respond to human emotion, providing clients with direct and immediate behavioral feedback (Trotter and Baggerly, 2019). Horses may provide real-time, accurate feedback to individuals who struggle with emotion regulation or social cues. The behavioral response from horses may help clients recognize their own presentation and recalibrate their approach.

Psychotherapy Incorporating Equine (P-EQ) is an intervention employed by licensed, credentialed mental health professionals who incorporate equines as part of a client's treatment plan (Wood *et al.*, 2021). P-EQ involves both an experiential and processing component that may be used with individuals, couples, families, or groups (Masini, 2010). Multiple researchers have suggested that the incorporation of the horse into psychotherapy fosters healthy attachments (Chardonnens, 2009; Yorke *et al.*, 2013; Trotter and Baggerly, 2019). There is overlap in the tenets of attachment theory (Bowlby, 1973) and core aspects of P-EQ, including the importance of establishing safety and physiological co-regulation (Bachi *et al.*, 2012). Some researchers have suggested that equine interventions may be best termed 'equine partnered therapy' (Earles *et al.*, 2015, p. 151). Others have suggested that horses in psychotherapy actually become co-therapists and sources of healing attachment for participants engaged with them (Klontz *et al.*, 2007; Sheade, 2021). A consensus document by Wood *et al.* (2021) emphasizes the necessity of leading with the therapy, hence the term 'psychotherapy incorporating equine' vs 'equine psychotherapy.' Additionally, industry best practice stresses the importance of therapy professionals obtaining specialized training on how to safely incorporate equines into treatment sessions.

P-EQ programs have shown promise with individuals experiencing anxiety and depression (Alfonso *et al.*, 2015; Earles *et al.*, 2015; Wilson *et al.*, 2017; Acri *et al.*, 2021), post-traumatic stress disorder (PTSD) (Johnson *et al.*, 2018; Romaniuk *et al.*, 2018; Shelef *et al.*, 2019), and attachment concerns (Burgon *et al.*, 2018). Similar to individuals in prior P-EQ studies, survivors of CSEC experience high rates of comorbid disorders, including complex PTSD, severe depression, personality disorders, anxiety (Ravi *et al.*, 2017), and attachment difficulties (Cecchet and Thoburn, 2014; O'Brien, 2018). Despite these mental health needs for CSEC, there is no known study that has investigated the use of a P-EQ program for the therapeutic healing of youth who have been trafficked for sex.

Consequently, the purpose of the study is to: (a) explore the experience of CSEC both uniquely and as a group in a P-EQ program, (b) identify the impacts of a P-EQ program for CSEC, and (c) develop implications for P-EQ programs to CSEC.

Methods

Interpretative phenomenological analysis (IPA) was utilized to capture CSEC experiences of P-EQ and explore the convergence and divergence of individual and group experiences (Smith *et al.*, 2012).

Participants

After IRB approval, participants were recruited via convenience sampling at a therapeutic working ranch in the Midwestern United States. Inclusion criteria required participants to: (a) assent to participation in a P-EQ Rhythmic Riding program at a working ranch, (b) be identified as CSEC, (c) be between the ages of 12–18 years, (d) assent to engage in an interview, (e) assent to journal about their lived experiences, (f) assent to allow the researcher to observe P-EQ group sessions and take field notes, and (g) engage in one focus group at the conclusion of the program. Parental consent was also obtained.

Because the horses were not the participants of the study, we did not submit an IACUC (Institutional Animal Care and Use Committee) protocol specific to this study. However, the same horses were under an IACUC protocol for a different study aimed at measuring equine heart rate variability during the sessions.

Data collection

Data collection consisted of semi-structured individual interviews, reflective journals, observational field notes, and one focus group at the conclusion of the study. Data was collected over the course of 3 months. Interview lengths varied between 10 and 36 min and were recorded, transcribed, and reviewed. Observational field notes were taken on seven different occasions, including before and after P-EQ sessions, semi-structured interviews, and the concluding focus group. Participants engaged in seven P-EQ sessions, once per week, with each lasting approximately 90–120 min. Before the groups, participants received guidance from the horse handler to maintain their safety. During the groups, participants were assigned a horse and only worked with that particular horse during the experience. The participants caught, groomed, bridled, and mounted the horses. In addition, the participants rode the horses bareback around the oval-shaped arena while following therapeutic prompts from the facilitator and listening to music. Riding bareback was opted to limit additional influences on participant experience and help participants more easily connect with the Rhythmic Ride and the movement of their horse. The therapeutic prompts asked participants to become aware of their experience (their thoughts, their feelings, their body), the horses' experience, and the interactive experience between participants and their horses. The only stated goal of the group was for participants to match the rhythm of the music, the rhythm of their bodies, the rhythm of their horses. One focus group was conducted at the conclusion of the program. Participants were encouraged to select the music as long as it was not therapeutically counterproductive. Throughout the entire experience, the horse handler was there to monitor the well-being of the horses.

Positionality statement

The primary researcher (first author) is a white, cisgender, female-identifying individual who has worked with trafficked individuals as well as former traffickers and has experience in crisis, family, couple, group, and individual counseling with specialties in trauma and sexual abuse. The second author is a white, cisgender, female-identifying Associate Professor in the Equine-Assisted Services Program at a Midwestern University. The third author identifies as a white, cisgender, female-identifying Associate Professor in the Counselor Education program at a Midwestern University. The final two authors work with horses.

Trustworthiness

Reflexive field notes were utilized before, during, and after the interviews and observations of the group sessions. Bracketing of researcher assumptions was used to support confirmability. Triangulation of qualitative methods in this study included the use of interviews, field notes, and journal prompt responses. Within this framework of triangulation, observations from field notes, interviews, and journals were analyzed for patterns of convergence and divergence. One aspect of the analysis paid attention to similarities and differences between what participants shared privately in their journals vs during interviews with the researcher. The triangulation of qualitative sources within the same method was useful to check for consistency and strengthen quality and credibility (Patton, 2002). An audit record was used along with reflexive journaling to further support researcher's credibility. In reporting results, thick descriptions were used to support the transferability of themes.

Analysis

IPA was used to identify, analyze, and synthesize themes. In keeping with the layered analysis process required by IPA, themes were analyzed from individuals and also across the entire data set. Primary analysis entailed the identification of themes for each individual participant. Secondary analysis entailed the identification of themes across all six participants as a group. Following the primary and secondary analyses, themes were collectively analyzed to explore convergence and divergence. Pseudonyms were used to protect the identities of the individuals.

Results

The purpose of this study was to explore the lived experiences of CSEC in a residential P-EQ Rhythmic Riding program. A total of six participants were included in the study. All six identified as female within the age range of 13–16 years. Two identified as Native American and four identified as white. None of the participants had experienced Rhythmic Riding in the past, though several had had prior exposure to horses. Table 1.1 provides an outline of the participant sub- and super-ordinate themes. Each participant identified components of the super-ordinate themes; however, there was some variation in the subthemes.

Shifting relationship with self and others

All six participants spoke about changes related to their relationship with themselves and others. Four of the participants spoke about how they began to develop a sense of mutual empathy for participants. Within this superordinate theme, two subthemes were identified: (a) self-confidence and sense of worth, and (b) interpersonal empathy.

Self-confidence and sense of worth

Each participant highlighted a growth experience related to their sense of confidence and worth due to their engagement in P-EQ. Participants spoke about feeling accomplished, more connected, and at ease with themselves and others, as well as gaining back aspects of themselves they felt were lost due to their trauma experiences.

Jas' growing self-confidence and sense of worth was articulated as an ability to influence those around her for the better. Despite reporting in earlier sessions, 'I don't like people,' Jas shifted into valuing her positive potential:

> [Rhythmic Riding] just makes an impact on me and my heart...I just want to make my mark on the world and make something better, because you can make a good impact and you can make a bad impact. And I want my positivity to impact on the world.

Jay shared that P-EQ made deep impacts on her sense of self:

> I never really liked myself. I never loved myself. I didn't care about myself and then let other people show me that's really what I was like. They treated me, not like a person. They treated me like something you find on the floor and throw it away. Now, today, I love myself and I care for myself. I care about my future and the people that come into my life.

In addition, Jay highlighted another aspect of her sense of worth that had been restored as a result of P-EQ despite being lost due to sexual trauma:

> What Moose has taught me, what he's brought back, I didn't know I could get back. There's one big thing that he's brought back, and that's my pureness within myself. Because I'm pretty sure there's other girls like me out there that didn't get a chance to give away their pureness freely, and it was taken from them. Moose taught me that I still have it. I still have all of it. That's the biggest gift anybody could give to me.

Table 1.1. Individual themes and superordinate themes.

Participant	Individual themes	Superordinate themes
Les	Self-confidence and sense of worth.	Shifting relationship with self and others.
	Interpersonal empathy.	
	Mirroring and identification.	Equine therapeutic factors.
	Understanding.	
	Sense of family and belonging.	
	Trust.	
	Equine competence.	Learning new skills.
	Increased distress tolerance.	
	Increased emotion regulation.	
Mary	Self-confidence and sense of worth.	Shifting relationship with self and others.
	Sense of family and belonging.	Equine therapeutic factors.
	Equine competence.	Learning new skills.
	Increased distress tolerance.	
	Increased emotion regulation.	
Jas	Self-confidence and sense of worth.	Shifting relationship with self and others.
	Interpersonal empathy.	
	Mirroring and identification.	Equine therapeutic factors.
	Understanding.	
	Sense of family and belonging.	
	Trust.	
	Equine competence.	Learning new skills.
	Increased distress tolerance.	
	Increased emotion regulation.	
Alice	Self-confidence and sense of worth.	Shifting relationship with self and others.
	Mirroring and identification.	Equine therapeutic factors.
	Understanding.	
	Equine competence.	Learning new skills.
	Increased distress tolerance.	
	Increased emotion regulation.	
Jackie	Self-confidence and sense of worth.	Shifting relationship with self and others.
	Interpersonal empathy	
	Mirroring and identification.	Equine therapeutic factors.
	Understanding.	
	Trust.	
	Equine competence.	Learning new skills.
	Increased distress tolerance.	
	Increased emotion regulation.	
Jay	Self-confidence and sense of worth.	Shifting relationship with self and others.

Continued

Table 1.1. Continued

Participant	Individual themes	Superordinate themes
	Interpersonal empathy.	
	Mirroring and identification.	Equine therapeutic factors.
	Understanding.	
	Sense of family and belonging.	
	Trust.	
	Equine competence.	Learning new skills.
	Increased distress tolerance.	
	Increased emotion regulation.	

Interpersonal empathy

Four of the participants demonstrated a growing awareness of others and an empathetic stance. Participants spoke about a dynamic process of giving and receiving mutual empathy during the program. In addition, the researcher observed participants helping others in need while they engaged in the therapeutic prompts of the sessions and with the horses.

Jas, who often described herself as having a difficult relational life, reported a growing awareness of other participants, their emotional experiences, and common humanity:

> I learned all the girls have their own emotions and own feelings, and some of them try to hold it in and they don't really show it. But when they're doing it, they smile or just calm, they go to sleep on their horse. It just shows that we're all human in a way. Well, we are all human, but it shows that we're all human and we all have emotions and that nobody's the same. And that's kind of what it shows to me.

During her interview, Jackie shared awareness of her impact on others, 'my mood affects other people's mood,' as a growing sense of interpersonal empathy. Furthermore, Jackie demonstrated empathy in her actions toward others. For example, Jackie assisted Alice in mounting her horse during a later equine session by repeatedly moving the large mounting block and helping Alice steady her horse. Jackie had not demonstrated this empathetic, collaborative attitude in previous sessions. During the next session, Jackie was able to ask for and receive help from others in ways she had not experienced before.

Jay described a relational shift from engaging in strategies of disconnection to authentically connecting with others:

> Everybody tried, and they succeeded at making a connection with me. My own parents couldn't even make a connection with me...I wasn't a kid that talked. I wasn't a kid that came out of my room. I don't know what kind of kid I was. I was just lost. As much as I didn't want any relationship with anybody and how much hate I carried for people, because of the other people in my life and what they did to me, it was really hard for staff and residents. But the more time they put into the relationship, and I saw the time that they were putting in, I started putting my time in. Because they were showing that they wanted a healthy person in their life and for me to become a healthy person, and I've grown so much from where I first was before I got here. I don't even, I barely remember who I was when I first got here. All I remember is this person that I am now because of the help I got and the people I had by my side.

Equine therapeutic factors

Each participant described and demonstrated ways in which the horses uniquely contributed to their therapeutic experience in the program. Five of the participants spoke about how the horses enabled them to identify their own experiences and relational patterns as well as facilitated a sense of emotional validation and experiential understanding. Four of the participants described how greater trust and a sense of belonging were created by being with the horses. Within this superordinate theme, four subthemes were identified: (a) mirroring and identification, (b) understanding, (c) sense of family and belonging, and (d) trust.

Mirroring and identification

Participants spoke about how connection with the horse allowed them to identify both with the horse and with their own experiences in real time. This connection allowed participants to identify relational patterns, recalibrate problematic behavior, and gain insight.

Les directly spoke to the connection she experienced with her horse through deep breathing. She explained:

> And this was my first day ever riding a horse. And at first I felt wobbly, like I was going to fall off, and I wanted to get off...but I just, I really liked how the way that when I breathed, the horse took a deep breath. And so I just stayed on because I liked that, just being connected with the horse. And I feel like our thoughts were connected with because Big Sam is so gentle and I love animals, or just being gentle like that.

In addition to connecting with Big Sam through deep breathing, Les also described other physical connections she experienced while riding at the end of one of the sessions: 'He breathes with me. I move my hips with him. It's like Bluetooth. I felt like I was in rhythm with him.'

Alice and her horse, Bear, may have demonstrated some of the most problematic behaviors. Bear tried to buck Alice off multiple times during the course of rhythmic riding. During one session, the facilitator asked Alice what she thought Bear was trying to tell her when he attempted to buck her off. Alice responded, 'I think he's trying to tell me he's sick of my shit.' In her journal that day, Alice similarly wrote, 'he remindz me I need to quit my shit. he's probably annoyed of how touchy I am ?.' Alice frequently displayed oppositional and contradictory behavior during the sessions. During the focus group at the conclusion of the program, Alice admitted that she would covertly and frequently kick Bear to make him bolt or trot but would then pull tight on the reins to hold him back. During her interview, Alice described how observing her horse helped her to identify her own problematic behavior and also become more accepting of Bear and his needs:

> I don't know, Bear doesn't do what I would like him to do, but then it's okay because I don't do things that people like me to do all the time, so it's okay, I let it go. It's not that big of a deal.

(Throughout this exchange, the horse handler was present to ensure the horse's safety.)

Similarly, Jas shared in her interview how she recognized some of the more difficult aspects of herself in her horse: 'Well, my horse is a dick. He's mean, but he's kind of like me, because I don't like people either.' In a journal entry, she described how her horse would often attempt to bully or bite other horses, adding that despite this behavior, 'I love him he's a lot like me.' In response to another prompt regarding how she thought the horses experienced her, Jas shared, 'cause my horse fights with people and is a dick like me and I have to calm him down which I relate to a lot.' Jas was able to identify behaviors in her that were problematic given the externalized behavior she observed in her horse.

Understanding

Participants frequently spoke about feeling understood by the horses and having their emotions validated. For most of the participants, there was a stated belief that the horses demonstrated a level of understanding that had not been felt in other therapy experiences or with other therapists. Les explained:

> I feel like talking to the horse was the best because, yeah, talking to him through body language. I feel like that was very exciting because I feel like he just understood me, you know?... Like a lot of the times people don't understand you when you say something and that's why sometimes people try to do it through body language, whether it's anger or sadness or trying to control something. I felt like when I was speaking to him through his language, he just understood me. And I just felt heard kind of.

For Alice, it was important to assert that the experience of being understood by the horses was facilitated without words. Alice shared in her interview:

> I don't know. Okay, not saying it, like you're not expressing how you feel. Like in therapy, you express how you feel, you use your words. You

> don't have to use your words, you don't have to do really anything to let everything out because I don't know, the horses also understand your emotions and understand your feelings so that's a big... yeah.

Jay's experience of understanding was related to not only emotional validation but also a sense that the horses knew about her trauma in ways that others did not. Jay shared:

> My life, I don't even know why my life comes up. Because in therapy group with people, I don't want to talk about my life. I don't want to remember my life, anything. Working with the horses, it's a different type of listening. You don't have to say anything, but they already know.

Sense of family and belonging

Four of the participants highlighted how being part of the P-EQ program gave them a sense of family and belonging that was deeply meaningful.

In her journal, Les wrote, 'Today in rythmic ride I felt chill. My horse was calm until another horse bit my babies ass. I got mad at first and wanted my horse bite em back.' Les described here a protective, even maternal, stance toward her 'baby,' Big Sam. The researcher observed when Big Sam was bitten from behind by Spook. In response, Les moved her horse away from Spook and attempted to soothe him by petting his neck and telling him repeatedly, 'You're okay.' After this happened, the facilitator led a discussion of what had happened and how Les had responded to the situation by caring for her 'baby' as well as how the other participants and horses were all connected as one big group.

Jas expressed a similar sentiment as Les in that she described her relationship with Spook as maternal. Jas wrote in her journal, 'I've always rode Spook and just always felt a connection with him. That's why I call him my son.' Jas also highlighted how the P-EQ group broadened her sense of belonging with other participants:

> And we just chill and everyone has a good week after that. Saturday, Sunday, we're just all friends in all the group. We're like a big family kind of. And when they're calm and everything, we all get along because I think of them as like my sisters and it's actually pretty cool.

In several journal entries and her interview, Jay described the experience of choosing and creating her own family as a result of P-EQ. She also underlined an experience of feeling safe and protected by the horses in a way she had not experienced from other males in the past:

> Now that I'm here, I'm doing really good. I'm happy. I created a family here, not just with people, but with the animals. The horses, to me, are human beings. That's why I call Big Sam and Moose my big brothers, because that's just the connection I feel. I could talk to them, and I could hear them say something. That's how strong the connection is with me and my brothers, because I've never had brothers growing up, older brothers. This is the first time that I ever feel a male person is trying to help me and not trying to do whatever else. That respect grew, the loyalty grew, the love, just everything. I don't think I'd be the same without Moose and Big Sam.

She further explained how she perceived her own role in the family she created with the horses:

> [The horses were] treating me like a princess, like a little girl. Made me happy and gave me memories that I thought I'd never get to have. I never got to be a kid, but he let me be a kid around him.

Trust

Four of the participants displayed and spoke of a growing sense of trust as a result of working with the horses. Participants frequently distinguished their ability to trust the horses as much easier and natural than learning to trust other people. At the same time, growing trust with the horses translated into more trusting relationships with others.

Les described how exciting it was to feel a sense of trust in the horses when it had been difficult to trust other people in the past: 'Like it was too good to be true almost because the horses, you can just trust them and you don't even know them, you know?' In her interview, Les shared that the ease with which she felt she could trust the horses almost seemed 'unreal' to her. Les shared that as she grew to trust the horses, she was also able to trust other participants:

> But I was just walking side by side with another pair, and I trusted the other horse that he wasn't going to do anything. And I trusted my horse and I trusted the peer that was on the other horse. And so neither of them was anxious, or it was me - I was the only one who was anxious, but not even really anxious because I was on the horse.

Jas' experience of growing trust in P-EQ shared similarities with Les' in that connection with the horses was fundamental to developing trust in other relationships. Jas spoke about how trusting a large and powerful animal was paradoxically easier than trusting other people.

> The best part for me is getting to have trust, because I have a really big struggle with that and just being able to trust a horse out of everything, an animal, is really crazy because you don't know if he's going to buck you off or just do all that. So the big thing for me was trust. And bareback you're not in support. And that's just the main thing.

She further described the level of physical and emotional vulnerability she was able to experience within the trusting relationship with her horse: 'And the trust is, he knows the path and he knows what he's doing. So then I can just close my eyes and let go of the ropes and just hold onto his back and just calm down.'

In response to the journal prompt regarding her relationship with the horses, Jackie wrote, 'It is good I love my horse we match and I just trust him.' During one session, the facilitator asked Jackie how the group had gone for her that week. Jackie shared, 'Different. I had more trust.' When asked by the facilitator about how the trust was built, Jackie replied, 'I listened to him.'

Learning new skills

All six participants displayed and spoke about the ways in which they learned new skills. Each participant identified new and concrete skills related to working with the horses, including riding and caring for them. Within this superordinate theme, three subthemes were identified: (a) equine competence, (b) increased distress tolerance, and (c) increased emotion regulation.

Equine competence

All six of the participants spoke about new abilities related to riding and caring for the horses. This equine competence was also observable in the sessions as participants developed new skill sets related to working with the horses.

In addition to the challenge presented by learning new skills, Les described some of the mechanics of riding her horse and maneuvering in the arena:

> The horse, at first when I wanted the horse to turn, it wouldn't turn, and then I was thinking I did something wrong. And so I was trying to experience with it and learn about their kind of signals. And so I bumped him with my right leg and pulled with my left hand. And so he turned right away and then we got back into the rhythm and he was on the track.

Although she felt concerned initially about riding the 'right' or 'wrong' way, Les learned how to implement her newly taught riding skills to successfully guide her horse. During the sessions, the researcher observed Les grooming, guiding, riding, and mounting her horse with greater skill over time.

Mary highlighted one of the ways that equine competence led to new opportunities outside of the therapeutic working ranch. During her interview, Mary frequently stated that she was now more comfortable being around horses and riding them. Moreover, she shared, 'Yeah, and now when I go to my aunt's farm, I can ride her horses.' In addition, Mary highlighted other aspects of equine competence. Mary spoke about her enjoyment of the grooming process, which the participants engaged in at the beginning and end of every session: 'I like brushing around them, and I like petting them.' Not only did Mary enjoy these opportunities, she also learned how to care for the horses in ways that demonstrated equine competence.

Jas spoke about some of the safety concerns that participants needed to keep in mind related to their horses. She shared how she had learned to monitor her horse's body language for signs of distress:

> Because if they're acting up, if his ears go back, he's going to attack somebody. And you got to know those things because he might bite another horse and then you might have to calm

> him down and stop him, take a few breaths so he could take breaths with you.

Increased distress tolerance

Distress tolerance facilitates change by helping individuals to accept the crises and unavoidable suffering of life through an empowered stance (Linehan, 2018). P-EQ provided a natural context within which participants were exposed to challenging situations in working with others and their horses.

Les reflected on grappling with discomfort or what she identified as the most challenging aspect of P-EQ during her interview,

> I feel like the hardest part was dealing with being uncomfortable. And I know there's certain situations where, if you're uncomfortable, then you should get out of. But there's going to be sometimes in your life where you're uncomfortable... And so if you're just always comfortable in your life, you're going to stay the same, in the same spot.

In her response, Les displayed knowledge of when to utilize distress tolerance skills vs when to leave an unsafe situation.

In one of her journal entries, Jackie shared how she learned to tolerate the distress of new and challenging situations with her horse:

> I thought it was cool and there was a couple of times where I got nervous but I did deep breaths and I felt like I was about to fall off at some points but I was able to calm myself down and it was fun and I felt that created a bond with my horse.

During her interview, Jay stated the importance of responding to challenging situations with distress tolerance skills. Her analogy of the crises and suffering of life to being 'stuck' in the 'dirt' provided context for her exploration of how to respond in ways that lead to empowerment. Jay shared:

> I just hope that the kids around, the kids everywhere know that there's still time to revive from everything. There's still time to regrow, work through everything. You just got to tell yourself that you could do it. It's going to be hard for a little bit, but there's always a rainbow after every storm. It'll get bad before it gets better. I want them to know...that you could stay in the dirt for a little bit and take your time for the little bit that you're sitting in the dirt or laying in the dirt. You got to realize that time doesn't stop, and don't be stuck on that with the time that you're losing, cause you could be doing something else with. You're going to be pushed to the dirt for a little bit, but it's your choice to get back up.

Jay highlighted the importance of personal empowerment as a result of utilizing distress tolerance skills in difficult situations.

Increased emotion regulation

Emotion regulation skills involve identifying and responding to primary and secondary emotions in ways that do not increase suffering (Linehan, 2018). Participants named a number of often conflicting and challenging feelings as well as the ways they learned to respond to them with emotion regulation skills.

Jas shared the ways in which she grew in her ability to identify and regulate emotions throughout the sessions as well as how she observed others developing these skills:

> I learned that I'm a lot different from other people. I have different values and just different emotions. And during rhythmic riding, I just let my emotions out and it actually helps a lot...So I don't know, [Rhythmic Riding] just keeps me calm and it helps me let go of my emotions...And then as I already know myself, I'm just learning about myself more and more.

Jas was able to draw distinctions between her emotions and others' feelings, as well as learn to experience her emotions more fully. Jas shared how she observed other participants growing in their ability to emotionally regulate with the horses. She reflected, 'I feel like some of the girls are able to let go of their pain and just put it onto the horse, and they are a lot calmer once they get out, and they're not as bitchy.'

Jackie also named a number of mixed emotions that came up for her during the sessions:

> Mad, whenever [my horse] wouldn't listen to me. And calm too, I guess...I don't know. Whenever he would... Kind of excited too, and happy, whenever he would do what I wanted him to do. And then it was exciting, and I was happy.

Jackie spoke about the ways in which primary and secondary emotions impacted her during the sessions. Although she largely described her emotions as linked to her horse's behavior, she also highlighted how choosing to regulate allowed her to move from more difficult emotions to positive ones during the sessions, thereby eliminating further personal pain.

In her journal, Jay discussed how learning emotion regulation skill is an ongoing practice. She shared how she had learned more about her feelings and a growing ability to be patient:

> I can be calm to an extent. I know I still got learning to do & I see others do too. Some others need more learning, some need a little extent. Overall the rythmic ride was beneficial to me because it taught me more patience.

In her interview, Jay also discussed how this growing patience was related to her emotion regulation by learning to 'feel and not hide my emotions.'

Discussion

This study is unique in that there is no known literature specifically investigating the use of P-EQ for CSEC. Overall, participants' perceptions of a P-EQ program were positive. Participants described overcoming challenges, with identification of personal growth in several areas at the conclusion of the program. Participants also reported that the P-EQ program provided new opportunities, such as equine competence. This finding is consistent with the existing literature related to therapies incorporating equines in that participants report the development of new equine skills and a sense of responsibility for their horses (Chardonnens, 2009; Burgon, 2011; Shelef *et al.*, 2019). Despite the challenges presented by prior trauma and commercial sexual exploitation, participants frequently spoke about the healing shift that resulted from experiencing connection and identification, understanding, sense of family and belonging, and trust with the horses in the P-EQ sessions. Participants described the healing connection with the horses as the 'best' part of their experience. These findings align with similar research that has highlighted the helpfulness of equine-facilitated interventions in experiences of empathic understanding, emotional validation, and non-judgment (Dell *et al.*, 2011; Buzel, 2016).

To answer the question of the impact of a P-EQ program on this population, the perceptions of CSEC survivors regarding their relationships with themselves and others while in the program were explored. Like the extant literature, participants spoke about growing ease with themselves, as well as self-confidence and sense of worth (Bizub *et al.*, 2003; Ewing *et al.*, 2007; Burgon, 2011; Burgon *et al.*, 2018). Just as participants in this study spoke of regaining a sense of trust in others and in them, the literature regarding participants in equine therapeutic programs highlighted the healing of betrayal and trauma (Burgon, 2011; Burgon *et al.*, 2018). Experiences of interpersonal empathy were also frequent in the participant responses and aligned with outcomes from other equine studies (Burgon, 2011; Lanning *et al.*, 2017; Burgon *et al.*, 2018).

In addition, participants in the P-EQ program described and demonstrated an increased ability to respond to distress and regulate emotions. These skills were frequently cited as positive changes in participants' thoughts, feelings, and behaviors. Moreover, increased distress tolerance and emotional regulation are crucial in responding to mental health needs. Like other individuals in P-EQ programs, the participants in this study cited therapeutic benefits related to reduction in problematic mental health symptoms (Burgon, 2011; Burgon *et al.*, 2018; Johnson *et al.*, 2018), such as PTSD (Johnson *et al.*, 2018; Romaniuk *et al.*, 2018; Shelef *et al.*, 2019), anxiety and depression (Alfonso *et al.*, 2015; Earles *et al.*, 2015; Wilson *et al.*, 2017; Acri *et al.*, 2021), and substance use (Dell *et al.*, 2011), among other diagnoses (Schultz *et al.*, 2006; Chardonnens, 2009). Like other equine intervention research findings, participants spoke about feeling calmer when encountering challenges in the sessions (Sheade, 2021).

Finally, to determine the implications of a P-EQ program for CSEC in residential treatment, participants were asked to compare and contrast their experiences in other therapy programs with their experience in the P-EQ program. Participants spoke about the equine therapeutic factors and ways the horses uniquely facilitated their treatment process. Overall, participants spoke more highly of the P-EQ program than other, more traditional therapies they had tried. Participants spoke about the ways in which the horses demonstrated understanding and patience in ways previously unexperienced by participants in other therapy formats.

At the same time, participants frequently spoke about safety concerns related to riding the horses and problematic behavior from the horses. Although no participants experienced a critical incident with the horses during this study and frequently reframed uncooperative behavior from the horses as overcoming a challenge, real risks related to participant safety remain a consideration in the use of P-EQ. Additionally, the welfare of the equines doing this work is paramount, and thus specific training with regard to equine behavior and horsemanship, as well as social competencies should be required for counselors utilizing P-EQ (Hallberg, 2018; Trotter and Baggerly, 2019). Those involved in facilitating P-EQ programs for CSEC should carefully evaluate how to maintain the safety of the participants and the well-being of the horses.

Overall, the qualitative themes in this study indicate a variety of positive therapeutic outcomes unique to partnering with horses for CSEC. As advocates for survivors, those interested in the healing of CSEC should take into consideration the benefits of P-EQ.

Limitations and directions for future research

There are several limitations related to the present study. The sample size in the study is relatively small. The sample is also reflective of female experiences and does not include the experiences of male survivors in the same context. Additionally, the sample does not fully represent the diverse identities and wide range of experiences within the commercial sexual exploitation of youth. Finally, the sample included children, which may pose additional limitations (Smith and Dunworth, 2003). To date, there is no research regarding the use of P-EQ for CSEC. Further exploration of the qualitative themes from this study could support additional qualitative, mixed-methods, and quantitative studies.

Acknowledgement

JMH received USD $250 from the NDSU School of Education and $250 from the NDSU College of Human Sciences and Education to complete this research.

Funding

First author received $250 from the School of Education and $250 from the College of Human Sciences and Education to complete this research.

Conflict of Interest

We do not have a conflict of interest to report with this manuscript.

Author Contributions

This was doctoral dissertation of JMH. ELB provided guidance on the animal-assisted part of the process, and JLT served as dissertation chair for JMH. LF and LB provided the equine therapy space for the study.

Ethics Statement

We complied by appropriate ethics (IRB and assent) in completing this study.

References

Acri, M., Morrissey, M., Peth-Pierce, R., Seibel, L., Seag, D. *et al.* (2021) An equine-assisted therapy for youth with mild to moderate anxiety: Manual development and fidelity. *Journal of Child and Family Studies*. DOI: 10.1007/s10826-021-02011-4.

Alfonso, S.V., Alfonso, L.A., Llabre, M.M. and Fernandez, M.I. (2015) Project stride: An equine-assisted intervention to reduce symptoms of social anxiety in young women. *Explore* 11(6), 461–467. DOI: 10.1016/j.explore.2015.08.003.

Bachi, K., Terkel, J. and Teichman, M. (2012) Equine-facilitated psychotherapy for at-risk adolescents: The influence on self-image, self-control and trust. *Clinical Child Psychology and Psychiatry* 17(2), 298–312. DOI: 10.1177/1359104511404177.

Banks, D. and Kyckelhahn, T. (2011) Characteristics of suspected human trafficking incidents, 2008–2010. Special Report (NCJ 233732). U.S. Department of Justice, Office of Justice Programs, Bureau of Justice Statistics, Washington, DC. Available at: https://bjs.ojp.gov/content/pub/pdf/cshti0810.pdf (accessed 5 December 2021).

Bizub, A.L., Joy, A. and Davidson, L. (2003) It's like being in another world:" Demonstrating the benefits of therapeutic horseback riding for individuals with psychiatric disability. *Psychiatric Rehabilitation Journal* 26, 377–384. DOI: 10.2975/26.2003.377.384.

Burgon, H.L. (2011) Queen of the world': Experiences of 'at-risk' young people participating in equine-assisted learning/therapy. *Journal of Social Work Practice* 25(2), 165–183. DOI: 10.1080/02650533.2011.561304.

Bowlby, J. (1973) *Attachment and loss, vol II Separation*. Basic Books, New York, NY.

Burgon, H., Gammage, D. and Hebden, J. (2018) Hoofbeats and heartbeats: Equine-assisted therapy and learning with young people with psychosocial issues – theory and practice. *Journal of Social Work Practice* 32(1), 3–16. DOI: 10.1080/02650533.2017.1300878.

Buzel, A. (2016) *Beyond words: The healing power of horses: bridging the worlds of equine assisted therapy and psychotherapy*. AuthorHouse, Bloomington, IN.

Cecchet, S.J. and Thoburn, J. (2014) The psychological experience of child and adolescent sex trafficking in the United States: Trauma and resilience in survivors. *Psychological Trauma: Theory, Research, Practice, and Policy* 6(5), 482–493. DOI: 10.1037/a0035763.

Chardonnens, E. (2009) The use of animals as co-therapists on a farm: The child–horse bond in person-centered equine-assisted psychotherapy. *Person-Centered and Experiential Psychotherapies* 8, 319–332. DOI: 10.1080/14779757.2009.9688496.

Cole, J. and Sprang, G. (2015) Sex trafficking of minors in metropolitan, micropolitan, and rural communities. *Child Abuse and Neglect* 40, 113–123. DOI: 10.1016/j.chiabu.2014.07.015.

Dell, C.A., Chalmers, D., Bresette, N., Swain, S., Rankin, D. *et al.* (2011) A healing space: The experiences of First Nations and Inuit youth with equine-assisted learning (EAL). *Child and Youth Care Forum* 40(4), 319–336. DOI: 10.1007/s10566-011-9140-z.

Earles, J.L., Vernon, L.L. and Yetz, J.P. (2015) Equine-Assisted Therapy for anxiety and posttraumatic stress symptoms: Equine-Assisted Therapy. *Journal of Traumatic Stress* 28(2), 149–152. DOI: 10.1002/jts.21990.

Ewing, C.A., MacDonald, P.M., Taylor, M. and Bowers, M.J. (2007) Equine-Facilitated Learning for youths with severe emotional disorders: A quantitative and qualitative study. *Child and Youth Care Forum* 36(1), 59–72. DOI: 10.1007/s10566-006-9031-x.

Hallberg, L. (2018) *The clinical practice of Equine-Assisted Therapy: Including horses in human healthcare*. Routledge, New York, NY.

International Labor Organization [ILO] (2022) *Global Estimates of Modern Slavery: Forced Labor and Forced Marriage*. International Labor Organization, Geneva, Switzerland. Available at: https://

www.ilo.org/wcmsp5/groups/public/_---_ed_norm/ipec/documents/publication/wcms_854733.pdf (accessed 8 December 2022).

Johnson, R.A., Albright, D.L., Marzolf, J.R., Bibbo, J.L., Yaglom, H.D. *et al.* (2018) Effects of therapeutic horseback riding on post-traumatic stress disorder in military veterans. *Military Medical Research* 5(1), 3. DOI: 10.1186/s40779-018-0149-6.

Jordan, J., Patel, B. and Rapp, L. (2013) Domestic minor sex trafficking: A social work perspective on misidentification, victims, buyers, traffickers, treatment, and reform of current practice. *Journal of Human Behavior in the Social Environment* 23(3), 356–369. DOI: 10.1080/10911359.2013.764198.

Klontz, B., Bivens, A., Leinart, D. and Klontz, T. (2007) The effectiveness of Equine-Assisted experiential therapy: Results of an open clinical trial. *Society and Animals* 15(3), 257–267. DOI: 10.1163/156853007X217195.

Lanning, B.A., Wilson, A.L., Krenek, N. and Beaujean, A.A. (2017) Using therapeutic riding as an intervention for combat veterans: An international Classification of Functioning, Disability, and Health (ICF) approach. *Occupational Therapy in Mental Health* 33(3), 259–278. DOI: 10.1080/0164212X.2017.1283282.

Linehan, M.M. (2018) *Cognitive-behavioral treatment of borderline personality disorder*. Guilford Press, New York.

Masini, A. (2010) Equine-Assisted Psychotherapy in clinical practice. *Journal of Psychosocial Nursing and Mental Health Services* 48(10), 30–34. DOI: 10.3928/02793695-20100831-08.

O'Brien, J.E. (2018) Sometimes, somebody just needs somebody– anybody – to care: The power of interpersonal relationships in the lives of domestic minor sex trafficking survivors. *Child Abuse and Neglect* 81, 1–11. DOI: 10.1016/j.chiabu.2018.04.010.

Patton, M.Q. (2002) *Qualitative research methods*, 3rd edn. Sage, Thousand Oaks, CA.

Rafferty, Y. (2018) Mental health services as a vital component of psychosocial recovery for victims of child trafficking for commercial sexual exploitation. *The American Journal of Orthopsychiatry* 88(3), 249–260. DOI: 10.1037/ort0000268.

Ravi, A., Pfeiffer, M.R., Rosner, Z. and Shea, J.A. (2017) Trafficking and trauma: Insight and advice for the healthcare system from sex-trafficked women incarcerated on Rikers Island. *Medical Care* 55(12), 1017–1022. DOI: 10.1097/MLR.0000000000000820.

Roe-Sepowitz, D.E., Hickle, K.E., Dahlstedt, J. and Gallagher, J. (2014) Victim or whore: The similarities and differences between victim's experiences of domestic violence and sex trafficking. *Journal of Human Behavior in the Social Environment* 24(8), 883–898. DOI: 10.1080/10911359.2013.840552.

Romaniuk, M., Evans, J. and Kidd, C. (2018) Evaluation of an equine-assisted therapy program for veterans who identify as 'wounded, injured or ill' and their partners. *PLOS ONE* 13(9), e0203943. DOI: 10.1371/journal.pone.0203943.

Schultz, P.N., Remick-Barlow, G.A. and Robbins, L. (2006) Equine-assisted psychotherapy: A mental health promotion/intervention modality for children who have experienced intra-family violence: Equine-assisted psychotherapy. *Health and Social Care in the Community* 15(3), 265–271. DOI: 10.1111/j.1365-2524.2006.00684.x.

Sheade, H. (2021) *Equine-Assisted Counseling and Psychotherapy: Healing through horses*. Routledge, New York, NY.

Shelef, A., Brafman, D., Rosing, T., Weizman, A., Stryjer, R. *et al.* (2019) Equine Assisted Therapy for patients with Post Traumatic Stress Disorder: A case series study. *Military Medicine* 184(9–10), 394–399. DOI: 10.1093/milmed/usz036.

Smith, J.A. and Dunworth, F. (2003) Qualitative methods in the study of development. In: Connolly, K. and Valsiner, J. (eds) *The Handbook of Developmental Psychology*. Sage, London.

Smith, J.A., Flowers, P. and Larkin, M. (2012) *Interpretative phenomenological analysis: Theory, method and research*. Sage, Thousand Oaks, CA.

The Trafficking Victims Protection Act (TVPA) (2000). Available at: https:// www.govinfo.gov/content/pkg/PLAW-106publ386/pdf/PLAW-106publ386. pdf (accessed 20 February 2020).

Trotter, K.S. and Baggerly, J.N. (2019) *Equine-assisted mental health for healing trauma*. Routledge, New York, NY.

Waring, G.H. (1983) *Horse behavior: The behavioral traits and adaptations of domestic and wild horses, including ponies*. Noyes Publications, Park Ridge, NJ.

Wilson, K., Buultjens, M., Monfries, M. and Karimi, L. (2017) Equine-assisted psychotherapy for adolescents experiencing depression and/or anxiety: A therapist's perspective. *Clinical Child Psychology and Psychiatry* 22(1), 16–33. DOI: 10.1177/1359104515572379.

Wood, W., Alm, K., Benjamin, J., Thomas, L., Anderson, D. *et al.* (2021) *Journal of Alternative and Complementary Medicine* 27(1), 88–95. DOI: 10.1089/acm.2020.0415.

Yorke, J., Nugent, W., Strand, E., Bolen, R., New, J. *et al.* (2013) Equine-assisted therapy and its impact on cortisol levels of children and horses: A pilot study and meta-analysis. *Early Child Development and Care* 183(7), 874–894. DOI: 10.1080/03004430.2012.693486.

2 Experimental Trial Demonstrates Positive Effects of Equine Facilitated Learning on Child Social Competence

Patricia Pendry* and Stephanie Roeter
Department of Human Development, Washington State University, Pullman, Washington

Abstract
Although equine facilitated programs have gained in popularity over the last decade, virtually nothing is known about the causal effects of equine facilitated interventions on human development and wellbeing. To address this gap in the literature, researchers conducted a randomized controlled trial to determine if an 11-week equine facilitated learning program enhanced 5th-8th grade children's social competence. Children were recruited for program participation through referral by school counselors and recruitment in schools and community agencies. Researchers then randomly assigned 64 physically and mentally able children to an experimental group or waitlisted control group. Children in the experimental group participated in an 11-week equine facilitated learning program designed to increase social competence through a series of once-weekly, 90-minute sessions of individual, team, and group-focused equine facilitated activities, whereas children in the control group did not until 16 weeks later. Parents of children in both groups provided ratings of child social competence at the beginning and again at the end of the 11-week program. Results indicated significant group differences in mean levels of child social competence at posttest ($p = 0.020$), suggesting a moderate positive effect of program participation ($d = 0.61$). Waitlisted children in the control group who completed the program at a later date demonstrated significantly higher posttest levels of social competence after program completion ($p = 0.000$), compared to their own pretest scores. Using a lagged dependent variable approach, program effects were robust ($p = 0.026$) when simultaneously considering children's pretest levels of social competence, age, gender, and referral status.

Keywords: child development, equine facilitated intervention, experimental trial, HAI, social competence

Introduction

Although there is growing evidence that promotion of behavioral and social-emotional competence in children and youth is an effective strategy to prevent mental, emotional, and behavioral disorders in adulthood (O'Connell *et al.*, 2009), identifying and expanding evidence-based prevention programs that promote these competencies in children and youth constitutes a significant challenge (Greenberg *et al.*, 2000; Huang *et al.*, 2005). One type of approach that has caught the attention of researchers and mental health professionals is equine facilitated

*Corresponding author: ppendry@wsu.edu

learning, which combines the experience of equine human interaction with counseling-based processing skills to increase participants' awareness and control of their emotions, cognitions, and behaviors. Although equine facilitated programs have gained popularity over the last decade, virtually nothing is known about their causal effects on human development. To help alleviate this gap in the literature, researchers conducted a randomized controlled trial to determine the effects of an 11- week equine facilitated learning program on the social competence of 5th-8th grade children.

Although research on the broader field of equine assisted interventions is in its infancy, a small body of pioneering work has shown promising results. Findings suggest significant associations between involvement in equine assisted intervention and adjustment in various socio-emotional, cognitive, and behavioral domains, including depression (Bowers and MacDonald, 2001), positive behavior (Chandler, 2005), behavioral and mood disorders (Mann and Williams, 2002), self-esteem (Brown and Alexander, 1991; Katcher and Wilkins, 1994; Gatty, 2001), feelings of social acceptance and peer popularity (MacDonald and Cappo, 2003), interpersonal communication (MacDonald, 2004), sensitivity towards others (Vidrine *et al.*, 2002), anger (Kaiser *et al.*, 2004), speech and language ability (Macauley and Guierrez, 2004), as well as more global measures of functioning and adjustment (Schultz *et al.*, 2007). Also, compared to more traditional classroom- based counseling activities, equine assisted interventions were more strongly associated with lower levels of internalizing and externalizing behavioral problems (Trotter *et al.*, 2008). Other researchers, in a study on therapeutic riding targeting boys with emotional disturbances in a residential treatment program, found no improvement was seen in depression or anxiety (Greenwald, 2001). While current evidence regarding the efficacy of equine assisted approaches is promising indeed, more causal evidence is needed to demonstrate the effects of equine facilities programs.

Causal pathways

Although there is not one specific human animal interaction theory that postulates how equine facilitated learning programs may affect child development, Bandura's social cognitive theory (1986) provides a nice framework (Serpell, 2000; Kruger and Serpell, 2010). According to this perspective, children learn by observing others, with the environment, behavior, and cognition all as important factors in influencing development. These three factors are not static or independent; rather, they are all reciprocal. For example, each behavior witnessed can change a person's way of thinking (cognition). Similarly, the environment one is raised in may influence later behavior.

This theory has implications for equine facilitated learning because equines possess several intrinsic and physical attributes that strongly influence the cognitions and environments of those interacting with them (McGreevy and McLean, 2010). For example, horses and mules are prey animals, characterized by a highly attuned awareness of their physical surroundings and a strong, instinctual inclination for a fight-or-flight response. This characteristic necessitates that program participants become more self-aware of their behavior, thoughts, emotional intent, and verbal and non-verbal communication to ensure their personal safety, and to enhance their ability to communicate with the equine. By interacting with equines in the context of horsemanship activities, children learn to provide and recognize non-verbal communication cues to their equine partner (e.g. pressure, release, posture, intent, physiological arousal), which may provide frequent opportunity to reflect on the meaning of their own behavior and intent. Since equines provide immediate, meaningful feedback - pinning ears, swishing tail, licking lips, blinking - children simultaneously gain experience in observing and interpreting non-verbal cues. As children – with the help from program facilitators - attempt to manage and direct their own behavior to evoke desirable responses from the horse (e.g. licking lips, blinking, resting foot), rather than less desirable responses (e.g. swishing tail, pinning ears, tuning out), reciprocal interactions between the child's behavior and cognitions take on an important role. Using human equine interactions as an analogy, program facilitators encourage children to reflect on their behavior, thoughts, feelings and communication skills and apply these to communication and collaboration with a *human* partner. Practicing of pro-social behaviors, reflecting on thoughts and feelings, and feedback – from the equine and facilitator - may thus improve self and social awareness, personal responsibility, goal-directed behavior, and communication skills.

Social role-activity theory (Brickel, 1982) also provides a useful perspective to explain effects of horse-human interactions on child development. Role theory proposes that when an individual assumes a new positive role (e.g. team leader), positive behavioral change may result (Lemon *et al.*, 1972). This relates nicely to equine facilitated activities, as participants are encouraged to take on a leadership role to facilitate positive interactions with equines, which rely heavily on the equine's instinctual expectations about the natural hierarchy of herd dynamics, leading them to oscillate between seeking and/ or assuming leadership, or accepting it from others (Budiansky, 1997).

During in-hand horsemanship activities, children are encouraged to engage in behaviors that assert leadership to the equine by moving the horse's body in a purposeful way through the use of progressive shaping techniques, or '... gradually increasing a behavior/pressure to elicit a specific behavior from the equine' (McGreevy and McLean, 2007, p. 114), and negative reinforcement to reward compliant responses from the horse. During these interactions children are frequently reminded by program facilitators that self-management, fostering a give-and-take relationship, quick decision making, and optimistic thinking are important skills that facilitate successful communication and thus completion of each task. Through these interactions with the horse, children thus learn to provide and interpret clear, non-verbal cues. In sum, equine characteristics may have a significant influence on the nature of interactions between participants and horses in ways that promote children's social and behavioral competencies through the practice of pro-social behavior, as well as reflection on participants' cognitions, behaviors, emotion and intent.

Description of current study and hypotheses

The overall objective of the current study was to conduct a randomized controlled trial to determine the efficacy of an 11 week, equine facilitated learning program in improving the social competence of 5th-8th grade children. Social competence is the ability of the child to successfully interact with other children and adults in a way that demonstrates an awareness of and ability to manage emotions in an age- and contextually appropriate manner. Social competence is comprised of social, emotional, and cognitive skills and behaviors that enhance social adaptation such as social awareness, self-awareness, goal-directed behavior, relationship skills, personal responsibility, and control of behavior. The main research question was: Does participation in an 11-week equine facilitated learning program increase children's social competence and/or components thereof?

The following hypotheses guided our analyses. The first hypothesis stated that children randomly assigned to participate in an 11-week equine facilitated learning program would have higher levels of social competence at posttest, compared to children assigned to a waitlisted condition. The second hypothesis stated that children in the waitlisted group would demonstrate significant within-person gains in social competence after program participation (i.e. *after* the experimental group completed the program). The third hypothesis stated that the positive effects of program participation on posttest levels of child social competence would be independent of children's pretest levels of social competence, gender, age, and referral status.

Social competence was the main outcome of interest to program designers and researchers, because it is considered a central domain of child development that plays a critical role in later academic achievement, mental health, and overall wellbeing (Shonkoff and Phillips, 2000). Lower social competence is positively associated with the development of anti-social behavior (Sørlie *et al.*, 2008), conduct problems and depression (Rockhill *et al.*, 2009), substance use (Griffin *et al.*, 2001) and lower academic competence and performance (Wentzel, 1991; Sørlie and Nordahl, 1998). Children with lower social competence are also more likely to experience difficulty regulating negative emotion and arousal, which can interfere with initiating and maintaining positive peer interactions, leading to peer rejection (Dodge and Somberg, 1987). Relationship problems with peers and adults, difficulty with emotion regulation, attention, behavior, and challenges with learning and achievement contribute to the development of mental health issues and poor child wellbeing. As such, enhancing child social competencies that support the development of self-regulation, academic skills, and peer social skills is considered a suitable target for preventive intervention programs (Greenberg *et al.*, 2000).

Method

The program

The program under study was PATH to Success[1] conducted at a Premier Accredited Center of the Professional Association of Therapeutic Horsemanship, International[2] (PATH, Intl.) in a university setting. The stated goal of the program under study is to enhance child social competence through an 11-week session of once-weekly, 90 min lessons of individual, team, and group-focused equine facilitated activities. The program was designed, implemented, and consistently supervised by a PATH, Intl. certified instructor in equine assisted mental health, and a doctoral level therapist. The curriculum contained standardized lessons with weekly goals and step-by-step instructions and procedures for each activity, which were piloted for 3 years before this study on its effects was conducted. The program was implemented by a team of volunteer horse specialists and facilitators, that included PATH certified instructors, undergraduate students in child-development, education, and animal science, professional counseling psychologists, and graduate-level counseling students. Program staff received extensive training through printed lessons, a handbook, hands-on program training with horses and child actors, as well as training in theoretical rationale of the program, principles of cognitive behavioral counseling, and child development through video and slide presentations. Weekly activities - based on principles of equitation science and natural horsemanship (McGreevy, 2007; McGreevy and McLean, 2007, 2010) - featured a combination of mounted and unmounted activities and horse-human interactions, including observation of equine behavior, engagement in equine management (e.g. grooming), in-hand horsemanship activities, some riding, and personal and group reflection activities. A description of weekly program objectives and a selection of sample activities are described in Table 2.1.

Children in the experimental group were assigned to the same facilitation team for the entire 11-week session that included one equine, two child participants, an equine specialist, and facilitator. Each day, a total of eight child participants - divided across four equine teams headed by four experienced equines – participated. The program was conducted four times a week, serving 32 children each week per 11-week session. The weekly lessons were conducted in the afternoons on weekdays and included transportation of participants from school to the program site and back immediately following their regular school day. Children who were waitlisted participated after children in the experimental group completed the program.

Procedures

The study was approved by the University Committee on Research Involving Human Subjects at Washington State University. Equine welfare of horses in the program was carefully monitored under the university's Institutional Animal Care and Use Committee (IACUC) regulations and campus veterinarian. In addition, equines were evaluated daily by a PATH, Intl. certified instructor for physical, mental, and emotional well-being before, during and after each session. Although equine participation in the program never elicited such behavior, program policy required that equines would be immediately released from program participation if they were to display stress behavior to ensure their wellbeing. In addition, any inappropriate or potentially harmful behavior of humans interacting with equines required

Table 2.1. Outline of Lesson Objectives by Week.

Week	Lesson Objective	Sample Activity
1	Basic safety: Meet horses and staff	Observing horse behavior and herd dynamics
2	Respect: Self, others and horses	Moving horses using 4 phases of direct or indirect pressure[3]
3	Communication: Verbal and non-verbal	Leading horses, interpreting horse body language
4	Leadership: Assertive and aggressive cues	Driving[4] activity using body language and phases
5	Trust: Coping with perceptions of stress	Riding and leading
6	Boundaries	Driving activity using indirect pressure
7	Overcoming challenges and building confidence	Desensitizing[5] horses
8	Enhancing self-regulation and relaxation	Horse massage[6], riding
9	Prepare for parents/visitors day	Incorporating horsemanship skills for team challenge
10	Parents/visitors day	Participants 'teach' parents horsemanship skills
11	Program wrap up	Obstacle course, riding and reflection

immediate intervention to prevent undue distress by the equine.

Program participants were recruited by program staff through referral by school counselors[7], distribution of flyers and advertisements, and information meetings in schools and the community. Criteria for program participation were that 1) parents and children had to be able to communicate effectively in English, 2) the child did not have a serious physical or mental disability and 3) the child had to attend the 5th through – 8th grade. Recruitment for the study occurred at the informational meetings, during which parents and children received information from the research team about the study, its screening, assessment, and assignment procedures as well as information about the process of consent and assent[8]. Interested parents were asked to complete a standardized measure of child social competence for screening and sample selection purposes, for which they were paid five dollars (see measures section for description of social competence measure). Children in each grade were subsequently rank-ordered by their social competence score with the aim of recruiting approximately equal numbers of boys and girls per grade, as well as selecting children with relatively lower social competence scores. Participation in the study was voluntary and children could attend the program without enrolling in the study.

Selected study participants were randomly assigned to a treatment group or wait-listed control group. Children assigned to the experimental group started program participation 2 weeks later, whereas children assigned to the waitlisted control group were offered program participation in the following session – 16 weeks later. Parents of both groups provided ratings of child social competence again at the posttest (i.e. at the end of the 11- week program). Parents of waitlisted children were asked to report on their child's social competence a third time, after their child completed their 11-week session.

Sample recruitment, screening and assignment

In the first year of this two-year study, program staff recruited a total of 78 interested children and their families from 10 different schools in a rural university town in the Pacific Northwest area of the United States. Sixty-four children were selected (N_{boys} = 37; N_{girls} = 27; M_{age} = 10.93 yrs) for study participation and random assignment to an experimental group (*N*=33) or waitlisted control condition (*N*=31). There were no significant differences between participants' age (*p*=0.988), gender (*p*=0.143), referral status (*p*=0.300), grade level (*p*=0.874), or pretest social competence (*p*=0.599) by group status (e.g. between experimental or waitlisted). Sample characteristics are reported in Table 2.2.

Sample enrollment, selection, randomization, follow up and analyses counts are described in a flow diagram (Fig. 2.1). A total of 54 participants completed data collection at pretest and posttest. Attrition in the experimental group was due to two active withdrawals immediately following assignment, one participant moving out of the area after 6 weeks of program participation, and one active withdrawal in week four due to allergy issues. Attrition in the control group occurred due to active withdrawal by three participants following random assignment, and three passive withdrawals as evidenced by failure to complete the posttest assessment. Of all children originally assigned to the waitlist control condition, a total of 20 eventually participated in the active program, which was conducted 4 weeks after children in the experimental group completed the program. There were no significant group differences (comparing waitlisted or experimental group) by age (*p*=0.637), referral status (*p*=0.581), grade level (*p*=0.639), social competence (*p*=0.548), or gender (*p*=0.089) between participants who completed the study and those who withdrew.

Table 2.2. Sample Descriptive and T-Tests Comparing Group Differences.

	Assigned (*N*=64) M (SD)	Treated (*N*=54) M (SD)
Social Competence	10.95 (1.06)	10.93 (1.08)
Age		
Male	26	25
Female	38	29
Referred	10	8
Grade		
5th	30	26
6th	21	18
7th	5	3
8th	8	7

Measures

As part of the screening and pretest measurement process, parents were asked to complete the parent report form of the Devereux Student Strength Assessment (DESSA), a measure of child social competence (LeBuffe *et al.*, 2009). The DESSA is a behavior rating scale that assesses the social- emotional competencies that serve as protective factors for children in kindergarten through eighth grade. In keeping with the prevention perspective of the program designers, an assessment tool was chosen that was strength- based and did not assess risk factors or maladaptive behaviors. The measure contains 72-items asking respondents to indicate how often various child behaviors occurred based on a 5 point Likert scale ranging from 0 (*never*), 1 (*rarely*), 2 (*occasionally*), 3 (*frequently*), to 4 (*very frequently*) over the past 4 weeks. The DESSA features a social competence composite score derived from 8 subscales. Examples of items and subscales include Optimistic Thinking (α=0.87; e.g. *How often did the child say good things about the future?*), Self-Management (α=0.86; e.g. *How often did the child focus on a task despite a problem or distraction?*), Goal- Directed Behavior (α=0.89; e.g. *How often did the child keep trying when unsuccessful?*), Self- Awareness (α=0.82; e.g. *How often did the child show an awareness of his/her personal strengths?*), Social-Awareness (α=0.81; e.g. *How often did the child respect another person's opinion?*), Personal Responsibility (α=0.87; e.g. *How often did the child prepare for school, activities, or upcoming events?*), Decision Making (α=0.91; e.g. *How often did the child accept responsibility for what she/he did?*), and Relationship Skills (α=0.93; e.g., *How often did the child greet a person in polite way?*). DESSA (LeBuffe *et al.*, 2009) raw scores were converted to T-scores using the DESSA coding procedure.

T-scores were used as dependent variables to analyze group differences and estimate regression models. T-scores were also used to categorize children into 3 groups ranging from low to high (0–40=*needs instruction*,

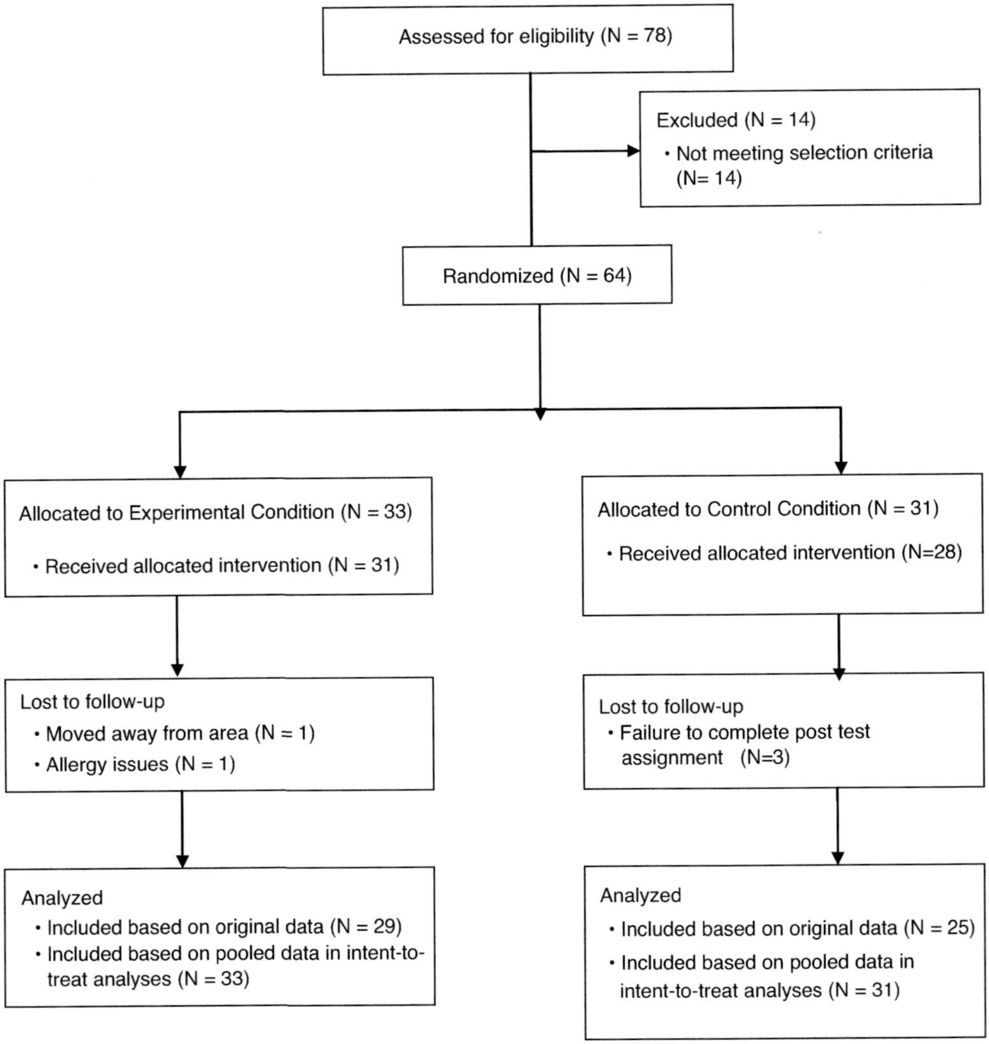

Fig. 2.1. Flow diagram describing sample enrollment, selection, randomization, follow up and analyses.

41–59 = *typical,* and 60–72 = *strength)* to enable interpretation of causal changes for intervention and prevention purposes. The DESSA has excellent internal reliability and shows significant, moderate-to-high correlations with widely used measures with good psychometric properties (LeBuffe *et al.*, 2009), including the BERS–2 and the Behavior Assessment System for Children–2 (Reynolds and Kamphaus, 2004). Completion of the DESSA by parents occurred at the beginning ($\alpha_1 = 0.98$) and end ($\alpha_2 = 0.98$), of the first 11 week program, and parents of waitlisted children reported once more (posttest2) at the end of the second 11 week program ($\alpha_3 = 0.98$).

Results

Descriptive analyses

Descriptive analyses on observed data revealed that overall pretest social competence levels of children in the study group were in the normal range (M = 41.41; SD = 10.16). The distribution in the sample suggests that the goal of recruiting children with relatively lower levels of social competence was met, as approximately half of the study sample was classified as needing instruction in social competence ($N_{needs\ instruction} = 30$, $N_{typical} = 31$, $N_{strengths} = 3$). A series of one-way ANOVAs revealed no statistically significant differences in pretest levels of child social competence between children in the control and experimental group composites scores ($M_{control} = 40.71$ (9.55), Min = 28, Max = 69; $M_{experimental} = 42.06$ (10.81), Min = 28, Max = 72) or any subscale scores (Table 2.3). Also, there was no evidence to suggest that the distribution of social competence classification groups (i.e. needing instruction, typical or strength) was different (p=0.649) between the control ($N_{needs\ instruction} = 13$, $N_{typical} = 7$, $N_{strengths} = 1$) and experimental group control ($N_{needs\ instruction} = 17$, $N_{typical} = 14$, $N_{strengths} = 2$). There were significant differences in pretest child social competence by gender showing higher social competence scores for girls (p=0.010; $M_{boys} = 37.63$ (9.03); $M_{girls} = 44.16$ (10.16)) and referral status (p=0.009; $M_{referred} = 34.45$ (6.85); $M_{unreferred} = 43.08$ (10.14)), although boys and girls were equally likely to be referred (p=0.851).

To determine whether participation in the program significantly affected child social competence we conducted a series of intent-to-treat analyses. This is a strategy for the analysis of randomized controlled trials that prevents overestimation of treatment effects by comparing participants in the groups to which they were originally randomly assigned, regardless of whether they subsequently withdrew or deviated from the protocol (Fisher *et al.*, 1990).

Table 2.3. Experimental Effects of Equine Facilitated Learning Program on Child Social Competence.

Pretest	Control M T-scores (SD)	Experimental M T-scores (SD)	ANOVA p-value
Social Competence COMP	40.71 (9.55)	42.06 (10.81)	0.599
Personal Responsibility	42.31 (8.90)	48.52 (9.12)	0.441
Optimistic Thinking	43.54 (11.07)	47.72 (11.23)	0.957
Goal Directed Behavior	37.85 (8.77)	43.64 (8.60)	0.572
Social Awareness	43.88 (11.40)	48.00 (10.13)	0.873
Decision Making	42.85 (10.99)	50.00 (10.60)	0.864
Relationship Skills	43.92 (11.00)	49.25 (10.27)	0.830
Self Awareness	41.92 (10.40)	50.65 (9.07)	0.823
Self Management	41.65 (11.75)	51.07 (9.10)	0.168
Posttest	**Control T-Scores**	**Experimental T-Scores**	**ANOVA p-values**
Social Competence COMP	41.45 (10.71)	47.85 (10.41)	0.020[a]
Personal Responsibility	42.72 (8.60)	48.34 (8.87)	0.009[b]
Optimistic Thinking	43.59 (10.50)	47.14 (11.30)	0.317
Goal-Directed Behavior	37.89 (8.38)	43.39 (8.15)	0.005[b]
Social Awareness	44.28 (11.70)	48.30 (10.31)	0.227
Decision Making	43.93 (10.70)	49.38 (10.01)	0.039[a]
Relationship Skills	43.84 (10.39)	48.99 (9.90)	0.041[a]
Self Awareness	42.13 (11.04)	50.98 (8.87)	0.001[b]
Self Management	42.33 (11.56)	51.35 (8.84)	0.000[c]

[a] $p < 0.05$
[b] $p < 0.01$
[c] $p < 0.001$

To accommodate this approach, missing data on posttest assessments on social competence (i.e. subscale and composite T- scores of child social competence measures at posttest) for 10 participants were thus imputed using five imputations according to multiple imputation procedures (Rubin, 1987; Rubin, 1996). Doing so retains the full sample for calculating intention-to-treat estimates and, under appropriate assumptions, should yield unbiased estimates of the intention to treat. The posttest results described in the next section thus reflect pooled estimates across five imputed datasets that were generated.

Hypothesis 1

To test the first hypothesis - participation in the program will improve child social competence - researchers conducted a one-way ANOVA comparing child T-scores on social competence measured at posttest between experimental and control groups. The within-subjects variable is the posttest, and the between-subjects variable constitutes the condition (control vs experimental group). Results support the first hypothesis (Table 2.3) by demonstrating a significant treatment effect of program participation as illustrated by significant group differences between experimental and control groups' social competence composite scores (p=0.020). Based on calculations using pooled variances, this represents an effect size of d=0.61, which is a moderate effect (Cohen, 1988). Researchers explored which aspects of social competence appeared most significantly improved by conducting a series of ANOVAs comparing T- scores on all subscales by treatment group. Using Tukey's method to correct for multiple comparisons, results revealed that self- awareness (p=0.000) and self-management (p=0.001) had most significantly - and positively - responded to the program participation (Table 2.3).

Hypothesis 2

Next, we tested the second hypothesis which stated that program participation by waitlisted control group children will increase their levels of social competence. Using t-tests, we compared social competence scores of waitlisted children obtained *after* they completed the 11 week session to their own pretest scores obtained during program screening. We found that waitlisted children in the control group who completed the program at a later date demonstrated significantly higher posttest levels of social competence (M=43.71; SD=11.07) after program completion compared to their own pretest scores (p=0.000).

Hypothesis 3

Testing the third hypothesis- positive effects of program participation on posttest levels of child social competence are independent of children's pretest levels of social competence, gender, age, and referral status - required us to isolate unbiased estimates of the influence of program participation on child social competence using a longitudinal lagged regression approach. Child social competence at posttest was expressed as a function of program participation and child social competence at pretest. The same measure used to estimate child social competence at posttest was included as an additional Time-1 covariate in the model to reduce omitted (unmeasured), time-invariant differences in children that were present at the beginning of the program (Cain, 1975). To reduce the threat that measured individual differences in program effects on child social competence at posttest were related to social competence levels at pretest, important factors potentially related to social competence at both time points (e.g. gender, age, referral status) were also included in the model as covariates. Analyses were conducted using the observed data (Table 2.4, Model 1) and using imputed data for missing posttest variables, resulting in pooled estimates (Table 2.4, Model 2).

Results support the third hypothesis by showing that models derived from observed and imputed data both show robust effects of program participation on posttest social competence scores. Based on imputed data and pooled estimates, results reveal a significant program effect on children's social competence ($\beta = 2.31$; $p = 0.026$). Results are independent of contributions of children's pretest levels of social competence, which are also positive and significant ($\beta = 6.43$; $p = 0.000$). There were no significant main effects of children's referral status, age, or gender when pretest levels of social competence were controlled. Although no hypotheses were stated regarding follow-up analyses to examine differences in program effects by group status of social competence, we found that children in the experimental group were much more likely to be reclassified from *needing instruction* to *typical* (Table 2.5) in response to program participation ($p = 0.014$) compared to children in the control group, the majority of whom remained classified as *needing instruction*.

Discussion

The current study employed a randomized controlled experimental trial to examine the causal effects of participation in an 11 week equine facilitated learning program on child social competence of 5th-8th grade children. This study responds to a call in the literature of human animal interaction to move the field of animal assisted intervention out of its 'fringe' status and establish credibility by means of carefully controlled trials and valid efficacy studies (Kruger and Serpell, 2010). Results indicated support for our first hypothesis suggesting a moderate, positive effect ($d = 0.61$; $p = 0.020$) on the social competence of 5th-8th grade children in response to program participation. Positive effects were observed on various aspects of

Table 2.4. Lagged Dependent Variable Model Predicting Posttest Social Competence.

Variable	Unstandardized B	SE	Beta	p-value
Model 1 (Original Data)				
Constant	25.39	12.85		0.054
Whether Experimental	4.84	0.220	2.015	0.021[a]
Social Competence Time 1	0.686	0.108	0.666	0.000[c]
Child Gender	−0.111	2.15	−0.041	0.673
Child Age	−1.00	0.959	−1.00	0.301
Whether Referred	−2.22	3.00	−0.073	0.463
Model 2 (Imputed Data/Pooled Estimates)				
Constant	27.594	12.55		0.030
Whether Experimental	4.97	2.15	2.31	0.026[a]
Social Competence Time 1	0.662	0.103	6.43	0.000[c]
Child Gender	−1.38	2.07	−0.666	0.507
Child Age	−1.08	0.933	−1.16	0.249
Whether Referred	−2.36	2.63	−0.898	0.370

[a]$p < 0.05$
[b]$p < 0.01$
[c]$p < 0.001$

Table 2.5. Causal Change in Social Competence Category by Treatment Group.

	Pretest		Posttest	
Group	Control	Experimental	Control	Experimental
Needs Instruction	13	17	17	7
Typical	17	14	12	20
Strength	1	2	2	6

social competence, including improvements in children's self-awareness, self- management, personal responsibility, decision making, goal-directed behavior, and relationship skills. In addition, children who were originally in the waitlisted control group showed similar gains in social competence after they completed the program, suggesting support for the second hypothesis, which predicted significant within-subject improvements for waitlisted children. Support for the third hypothesis demonstrated that treatment effects were independent of children's pretest social competence levels, referral status, gender, and age.

The results are important for the following reasons. First, the fact that participation in an equine facilitated program conducted as an after-school program effectively enhanced children's social competence is exciting, since research indicates that promotion of personal and social skills in after-school settings are also known to improve youths' feelings of self-confidence and self-esteem, school bonding (positive feelings and attitudes toward school), positive social behaviors, school grades, and achievement test scores (Durlak and Weissberg, 2007). As such, equine facilitated learning may thus have produced multiple benefits that pertain to youths' personal, social, and academic lives. Second, these results represent a novel contribution to the literature on equine assisted intervention, as there is no prior work that has employed an experimental design to examine *causal* effects of equine facilitated learning programs on child social competence. Third, results echo findings from prior correlational, anecdotal, and case study evidence, which suggest significant positive associations between participation in equine facilitated programs and various aspects of adjustment and wellbeing (Brown and Alexander, 1991; Katcher and Wilkins, 1994; Bowers and MacDonald, 2001; Gatty, 2001; Kaiser *et al.*, 2004; Macauley and Guierrez, 2004; Chandler, 2005; Klontz *et al.*, 2007: Vidrine *et al.*, 2002; Mann and Williams (2002); MacDonald and Cappo (2003); MacDonald (2004); Trotter (2005); Schultz *et al.* (2007); Trotter *et al.* (2008)). Fourth, the results give greater credence to the claim of therapeutic horsemanship professionals, participants, and parents who have reported and experienced significant positive effects of equine facilitated activities firsthand. Faced with skepticism about the efficacy of equine facilitated programs by potential funders and third party payers, therapeutic professionals and clients can now point to causal evidence. This may not only increase the public's confidence in equine programs' ability to positively affect child development, but also translate into increased structural support to increase accessibility to such programs. Equine facilitated learning programs may serve as viable alternatives to after-school programs focusing on athletic or academic achievement, to provide after-school opportunities to children with different interests and needs. Last, the results directly inform research questions for future research, particularly research focused on understanding of the mechanisms that may underlie these positive effects.

Interpretation of results

Although this study did not examine the pathways referred to in the introduction, we discuss several processes we think are important to consider in obtaining positive program effects. First, while we believe that equines possess key characteristics (e.g. prey animal, herd animal) that influence the behavior of humans interacting with them, we do not believe that equines' innate characteristics *alone* fostered gains in social competence. Program facilitators helped children recognize analogies between horse-human interactions to those between humans in various social contexts (e.g. school, family). For example, during one of the horse activities, children were taught to use gradually increasing amount of finger pressure on the horse's hair, skin, muscle or bone as a way of 'asking' the horse to move its front or hind quarters. When the child started the activity using 'muscle' instead of 'hair', facilitators asked the child to reflect on whether he or she would use the analogous behavior - shove or push - to 'ask' another person to move out of the way, or if he or she may have used a different communication tool to express the request. By engaging children in practice and reflection within the context of a horsemanship activity, children may have been more open to reflect on their behavior during interactions with non-equine partners.

Second, embedded in lessons targeting children's goal-directed behavior and social skills were several activities designed to reduce 'stress.' The rationale for including such activities was that children who experience stress - through exposure to stressful life events or by perceiving common events as stressful - tend to have distorted perceptions of the degree of threat present in social situations. They also lack effective coping skills to manage their internal distress. In addition to affecting peer relations, these tendencies lead children to be more susceptible to developing some form of psychopathology. Thus, social cognitions, especially regarding self-efficacy, personal agency, self-control, and cognitive distortion, as well as tools to help children regulate their physiological arousal and emotions, may have enhanced social competence through stress reduction. Examples of program activities that targeted stress reduction included silent grooming, rhythmic stroking of the horse, encouraging slow breathing, muscle relaxation and rhythmicity during mounted activity, as well as encouraging optimistic thinking, reflecting on emotions experienced during the lesson, and utilizing cognitive, behavioral, and physiological coping strategies towards adaptive self- management.

Third, apart from equine facilitation, children may have benefitted from experiencing an optimal social niche that enabled them to improve on social skills important to success in other, more complex social settings. For example, large, unstructured peer group settings (e.g. recess) can be particularly difficult situations for children with poorer social competencies, who may have difficulty initiating and maintaining normative peer interactions. The smaller, more structured peer interaction setting may have provided an appropriate and positive learning environment as equine team members (horse specialists and facilitators) provided support to guide positive peer interactions, friendship development, and modeling of social competence. In addition, given that the sample was not exclusively targeting referred children, program participants experienced normative peers as pro-social models. Participation in the program may have provided less competent children with opportunities to interact with more competent social partners, leading to positive social interactions with peers. These serve as models for subsequent interactions outside the context of the intervention. This small, structured format may have also counteracted the alienation experienced by children who are receiving special education services or targeted interventions that may make them vulnerable to labeling and the influence of less competent peers. In sum, we suggest that engaging in equine facilitated activities enhance social competence when conducted in an appropriate learning environment that provides optimal opportunities for promoting self-efficacy, performance accomplishment, and personal agency (Kruger and Serpell, 2010).

Strengths and limitations

There are several strengths and limitations to consider when interpreting results of this study. The main limitation of the study is that the results are based on parents' perception of child social competence. While parents tend to be appropriate reporters of children's social competencies as they intimately *know* the child, they are, by definition, not blind to their child's treatment status. As such, it is possible that the results reflect parents' positive expectations about the efficacy of the program. This limitation would have been of less concern if teacher reports on child social competence had been included, assuming that researchers would have been more successful in keeping the child's treatment status blind to teachers. Teacher reports were not collected as several teachers expressed concern about the fact that measures were collected at the beginning of the school year, when teachers had not yet interacted much with the children.

Second, although participation in the intervention was offered to all 5th-8th grade students in the geographical area, students and parents had a choice to enroll their child in the program. As such, the study cannot address whether this equine facilitated intervention would be effective for children who may have been less willing to interact with horses, or whose parents were less supportive of their child's program participation. While we informally observed individual differences in the extent to which children - and parents - expressed enthusiasm about the efficacy of equine assisted activities, we did not measure or model these differences. Hence, this limitation needs to be considered when determining whether equine facilitated growth and learning programs can be used as universal prevention programs. Similarly, because of the relatively short-term nature of the intervention and lack of longitudinal data at this time, researchers could not ascertain whether the program was of sufficient duration or intensity to potentially alter developmental pathways of children at significant risk for developing mental health problems. Last, equine-facilitated activities involve a great deal of human-human interaction, which the participants in the control group did not receive. The design of the study would have benefitted from an additional control group to compare effects on social competence from the equine facilitated activities to the effects on social competence from interventions that do not involve interactions with equines.

One of the main strengths of this study and its findings is that the results are based on an experimental design, which allows researchers to make causal inferences. This constitutes a novel contribution to research on human-animal interaction and child development generally, and equine facilitated interventions specifically. Second, results yielded a moderate effect size that was statistically significant despite the small sample size and being based on a calculation method using pooled variances. In addition, results were robust when children's pretest levels of social competence were controlled, along with other potential confounding variables. Last, although the sample was relatively small, participants were drawn from ten public and private schools across two different towns, and featured referred and non-referred children of various ages, somewhat alleviating concerns about the generalizability of the sample.

Conclusion

Given the causal design of this study, it provides the broader field of therapeutic horsemanship professionals with scientific evidence that can be used to generate support for expanding innovative, evidence-based approaches through equine assisted activities. That said, observed program effects are specific to the program under study, which is an established, equine facilitated learning program with a structured curriculum and a team of experienced, PATH, Intl. certified instructors, horse specialists, therapists, and facilitators. Until more is known about the mechanisms underlying its observed effects, prevention scientists should pair enthusiasm with caution before encouraging scaling up equine facilitated programs in school and community settings to serve as targeted prevention programs. In fact, future research should attempt to elucidate which components of equine facilitated learning are central to program success. For example, is it grooming horses, riding horses, or being in the presence of horses or other children that drove program effects? Or, was it the cognitive- behavior learning processes practiced during natural horsemanship activities that uniquely enhanced children's social competencies? Similarly, to what extent does the equine component of this type of program infer benefits above and beyond evidence-based prevention focusing on human to human interaction? Last, future research should examine effects of equine assisted interventions on other developmental outcomes.

Notes

1 PATH to Success was designed by Sue Jacobson, coordinator of People-Pet Partnership at the College of Veterinary Medicine and Dr Phyllis Erdman, Associate Dean of the College of Education, both at Washington State University.

2 The professional association was known as the North American Riding for the Handicapped Association (NARHA) until its renaming in the Spring of 2011.

3 Pressure refers to a directly physical or implied signal to the horse with fingers/hand, or leg until the desired behavior or the equine is fulfilled (e.g., moving away from the pressure).

[4] Driving refers to signaling the horse to move forward by pointing toward the desired space, and applying pressure behind the equine's shoulder to 'drive' them forward of the equestrian.
[5] Desensitizing refers to exercises to condition the equine to ignore a stimulus or object, thus avoiding 'spooking.'
[6] Horse massage refers to an exercise taught to participants in which they stroke the horse with slow, rhythmic strokes down the equines' body to 'relax' the equine before riding.
[7] Children referred by school counselors were children who were receiving school counseling services for academic and/or behavioral adjustment issues, or those whose parents had consulted with school counseling staff about the presence of stress in the home.
[8] Using a slide presentation, parents were given information about the consent and assent process, along with hard copies of both forms, and a typed script of the assent procedures, which described study procedures and children's rights using developmentally appropriate language (i.e., children 11 and younger, children 12 and older). The Principal Investigator (PI) assented all child participants in person.

Acknowledgements

The project described was supported by Grant Number 1R03HD066590-01, awarded to the first author, Dr. Patricia Pendry, from the Eunice Kennedy Shriver National Institute of Child Health & Human Development and Mars-Waltham. The content is solely the responsibility of the authors and does not necessarily represent the official views of the Eunice Kennedy Shriver National Institute of Child Health & Human Development, the National Institutes of Health, or Mars-Waltham. We acknowledge Sue Jacobson, program coordinator of People-Pet Partnership at the College of Veterinary Medicine and Dr. Phyllis Erdman, Associate Dean of the College of Education at Washington State University, who together designed the PATH to Success program. We also acknowledge Dan Bayly and Dr. Leanne Parker for providing counseling expertise and facilitation supervision, and PATH to Success Study volunteers, research assistants, participants, their parents and teachers.

References

Bandura, A. (1986) *Social foundations of thought and action: A social cognitive theory*. Prentice Hall, Englewood Cliffs, N.J.

Bowers, M.J. and MacDonald, P.M. (2001) The effectiveness of equine-facilitate psychotherapy with at-risk adolescents: A pilot study. *Journal of Psychology and Behavioral Sciences* 15, 62–76.

Brickel, C.M. (1982) Pet-facilitated psychotherapy: A theoretical explanation via attention shifts. *Psychological Reports* 50, 71–74. DOI: 10.2466/pr0.1982.50.1.71.

Brown, L. and Alexander, J. (1991) *Self Esteem Index examiner's manual*. Proed, Austin, TX.

Budiansky, S. (1997) *The nature of horses: Exploring equine evolution, intelligence, and behavior*. The Free Press, New York.

Cain, G. (1975) Regression and selection models to improve nonexperimental comparisons. In: Bennett, C.A. and Lumsdaine, A.A. (eds) *Evaluation and Experiment: Some Critical Issues in Assessing Social Programs*. Academic Press, New York.

Chandler, C. (2005) *Animal assisted therapy in counseling*. Routledge, New York, NY.

Cohen, J. (1988) *Statistical power analysis for the behavioral sciences*, 2nd edition. Erlbaum, Hillsdale, NJ.

Dodge, K.A. and Somberg, D.R. (1987) Hostile attributional biases among aggressive boys are exacerbated under conditions of threats to the self. *Child Development* 58, 213–224. DOI: 10.2307/1130303.

Durlak, J.A. and Weissberg, R.P. (2007) Meta-analysis of youth intervention programs. Collaborative for Academic, Social and Emotional Learning. Available at: http://casel.org/publications/the-impact-of-after-school-programs-that-promote-personal-and-social-skills (accessed 24 April 2012).

Fisher, L.D., Dixon, D.O., Herson, J., Frankowski, R.K., Hearon, M.S. *et al.* (1990) Intention to treat in clinical trials. In: Pearce, K.E. (ed.) *Statistical Issues in Drug Research and Development*. Marcel Dekker, New York, pp. 331–350.

Gatty, C.M. (2001) Psychosocial impact of therapeutic riding: A pilot study. *Equine- Facilitated Mental Health Association* 5(2), 8.

Greenberg, M.T., Domitrovic, C. and Bumbarger, B. (2000) Preventing mental disorders in school-age children: A review of the effectiveness of prevention programs. Retrieved from Pennsylvania State University, College of Health and Human Development. Available at: http://prevention.psu.edu/pubs/documents/MentalDisordersfullreport.pdf

Greenwald, A. (2001) The effect of a therapeutic horsemanship program on emotionally disturbed boys. *Dissertation Abstracts International: Section B: The Sciences and Engineering* 62(2-B), 178.

Griffin, K.W., Epstein, J.A., Botvin, G.J. and Spoth, R.L. (2001) Social competence and substance use among rural youth: Mediating role of social benefit expectancies of use. *Journal of Youth and Adolescence* 30, 485–498. DOI: 0047-2891/01/0800-0485.

Huang, L., Stroul, B., Friedman, R., Mrazek, P., Friesen, B. *et al.* (2005) Transforming mental health care for children and their families. *American Psychologist* 60, 615–627. DOI: 10.1037/0003-066X.60.6.615.

Kaiser, L., Spence, L.J., Lavergne, A.G. and Vanden Bosch, K.L. (2004) Can a week of therapeutic riding make a difference?—A pilot study. *Anthrozoos* 17, 63–72. DOI: 10.2752/089279304786991918.

Katcher, A.H. and Wilkins, G.G. (1994) The use of animal assisted therapy and education with attention-deficit hyperactive and conduct disorders. *Interactions* 12(4), 1–5. DOI: 10.1145/1070960.1070962.

Klontz, B.T., Bivens, A., Leinart, D. and Klontz, T. (2007) The effectiveness of equine-assisted experiential therapy: Results of an open clinical trial. *Society & Animals* 15, 257–267. DOI: 10.1163/156853007X217195.

Kruger, K.A. and Serpell, J.A. (2010) Animal-assisted interventions in mental health: Definitions and theoretical foundations. In: Fine, A. (ed.) *Handbook on Animal-Assisted Therapy*. Academic Press, New York.

LeBuffe, P.A., Shapiro, V.B. and Naglieri, J.A. (2009) *The Devereux Student Strengths Assessment (DESSA)*. Kaplan Press, Lewisville, NC.

Lemon, B.W., Bengston, V.L. and Peterson, J.A. (1972) An exploration of the activity theory of aging: Activity types and life satisfaction among in-movers to a retirement community. *Journal of Gerontology* 27, 511–523. DOI: 10.1093/geronj/27.4.511.

Macauley, B. and Guierrez, K. (2004) The effectiveness of hippotherapy for children with language-learning disabilities. *Communication Disorders Quarterly* 24, 205–217. DOI: 10.1177/15257401040250040501.

MacDonald, P.M. (2004) *The effects of equine-facilitated therapy with at-risk adolescents: A summary of empirical research across multiple centers and programs*. The Center for the Interaction of Animals and Society (CIAS). Philadelphia: University of Pennsylvania School of Veterinary Medicine. Available at: www2.vet.upenn.edu/research/centers/cias/pdf/CIAS_AAI_white_paper.pdf

MacDonald, P.M. and Cappo, J. (2003) Equine-facilitated therapy with '"at-risk"' youth: Does it work? *Strides* 9, 30–31. DOI: 10.1007/s10566-006-9031-x.

Mann, D. and Williams, D. (2002) Equine assisted family therapy for high-risk youth: Defining a model of treatment and measuring effectiveness. Unpublished manuscript.

McGreevy, P.D. (2007) The advent of equitation science. *The Veterinary Journal* 174, 492–500. DOI: 10.1016/j.tvjl.2006.09.008.

McGreevy, P.D. and McLean, A.N. (2007) Roles of learning theory and ethology in equitation. *Journal of Veterinary Behavior: Clinical Applications and Research* 2, 108–118. DOI: 10.1016/j.jveb.2007.05.003.

McGreevy, P.D. and McLean, A.N. (2010) *Equestrian Science*. Blackwell Publishing, United Kingdom.

O'Connell, M.E., Boat, T. and Warner, K.E. (2009) *Preventing Mental, Emotional, and Behavioral Disorders Among Young People: Progress and Possibilities*. The National Academic Press, Washington, D.C.

Reynolds, C.R. and Kamphaus, R.W. (2004) *Behavior assessment system for children*, 2nd ed. American Guidance Service, Circle Pines, MN.

Rockhill, C.M., Vander Stoep, A., McCauley, E. and Katon, W.J. (2009) Social competence and social support as mediators between comorbid depressive and conduct problems and functional outcomes in middle school children. *Journal of Adolescence* 32, 535–553. DOI: 10.1016/j.adolescence.2008.06.011.

Rubin, D.B. (1987) *Multiple Imputation for Nonresponse in Surveys*. J. Wiley & Sons, New York.

Rubin, D.B. (1996) Multiple imputation after 18+ year. *Journal of the American Statistical Association* 91, 473–489.

Schultz, P.N., Remick-Barlow, G.A. and Robbins, L. (2007) Equine-assisted psychotherapy: A mental health promotion/intervention modality for children who have experienced intra-family violence. *Health & Social Care in the Community* 15, 265–271. DOI: 10.1111/j.1365-2524.2006.00684.x.

Serpell, J.A. (2000) Creatures of the unconscious: Companion animals as mediators. In: Podberscek, A.L., Paul, E.S. and Serpell, J.A. (eds) *Companion Animals and Us*. Cambridge University Press, Cambridge, pp. 108–121.

Shonkoff, J. and Phillips, D. (2000) *From neurons to neighborhoods: The science of early childhood development*. National Academy Press, Washington, DC.

Sørlie, M.A. and Nordahl, T. (1998) *Problematferd i skolen: Hovedfunn, forklaringer og pedagogiske implikasjoner: hovedrapport fra forskningsprosjektet "Skole og samspillsvansker" [Problem behavior in school: Main findings, explanations and educational implications Main report from the research project "School and difficult social interactions]*. NOVA, Oslo.

Sørlie, M., Hagen, K.A. and Ogden, T. (2008) Social competence and antisocial behavior: Continuity and distinctiveness across early adolescence. *Journal of Research on Adolescence* 18, 121–144. DOI: 10.1111/j.1532-7795.2008.00553.x.

Trotter, K.S. (2005) The efficacy of equine assisted group counseling with at-risk children and adolescents. *Dissertation*.

Trotter, K.S., Chandler, C.K., Goodwin-Bond, D. and Casey, J. (2008) A comparative study of the efficacy of group equine assisted counseling with at-risk children and adolescents. *Journal of Creativity in Mental Health* 3, 254–284. DOI: 10.1080/15401380802356880.

Vidrine, M., Owen-Smith, P. and Faulkner, P. (2002) Equine-facilitated group psychotherapy: Applications for therapeutic vaulting. Issues in mental health nursing. *Special Issue: Child and Adolescent Mental Health* 23, 587–603. DOI: 10.1080/01612840290052730.

Wentzel, K.R. (1991) Relations between social competence and academic achievement in early adolescence. *Child Development* 62, 1066–1078. DOI: 10.1111/j.1467-8624.1991.tb01589.x.

3 Randomized Trial Examines Effects of Equine Facilitated Learning on Adolescents' Basal Cortisol Levels

Patricia Pendry*, Annelise N. Smith and Stephanie M. Roeter
Department of Human Development, Washington State University

Abstract

Although equine facilitated programs have gained in popularity over the last decade, virtually nothing is known about the causal effects of equine facilitated interventions on human development and well-being. Researchers conducted a randomized trial to determine the effects of an 11-week equine facilitated learning program on the activity of the Hypothalamic Pituitary Adrenal (HPA) axis of fifth through eighth graders through salivary cortisol levels. Children (N = 131) referred by school counselors and recruited from the community were randomly assigned to either an experimental (N = 53) or waitlisted condition (N = 60). Six samples of salivary cortisol were collected in participants' own home over two consecutive days at pre-test, and another set of six samples were collected at post-test in both groups of children. Children in the experimental group who participated in a series of once-weekly, 90-minute sessions of equine facilitated activities had lower afternoon cortisol levels ($F(1, 112) = 8.56, p = 0.017; d = 0.48$) and lower total cortisol concentration per waking hour ($F(1, 112) = 11.12, p = 0.017, d = 0.46$) at post-test, compared to waitlisted children. Multivariate regression analyses showed that program effects were independent from baseline levels of child cortisol, child gender, age, and referral status.

Keywords: randomized trial, adolescent, basal cortisol levels, equine facilitated intervention, HPA-axis

Introduction

Since the practice of incorporating Human-Animal Interaction (HAI) into therapeutic and educational settings has grown in popularity (Nimer and Lundahl, 2007), there has been a call to conduct evidence-based research on its effects on child physical and mental health, well-being, and development (Esposito *et al.*, 2011). Along with a call for clinical trials and examination of treatment effects in various populations, the importance of examining biobehavioral and physiological responses to HAI were emphasized (Esposito *et al.*, 2011). Incorporating these themes, the present study reports on a randomized trial on the effects of an 11-week, equine facilitated learning (EFL)[1] program on activity of the Hypothalamic Pituitary Adrenal (HPA) axis of adolescents, measured through salivary cortisol. In particular, we focus on examining the effects of EFL on adolescents' basal cortisol levels, which in healthy adolescents are typically highest in the morning soon after waking, drop rapidly in the first few hours

*Corresponding author: ppendry@wsu.edu

after waking, and then continue to drop more slowly, reaching a low point around midnight (Klimes-Dougan *et al.*, 2001). The main goal was to examine whether engagement in an 11-week EFL program strengthens adaptive patterns of cortisol production of adolescents, with the goal of preventing development of physical and mental health problems associated with deviations in basal cortisol patterns, such as higher cortisol levels in the afternoon and a higher concentration of cortisol levels per waking hour (Ruttle *et al.*, 2010; Zeiders *et al.*, 2011). This investigation thus responds to a recent call in the developmental literature to design and evaluate preventive interventions aimed at modifying basal activity of the HPA- axis (Cicchetti and Gunnar, 2008) and builds on the notion proposed in a recent HAI review that dampening of physiological stress parameters may represent a core mechanism in explaining positive effects of HAI (Beetz *et al.*, 2012).

Rationale for Strengthening Basal Cortisol Patterns in Adolescents

Strengthening adolescents' basal cortisol levels - optimizing the development of adaptive daily patterns of production of cortisol - through participation in EFL was the main goal of this study for several reasons. First, the transition to adolescence is thought to represent a developmental period during which the brain might be especially sensitive to exposure to elevated levels of basal cortisol. Based on the observed links between the emergence of psychopathology and periods of heightened stress - and elevations in cortisol – heightened HPA activity is hypothesized to play a role in the development of stress-related disorders in adolescence (Lupien *et al.*, 2009). Second, given research showing that increased HPA activity occurs as a result of puberty-associated neurobiological changes in systems underlying stress and emotional functioning (Stroud *et al.*, 2009), the onset of puberty in and of itself may be a transition point that represents a period of increased risks in adolescence. This may be problematic given that this developmental period is already associated with psychosocial factors of vulnerability. Third, researchers have posited that lowered HPA activity (especially later in the day) in a relatively lower-risk, normally functioning sample of early adolescents may be considered more adaptive and thus constitute a protective influence against the development of later psychopathology (Lupien *et al.*, 2009). Although associations between lower concentrations of cortisol levels and lower afternoon levels of cortisol and behavior and psychopathology are complex, evidence suggests that weaker diurnal patterns of cortisol are associated with various adjustment problems. For example, Tyrka *et al.* (2010) found that higher afternoon cortisol levels were linked to internalizing problems in a sample of 8–11 year-old boys. Furthermore, in a two-year follow-up, Tyrka *et al.* (2012) found that afternoon cortisol levels assessed at baseline were significant positive predictors of child depressive symptoms two years later. In addition, associations have also been found for average daily levels. Researchers (Scerbo and Kolko, 1994) found that elevated average child cortisol levels were positively related to child internalizing behavior, which has also been shown to relate to exclusion by peers at school (Peters *et al.*, 2011). In a sample of 9–14 year-olds, greater increases in average daily levels of cortisol measured across the span of a year predicted higher general and social anxiety in girls six months later at post-test (Schiefelbein and Susman, 2006). In sum, preventing or alleviating deviations from typical circadian patterns (e.g. high instead of low cortisol levels in the afternoon or evening, constant high or low levels of cortisol throughout the day) may have positive implications for various indices of adolescent physical and mental health.

Rationale for Reducing Basal Cortisol Levels Through Human Equine Interaction

Participation in an 11-week EFL program may strengthen adaptive patterns of adolescents' basal cortisol for several reasons. First, EFL incorporates several approaches (e.g. human-animal interaction; cognitive-behaviorally-based stress management; and physiological relaxation techniques) are thought to lower basal cortisol activity in adults. Although based on correlational designs, studies show promising within-person changes of cortisol levels in response to interactions with animals (Odendaal and Meinjes, 2003; Barker *et al.*, 2005; Barker *et al.*, 2010), relaxation training (Pawlow and Jones, 2005), cognitive- behavioral stress management (Cruess *et al.*, 2000; Gaab *et al.*, 2003), yoga (Curtis *et al.*, 2011), reiki and imagery (Bowden *et al.*, 2010), transcendental meditation (Kamei *et al.*, 2000), and massage (Field *et al.*, 2004). Second, although not specifically examined in the context of human-equine interaction, a recent review on the psychophysiological implications of human- animal interaction suggests that human-animal interaction, and possibly human-equine interaction, may modulate physiological stress parameters through their connection with the oxytocin system (Beetz *et al.*, 2012). For example, several investigators (Odendaal, 2000; Odendaal and Meinjes, 2003; Handlin *et al.*, 2011) have noted significant increases in oxytocin levels in human plasma levels after as little as three minutes of physically interacting (e.g. stroking) with a dog, with effects depending on the relationship quality between the human-dog pair. Similarly, physical interactions that 'tap into' caregiving and grooming behaviors have also been associated with increases in oxytocin in human to human contact (Carter, 2003; White-Traut *et al.*, 2009). This is relevant, as the EFL program under study explicitly engages participants in physical interactions with the horse (i.e. grooming, stroking, petting, touching) for the purpose of down-regulating their physiological arousal. For example, in addition to the regular ten-minute grooming activity children engaged in during each session, children were at times instructed to use their hands to stroke (e.g. massage) the horse slowly and rhythmically for ten minutes, while mindfully engaging in relaxation behaviors. These behaviors included breathing slowly and rhythmically to match their breathing to the breathing of the horse, refraining from any verbal communication with others, and being mindful about the experience of touching the horse and feeling its warmth. Since increases in oxytocin are associated with decreases in cortisol levels in humans and non-human animals (Legros *et al.*, 1988; Petersson *et al.*, 1999; Neumann *et al.*, 2000), engagement in this type of EFL activity - and the hypothesized increases in oxytocin - may thus lead to suppressed HPA activity as evidenced by reduced cortisol levels.

Last, given that cortisol levels are extremely sensitive to social stressors and supports, (Gunnar and Donzella, 2002), participation in EFL may effectively lower basal cortisol production through adaptations in participants' stress-relevant cognitive appraisals. Specifically, basal HPA activity is dependent not only on the nature and frequency of daily stressors encountered, but also on appraisal processes (Herman and Cullinan, 1997). These include both the individual's interpretation of stressors, as well as the assessment of the availability – and adequacy – of coping resources. One key component of EFL that may alter participants' stress-relevant cognitive appraisals is that participants are provided with countless opportunities to experience positive human-equine interactions in a supportive, non- judgmental environment. Combined with observed improvements in social competence (Pendry and Roeter, 2013), the development of a trust-based relationship with the horse, individual counselors, and other participants, EFL may thus effectively increase participants' performance accomplishment and perception of social support, which is likely to affect how stressors are perceived and, physiologically, responded to.

Research Questions and Hypothesis

Given that correlational studies have demonstrated promising alterations towards more adaptive basal cortisol patterns in adults (e.g. lower average cortisol, lower afternoon and evening levels), along with findings suggesting that HAI may increase oxytocin (associated with suppressing HPA activity), and improve stress-relevant cognitive appraisals, the following hypotheses guided our analyses. The first hypothesis predicted that children randomly assigned to participate in an 11-week equine facilitated learning program would on average have lower levels of average daily cortisol levels at post-test, compared to children assigned to a waitlisted condition. The second hypothesis predicted that children randomly assigned to participate in an 11-week equine facilitated learning program would have on average lower levels of afternoon and bedtime cortisol levels at post-test, compared to children assigned to a waitlisted condition.

Method

The program

The equine facilitated learning program, PATH to Success, was conducted at a Premier Accredited Center of PATH, Int'l in a university setting. The stated goal of the program is to enhance child social competence and reduce stress through an 11-week session of once- weekly, 90 min lessons of individual, team, and group-focused equine facilitated activities. Activities were based on principles of equitation science and natural horsemanship (McGreevy and McLean, 2010) featuring a combination of mounted and un-mounted activities and human- equine interactions, including observation of equine behavior, engagement in equine management (e.g. grooming), in-hand horsemanship, some riding, and personal and group reflection. The program was implemented by a team of certified instructors, professional counseling psychologists, graduate-level counseling students, and undergraduate students in child development, education, and animal science. Children in the experimental group were assigned to the same facilitation team for the entire 11-week session that included one equine, two child participants, an equine specialist, and a facilitator. Each day, a total of eight child participants - divided across four equine teams headed by four experienced facilitators - participated. The program was conducted four times a week, serving 32 children each week per 11-week session. The weekly lessons were conducted on weekday afternoons and included transportation of participants from school to the program site and back immediately following their regular school day. Children who were waitlisted participated after children in the experimental group completed the program. A description of weekly program objectives and a selection of sample activities are described in Table 3.1.

Procedures

The study was approved by Washington State's University Committee on Research Involving Human Subjects. Equine welfare of horses in the program was carefully monitored under the university's Institutional Animal Care and Use Committee (IACUC) regulations and campus veterinarian. Program participants were recruited through referral by school counselors and distribution of flyers and advertisements throughout schools and the community. Children referred by school counselors were those receiving counseling services for academic and/or behavioral adjustment issues or those whose parents had consulted with counseling staff about concerns over their child's exposure to school and/or home-based stress. Criteria for participation were 1) parents and children were proficient English speakers, 2) the child did not have physical or mental disabilities, and 3) children attended the fifth through eighth grade. Parents provided information about the adolescent's use of steroid-based medication, health and behavior, and child's exposure to school and/or family based stress. Although this did not occur, children could have been excluded from participation if a history of animal abuse or suicidal behavior had been reported. Parents and children were instructed and consented/ assented by the PI, and received materials and demonstrations on in-home salivary sampling procedures. Parents completed a standardized measure of child social competence (see measures) for which they were paid five dollars. Given that a limited number of program and study slots were available, children who were referred by school counselors, those with greater exposure to current or recent school or home-based stressors, and those with lower social competence were given priority during study selection - but not randomization - procedures.

Sample recruitment, screening, and assignment

Program staff recruited a total of 131 interested children and their families from ten different schools in a rural university town in the Pacific Northwest area of the United States, of which 113 were selected ($N_{boys} = 41$; $N_{girls} = 72$; $N_{referred} = 16$; $M_{age} = 11.35$ yrs; $M_{social\ competence} = 42.99$) for study participation. Using a randomization schedule, study participants were randomly

Table 3.1. Outline of Lesson Objectives by Week.

Week	Lesson Objective	Sample Activity
1	Basic safety: Meet horses and staff	Observing horse behavior and herd dynamics
2	Respect: Self, others and horses	Moving horses using 4 phases of direct or indirect pressure
3	Communication: Verbal and non-verbal	Leading horses, interpreting horse body language
4	Leadership: Assertive and aggressive cues	Driving activity using body language and phases
5	Trust: Coping with perceptions of stress	Riding and leading
6	Boundaries	Driving activity using indirect pressure
7	Overcoming challenges and building confidence	Desensitizing horses
8	Enhancing self-regulation and relaxation	Horse massage, riding
9	Prepare for parents/visitors day	Incorporating horsemanship skills for team challenge
10	Parents/visitors day	Participants 'teach' parents horsemanship skills
11	Program wrap up	Obstacle course, riding and reflection

assigned to an experimental group (N=53) or waitlisted control condition to start program participation 16 weeks later (N=60). Participants were predominantly White (81.6%), and of non-Latino or Hispanic ethnicity (88.8%), with the remaining children reporting across racial categories that included more than one race (8%), Asian (3.2%), American Indian or Alaska Native (1.6%), or unknown race (5.6%). There were no significant differences between participants' age (p=0.60), gender (p=0.77), referral status (p=0.18), grade level (p=0.75), or pre-test levels of social competence (p=0.31) by experimental condition. Sample enrollment, selection, randomization, follow up, and analyses counts are described in a flow diagram (Fig. 3.1). Of participants originally assigned, 84% completed data collection at post- test (N_{boys} = 36; N_{girls} = 58; $N_{referred}$ = 13; M_{age} = 11.34 yrs). There were no significant group differences by age (p=0.96), referral status (p=0.91), grade level (p=0.90), baseline social competence (p=0.93), gender (p=0.25), or amount of previous experience working with horses (p=0.94) between participants who completed the study and those lost to attrition.

Measures

Child social competence

Parents completed the Devereux Student Strength Assessment (DESSA; LeBuffe *et al.*, 2009) which contains 72 items asking respondents to indicate how often various child behaviors occurred over the past four weeks based on a 5-point Likert scale ranging from 0 (*never*) to 4 (*very frequently*). The DESSA composite score (α_1 = 0.98) was derived from 8 subscales including Optimistic Thinking (α = 0.87), Self-Management (α = 0.86), Goal-Directed Behaviour (α = 0.89), Self- Awareness (α = 0.82), Social-Awareness (α = 0.81), Personal Responsibility (α = 0.87), Decision Making (α = 0.91), and Relationship Skills (α = 0.93). Based on comparisons of mean levels of pre-test social competence in our sample to mean levels obtained in a 'normal' (M = 48.20; SD = 10.01) and 'emotionally disturbed' (M = 37.10; SD = 7.90) population sample (LeBuffe *et al.*, 2009), it appears that the goal of recruiting children with relatively lower levels of social competence was met.

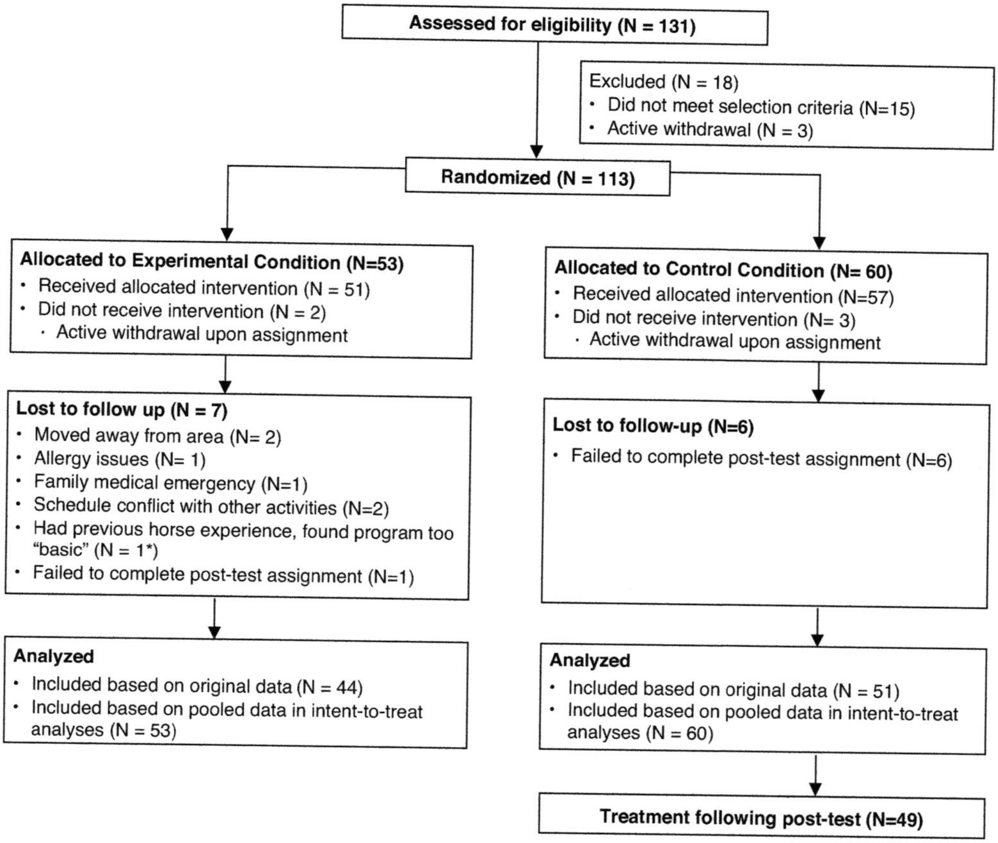

Fig. 3.1. Flow diagram describing sample enrollment, selection, randomization, follow up and analyses. *Participant withdrew after 4 weeks of participation but completed post-test assessments.

Salivary cortisol sampling

In-home cortisol sampling at pre-test was done by all participants over two consecutive days after participant consent/assent and selection, but before random assignment was communicated to participant, and before the start of the program. The in-home sampling paradigm conducted at pre-test was repeated at post-test i.e. after completion of the treatment period, twelve weeks later. During each sampling period, participants collected six salivary cortisol samples by spitting through a straw into a sterile 1.8 ml cryovial, three times a day, on each of two consecutive weekdays at prescribed events (immediately upon waking, immediately before bedtime) and at prescribed times (4:00 pm). Participants' recorded sampling time, use of steroid-based medication, timing of food/beverage intake, sleep and wake-times, as well as sampling confounds (e.g. sickness). Substantial efforts were made to impress upon participants the importance of compliance with the timing of sampling and reporting of sampling times. Payment of 5 dollars for each set of six samples was offered to each child upon completion at pre-test and again post-test. Compliance was high as 98% of children sampled provided all six samples as requested. Participants brought samples to the program site at the beginning and end of the program. Upon collection, samples were transferred by cooler to our laboratory-based freezer for storage at −20 degrees Celsius. Samples were analyzed by a laboratory specializing in the assaying of saliva by enzyme immunoassay. The test used for these assays had a range of sensitivity from 0.007 to 1.8 µg/dl, and average intra- and inter-assay coefficients of variation less than 3 and 7% respectively.

Data reduction of cortisol values

Cortisol parameters (area under the curve (AUC), morning, afternoon, bedtime) for each child were calculated at pre-test and six at post- test. To limit the influence of extreme cortisol values, values were first winzorised to 3 standard deviations above and below the mean. Six cortisol values across two consecutive days were used to calculate each child's average daily cortisol level (AUC) using the polygon method, at pre-test and post-test. The polygon AUC is a valuable measure, especially in establishing a link between cortisol levels and psychological functioning (Stewart and Seeman, 2000). Next, based on averaging day one and day two cortisol values for each time of day (e.g. morning, afternoon, bedtime), we calculated an average cortisol value for each parameter (e.g. morning, afternoon, bedtime). A natural logarithmic transformation for each cortisol parameter at each time point was used to reduce positive skewness of the data. Descriptives on cortisol levels for each two-day period and their associations for each time period are presented in Table 3.2.

Results

To determine whether participation in the program significantly strengthened child basal cortisol levels we conducted a series of intent-to- treat analyses (Fisher *et al.*, 1990), which prevents overestimation of treatment effects by comparing participants in the groups to which they were originally randomly assigned, regardless of whether they subsequently withdrew or deviated from the protocol. Missing data on post-test assessments were imputed using five imputations according to multiple imputation procedures (Rubin, 1996). The post- test results thus reflect pooled estimates across five imputed datasets that were generated.

A series of one-way ANOVAs revealed no statistically significant differences in pre-test levels of child cortisol levels by experimental condition for any of the cortisol parameters. To test the first hypothesis - participation in

Table 3.2. Descriptives by Day and Correlations Between Days of Original Untransformed Child Cortisol Data.

Variable	Day 1 M (SD)	Day 2 M (SD)	Correlation
Pre-test			
Morning Cortisol	0.290 (0.18)	0.322 (0.29)	0.273[a]
Afternoon Cortisol	0.085 (0.12)	0.116 (0.23)	0.149[b]
Bedtime Cortisol	0.065 (0.16)	0.049 (0.12)	0.540[a]
Morning Time	7.32 (1.00)	7.41 (1.17)	0.434[a]
Afternoon Time	16.40 (0.88)	16.31 (0.94)	0.247[b]
Bedtime Time	21.12 (0.90)	21.17 (0.92)	0.592[a]
Post-test			
Morning Cortisol	0.289 (0.16)	0.301 (0.16)	0.255[a]
Afternoon Cortisol	0.078 (0.06)	0.085 (0.08)	0.376[a]
Bedtime Cortisol	0.057 (0.09)	0.050 (0.08)	0.500[a]
Morning Time	7.61 (1.11)	7.48 (1.10)	0.565[a]
Afternoon Time	16.07 (1.25)	16.04 (1.13)	0.591[a]
Bedtime Time	21.41 (1.02)	21.33 (1.11)	0.473[a]

[a]Correlation is significant at the 0.01 level.
[b]Correlation is significant at the 0.05 level.

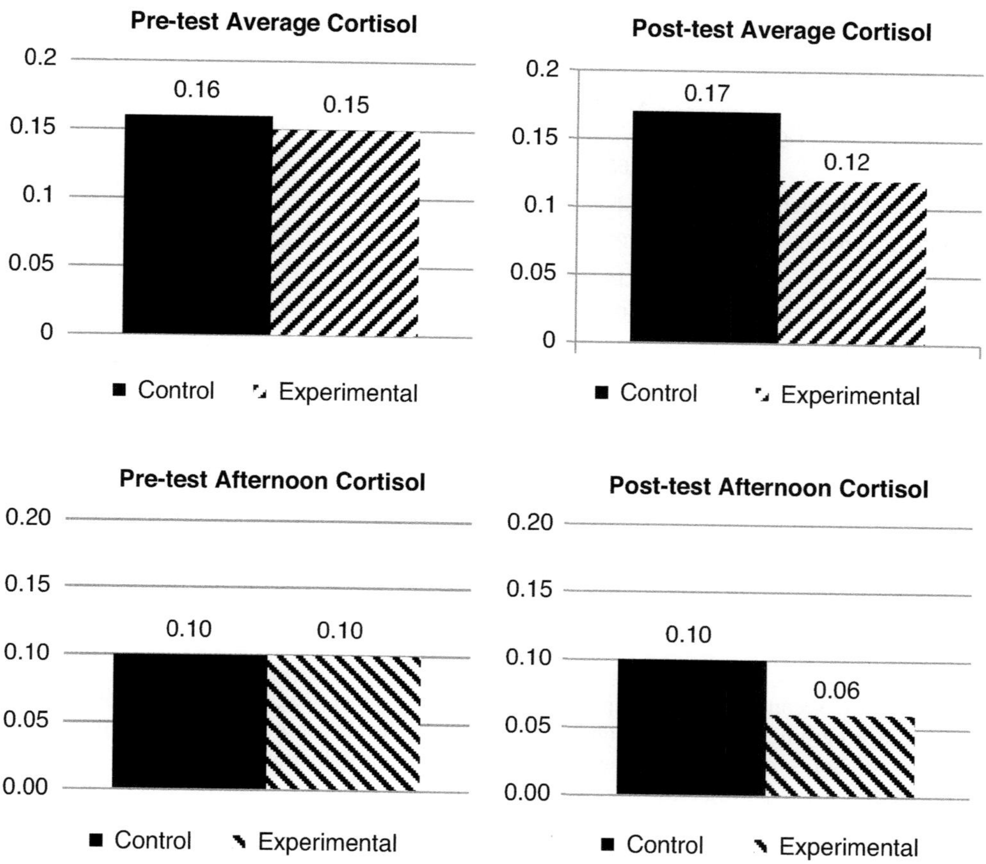

Fig. 3.2. Basal pattern of cortisol production demonstrating significant lower afternoon and daily average levels of cortisol at post-test for children assigned to experimental group.

the program would reduce average daily (AUC) levels of cortisol - researchers conducted a one-way ANOVA comparing average daily cortisol levels measured at post-test by experimental condition. Results support the first hypothesis indicating that children who participated in the intervention had significantly lower levels of cortisol per waking hour at post-test (F (1, 112) = 11.12, p=0.017), compared to children who were waitlisted (see Fig. 3.2). Based on calculations using pooled variances, the effect size of program participation on average daily cortisol level was 0.46, a moderate effect (Cohen, 1988).

Using a lagged regression approach, researchers next predicted adolescents' basal cortisol levels at post-test (Time −2), while including a pre-test (Time −1) covariate in the model. The aim of this approach was to examine program effects while considering omitted (unmeasured) time-invariant differences in children that may have been present at the beginning of the program. Important factors potentially related to adolescent basal cortisol levels at both time points (e.g. gender, age, referral status) were also included in the model as covariates. Results (Table 3.3, model 1) revealed a significant treatment effect on adolescents' average daily (AUC) cortisol levels ($\beta = -0.223$; p=0.019), which was independent of contributions of children's pre-test levels of average daily cortisol, which were positive and significant ($\beta = 0.263$; $p<0.006$). There were no significant main effects of children's referral status, age, or gender, nor did we observe that treatment effects were moderated by these child characteristics.

Table 3.3. Lagged Dependent Variable Models Predicting Post-test Basal Cortisol Levels by Experimental Condition.

Variable	Unstandardized B	SE	Beta	p-value
Average Daily (AUC) Cortisol				
Constant	−0.639	0.540		0.242
Experimental Condition	−0.209	0.087	−0.223	0.019
Pre-test AUC Cortisol	0.219	0.078	0.263	0.006
Child Gender	−0.143	0.095	−0.145	0.137
Child Age	−0.007	0.004	−0.159	0.103
Whether Referred	−0.054	0.132	−0.039	0.684
Afternoon Cortisol				
Constant	−1.586	0.784		0.046
Experimental Condition	0.314	0.119	−0.278	0.010
Pre-test Afternoon Cortisol	0.156	0.086	0.195	0.074
Child Gender	−0.158	1.33	−0.131	0.237
Child Age	−0.004	0.006	−0.075	0.486
Whether Referred	−0.065	0.176	−0.040	0.710

Testing the second hypothesis - participation in the program would reduce afternoon levels of cortisol - researchers conducted a one-way ANOVA comparing afternoon cortisol levels measured at post-test by experimental condition. Results support the second hypothesis (Fig. 3.2) by demonstrating a treatment effect of program participation as illustrated by significant group differences in afternoon cortisol levels, ($F(1, 112) = 8.56$, $p=0.017$) by experimental condition. Based on calculations using pooled variances, program participation resulted in an effect size of $d=0.48$, a moderate effect. Using a longitudinal lagged regression approach, researchers next predicted adolescents' afternoon basal cortisol levels at post-test (Time-2), while including a pre-test (Time-1) covariate in the model, as well as factors potentially related to adolescent afternoon cortisol levels at both time points. Results (Table 3.3, model 2) revealed a significant treatment effect on adolescents' afternoon cortisol level ($\beta=0.278$; $p=0.010$), which was independent of contributions of children's pre- test levels of afternoon cortisol ($\beta=0.195$; $p<0.074$). There were no significant main effects of children's referral status, age, or gender, nor did we observe that treatment effects were moderated by these child characteristics. We found no evidence suggesting that morning or bedtime cortisol levels were significantly affected by program participation.

Discussion

The current study reports on the results of a randomized trial examining causal effects of participation in an 11 week equine facilitated learning program on basal HPA activity of adolescents assessed through in-home sampling of salivary cortisol at pre-test (before program assignment) and post-test (12 weeks later). Results showed a moderate lowering effect on average daily cortisol levels (AUC) ($d=0.48$) and afternoon cortisol levels ($d=0.48$) of adolescents who participated in an 11 week EFL program, compared to those who were waitlisted. Findings also showed that program effects on afternoon and daily average cortisol were independent from baseline levels of cortisol, child gender, age, and referral status. Although associations between lower average daily cortisol/afternoon levels of cortisol and developmental psychopathology and/or adolescent health are by no means clearly understood, several researchers have suggested that lower basal cortisol levels in a relatively normal adolescent sample may constitute a protective influence against the development of psychopathology and health problems in certain populations (Lupien *et al.*, 2009). As such, this study suggests that EFL may be an effective approach to support positive development in adolescence. To our knowledge, this is the first randomized, controlled trial to demonstrate treatment effects of a prevention program on basal HPA activity in a sample of predominantly normal sample of adolescents. These results also represent the first causal evidence demonstrating significant effects of EFL on basal neuroendocrine functioning of adolescents. Our work complements findings based on recent randomized trials on EFL on social competence (Hauge *et al.*, 2013; Pendry and Roeter, 2013) and responds to calls in the growing field of Human- Animal Interaction (HAI) and Animal Assisted Intervention (AAI) (Esposito *et al.*, 2010) for evidence-based approaches examining effects on child development. This study also contributes to a relatively new and much needed body of literature of randomized trials on preventive interventions with adolescents, which have paid minimal attention to neurobiological and physiological systems in their evaluations of treatment efficacy (Cicchetti and Gunnar, 2008).

Human-animal interaction and cortisol

The observed treatment effects of equine facilitated learning on young adolescents' cortisol levels represent a novel finding that complements a small body of research linking interaction between humans and dogs to reduced momentary cortisol activity. For example, Barker *et al.* (2010) compared cortisol reactivity and recovery of therapy dog owners to non- therapy dog owners in the context of a laboratory-based stress task. In both groups, cortisol levels dropped below baseline after minute of interaction with the therapy dog, offsetting initial increases in cortisol levels in response to the stress task. Barker *et al.* (2005) also demonstrated that serum and salivary momentary cortisol levels of healthcare professionals were significantly lower after interacting with a therapy dog in a hospital setting. Lastly, interaction with a familiar dog in the home has been found to decrease human cortisol levels (Odendaal and Meinjes, 2003). While the use of therapy dogs has been shown to reduce cortisol levels in adults, interventions incorporating the use of therapy dogs with children have been less successful. For example, Kaminski *et al.* (2002) compared salivary cortisol levels of participants participating in child-life intervention sessions - one group incorporating the use of therapy dogs and the other without – and found that reductions in salivary cortisol in either group were not significant.

Implications for adolescent development

The implications of these results for adolescent development need to be interpreted thoughtfully. First, extremely little is known about the extent to which cortisol activity of young adolescents - and changes therein - may precede adaptive changes on the behavioral level in the long term. As such, it is not clear whether the observed changes in cortisol levels constitute an influential protective factor that promotes future adaptation and resilience (Masten, 2001) in the context of risk factors. Second, it is important to consider both the nature of results (lower AUC and afternoon cortisol), as well as the population in which they were obtained (i.e. a relatively normal sample of young adolescents, rather than a clinical sample of children characterized by a history of severe maltreatment, extreme rearing conditions, and/or existing developmental psychopathology). Specifically, it is important to consider that basal cortisol levels in our sample at pre-test were characterized by what are thought of as relatively normal patterns of cortisol production, although a wide range of individual differences in steepness of these patterns were observed. Effects of lower AUC and afternoon cortisol levels observed in our study should thus not be equated with altering the presence of maladaptive patterns of HPA activity, such as hyper or hypocortisolism. In sum, the results of this study reflect moderate treatment effects of equine facilitated learning on average daily (AUC) and afternoon cortisol levels within the normal range of basal HPA activity.

Limitations and implications

Future studies examining the effects of preventive equine facilitated learning programs on adolescent basal HPA activity will benefit from addressing the following limitations of this study. Since the program involved a significant amount of human interaction, the design of the study would have benefitted significantly from an additional control group to compare effects of equine facilitated activities to the effects of the intervention components that did not involve interactions with equines. Also, the use of additional control groups could have facilitated an examination of the role of particular types of equine interactions thought to be central to the intervention's success. For example, is it grooming horses, riding horses, or engaging in natural horsemanship activities that affect child HPA activity? What is the role of various other program features and activities possibly underlying the treatment effects (e.g. the role of cognitive behavioral approaches to affect cognitive appraisals about stress and coping vs relaxation activities aimed at reducing physiological arousal)? Second, although participation in the intervention was offered to all fifth through eighth grade students in the geographical area, students and parents had a choice to enroll their child in the program. As such, the study cannot address whether this equine facilitated intervention would be equally effective for children who may have been less willing or comfortable to interact with horses.

Our findings have several implications for practice. First, the experimental nature of these findings gives greater credence to the claim of therapeutic horsemanship professionals, participants, and parents who have reported significant positive effects of equine facilitated activities firsthand. Faced with skepticism about the efficacy of equine facilitated programs by potential funders and third party payers, therapeutic professionals and clients can now point to causal effects on physiological outcomes associated with current and future physical and mental health. Second, this study suggests that equine facilitated learning programs may serve as viable alternatives to after school programs focusing on athletic or academic achievement, providing after school opportunities to children with different interests and needs. Third, while this is only the first step in more fully understanding our ability to alter HPA activity of adolescents through the use of preventive interventions, the effect of a relatively short-term EFL program on diurnal HPA activity may serve as the impetus to conduct research on effects of EFL in other populations, including children with a history of stress exposure or those with diagnoses of stress- related adjustment issues, as well as closer examination of potential mediating variables, underlying program effects.

Note

[1] EFL combines mounted and unmounted, experiential learning, interaction with horses, ponies, miniature horses, donkeys, and mules, and counseling-based processing skills to increase adolescents' awareness and control of their emotions, cognitions, behaviors, and physiological arousal.

Acknowledgements

This work was supported through grant number 1R03HD066590-01 to the first author from the Eunice Kennedy Shriver National Institute of Child Health & Human Development and Mars-Waltham. The content is solely the responsibility of the authors and does not necessarily represent the official views of the Eunice Kennedy Shriver National Institute of Child Health & Human Development, the National Institutes of Health, or Mars-Waltham. We would like to acknowledge Sue Jacobson, program coordinator of People-Pet Partnership at the College of Veterinary Medicine at Washington State University and Dr. Phyllis Erdman, Associate Dean of the College of Education also at Washington State University, who implemented the PATH to Success program. We also acknowledge Dan Bayly and Dr. Leanne Parker for providing counseling expertise and facilitation supervision. We also thank all PATH to Success Study volunteers, research assistants, participants, their parents, and teachers.

References

Barker, S.B., Knisely, J.S., McCain, N.L. and Best, A.M. (2005) Measuring stress and immune response in healthcare professionals following interaction with a therapy dog: A pilot study. *Psychological Reports* 96, 713–729. DOI: 10.2466/pr0.96.3.713-729.

Barker, S.B., Knisely, J.S., McCain, N.L., Schubert, C.M. and Pandurangi, A.K. (2010) Exploratory study of stress- buffering response patterns from interaction with a therapy dog. *Anthrozoös* 23, 79–91. DOI: 10.2752/175303710X12627079939341.

Beetz, A., Uvnäs-Moberg, K., Julius, H. and Kotrschal, K. (2012) Psychosocial and psychophysiological effects of human- animal interactions: The possible role of oxytocin. *Frontiers in Psychology* 3, 1–15. DOI: 10.3389/fpsyg.2012.00234.

Bowden, D., Goddard, L. and Gruzelier, J. (2010) A randomised controlled single- blind trial of the effects of reiki and positive imagery on well-being and salivary cortisol. *Brain Research Bulletin* 81, 66–72. DOI: 10.1016/j.brainresbull.2009.10.002.

Carter, S.C. (2003) Developmental consequences of oxytocin. *Physiology & Behavior* 79, 383–397. DOI: 10.1016/S0031-9384(03)00151-3.

Cicchetti, D. and Gunnar, M.R. (2008) Integrating biological processes into the design and evaluation of preventive interventions. *Development and Psychopathology* 20, 737–743. DOI: 10.1017/S0954579408000357.

Cohen, J. (1988) *Statistical power analysis for the behavioral sciences*, 2nd ed. Erlbaum, Hillsdale, NJ.

Cruess, D.G., Antoni, M.H., Kumar, M. and Schneiderman, N. (2000) Reductions in salivary cortisol are associated with mood improvement during relaxation training among HIV seropositive men. *Journal of Behavioral Medicine* 23, 107–122. DOI: 10.1023/A:1005419917023.

Curtis, K., Osadchuk, A. and Katz, J. (2011) An eight-week yoga intervention is associated with improvements in pain, psychological functioning and mindfulness, and changes in cortisol levels in women with fibromyalgia. *Journal of Pain Research* 4, 189–201. DOI: 10.2147/JPR.S22761.

Esposito, L., McCune, S., Griffin, J.A. and Maholmes, V. (2011) Directions in human-animal interaction research: Child development, health, and therapeutic interventions. *Child Development Perspectives* 5(3), 205–211. DOI: 10.1111/j.1750-8606.2011.00175.x.

Esposito, L.E., McCardle, P., Maholmes, V., McCune, S. and Griffin, J.A. (2010) Introduction. In: McCardle, P., McCune, M., Griffin, J.A., Esposito, L. and Freund, L. (eds) *Animals in Our Lives: Human-Animal Interaction in Family, Community, & Therapeutic Settings*. Brookes, Baltimore, pp. 1–5.

Field, T., Hernandez-Reif, M., Diego, M., Feijo, L., Vera, Y. *et al.* (2004) Massage therapy by parents improves early growth and development. *Infant Behavior & Development* 27, 435–442. DOI: 10.1016/j.infbeh.2004.03.004.

Fisher, L.D., Dixon, D.O., Herson, J., Frankowski, R.K., Hearon, M.S. *et al.* (1990) Intention to treat in clinical trials. In: Pearce, K.E. (ed.) *Statistical Issues in Drug Research and Development*. Marcel Dekker, New York, pp. 331–350.

Gaab, J., Blättler, N., Menzi, T., Pabst, B., Stoyer, S. *et al.* (2003) Randomized controlled evaluation of the effects of cognitive-behavioral stress management on cortisol responses to acute stress in healthy subjects. *Psychoneuroendocrinology* 28, 767–779. DOI: 10.1016/S0306-4530(02)00069-0.

Gunnar, M.R. and Donzella, B. (2002) Social regulation of the cortisol levels in early human development. *Psychoneuroendocrinology* 27, 199–220. DOI: 10.1016/S0306-4530(01)00045-2.

Handlin, L., Hydbring-Sandberg, E., Nilsson, A., Ejdgeback, M., Jansson, A. *et al.* (2011) Short-term interaction between dogs and their owners—effects on oxytocin, cortisol, insulin and heart rate— an exploratory study. *Anthrozoös* 24, 301–316.

Hauge, H., Kvalem, I.L., Berget, B., Enders- Slegers, M. and Braastad, B.O. (2013) Equine-assisted activities and the impact on perceived social support, self-esteem and self-efficacy among adolescents—an intervention study. *International Journal of Adolescence and Youth*. DOI: 10.1080/02673843.2013.779587.

Herman, J. and Cullinan, W. (1997) Neurocircuitry of stress: Central control of the hypothalamo-pituitary-adrenocortical axis. *Trends in Neuroscience* 20, 78–84. DOI: 10.1016/S0166-2236(96)10069-2.

Kamei, T., Torui, Y., Kimura, H., Ohno, S., Kumano, H. *et al.* (2000) Decrease in serum cortisol during yoga exercise is correlated with alpha wave activation. *Perceptual and Motor Skills* 90, 1027–1032. DOI: 10.2466/PMS.90.3.1027-1032.

Kaminski, M., Pellino, T. and Wish, J. (2002) Play and pets: The physical and emotional impact of child-life and pet therapy on hospitalized children. *Children's Health Care* 31(4), 321–335.

Klimes-Dougan, B., Hastings, P.D., Granger, D.A., Usher, B.A. and Zahn-Waxler, C. (2001) Adrenocortical activity in at-risk and normally developing adolescents: Individual differences in salivary cortisol basal levels, diurnal variation, and responses to social challenges. *Developmental Psychopathology* 13, 695–719. DOI: 10.1017/S0954579401003157.

LeBuffe, P.A., Shapiro, V.B. and Naglieri, J.A. (2009) *The Devereux Student Strengths Assessment (DESSA)*. Kaplan Press, Lewisville, NC.

Legros, J.J., Chiodera, P. and Geenen, V. (1988) Inhibitory action of exogenous oxytocin on plasma cortisol in normal human subjects: Evidence of action at the adrenal gland. *Neuroendocrinology* 48, 204–206.

Lupien, S.J., McEwen, B.S., Gunnar, M.G. and Heim, C. (2009) Effects of stress throughout the lifespan on the brain, behaviour and cognition. *Nature Reviews Neuroscience* 10, 434–445. DOI: 10.1038/nrn2639.

Masten, A.S. (2001) Ordinary magic: Resilience processes in development. *American Psychologist* 56, 227–238. DOI: 10.1037/0003-066X.56.3.227.

McGreevy, P.D. and McLean, A.N. (2010) *Equestrian Science*. Blackwell Publishing, United Kingdom.

Neumann, I.D., Wigger, A., Torner, L., Holsboer, F. and Landgraf, R. (2000) Brain oxytocin inhibits basal and stress-induced activity of the hypothalamo-pituitary- adrenal axis in male and female rats: Partial action within the paraventricular nucleus. *Journal of Neuroendocrinology* 12, 235–243.

Nimer, J. and Lundahl, B. (2007) Animal- assisted therapy: A meta-analysis. *Anthrozoös* 20(3), 225–238. DOI: 10.2752/089279307X224773.

Odendaal, J.S. (2000) Animal assisted therapy—magic or medicine? *Journal of Psychosomatic Research* 49, 275–280.

Odendaal, J.S.J. and Meinjes, R. (2003) Neurophysiological correlates of affiliative behavior between humans and dogs. *Veterinary Journal* 165, 296–301. DOI: 10.1016/S1090-0233(02)00237-X.

Pawlow, L.A. and Jones, G.E. (2005) The impact of abbreviated progressive muscle relaxation on salivary cortisol and salivary immunoglobulin a (slga). *Applied Psychophysiology and Biofeedback* 30, 375–387. DOI: 10.1007/s10484-005-8423-2.

Pendry, P. and Roeter, S. (2013) Experimental trial demonstrates positive effects of equine facilitated learning on child social competence. *Human-Animal Interaction Bulletin* 1, 1–19.

Peters, E., Riksen-Walraven, J.M., de Cillessen, A.H. and Weerth, C. (2011) Peer rejection and HPA activity in middle childhood: Friendship makes a difference. *Child Development* 82(6), 1906–1920. DOI: 10.1111/j.1467-8624.2011.01647.x.

Petersson, J., Lundeberg, T. and Unvas-Moberg, K. (1999) Short-term increase and long-term decrease of blood pressure in response to oxytocin-potentiating effect of female steroid hormones. *Journal of Cardiovascular Pharmacology* 33, 102–108.

Rubin, D.B. (1996) Multiple imputation after 18+ year. *Journal of the American Statistical Association* 91, 473–489. DOI: 10.2307/2291635.

Ruttle, P.L., Shirtcliff, E.A., Serbin, L.A., Fisher, D.B., Stack, D.M. *et al.* (2010) Disentangling psychobiological mechanisms underlying internalizing and externalizing behaviors in youth: Longitudinal and concurrent associations with cortisol. *Hormones and Behavior* 59, 123–132. DOI: 10.1016/j.yhbeh.2010.10.015.

Scerbo, A.S. and Kolko, D.J. (1994) Salivary testosterone and cortisol in disruptive children: Relationship to aggressive, hyperactive, and internalizing behaviors. *Journal of the American Academy of Child & Adolescent Psychiatry* 33(8), 1174–1184.

Schiefelbein, V.L. and Susman, E.J. (2006) Cortisol levels and longitudinal cortisol change as predictors of anxiety in adolescents. *The Journal of Early Adolescence* 26(4), 397–413. DOI: 10.1177/0272431606291943.

Stewart, J. and Seeman, T. (2000) Salivary Cortisol Measurement. Available at: http://www.macses.ucsf.edu/research/allost atic/salivarycort.php

Stroud, L.R., Foster, E., Papandonatos, G.D., Handwerger, K., Granger, D.A. *et al.* (2009) Stress response and the adolescent transition: Performance versus peer rejection stressors. *Development and Psychopathology* 21, 47–68. DOI: 10.1017/S0954579409000042.

Tyrka, A.R., Kelly, M.M., Graber, J.A., DeRose, L., Lee, J.K. *et al.* (2010) Behavioral adjustment in a community sample of boys: Links with basal and stress-induced salivary cortisol concentrations. *Psychoneuroendocrinology* 35, 1167–1177. DOI: 10.1016/j.psyneuen.2010.02.002.

Tyrka, A.R., Lee, J.K., Graber, J.A., Clement, A.M., Kelly, M.M. *et al.* (2012) Neuroendocrine predictors of emotional and behavioral adjustment in boys: Longitudinal follow-up of a community sample. *Psychoneuroendocrinology* 37, 2042–2046. DOI: 10.1016/j.psyneuen.2012.04.004.

White-Traut, R.C., Schwertz, D., McFarlin, B. and Kogan, J. (2009) Salivary cortisol and behavioral state responses of healthy newborn infants to tactile-only and multisensory interventions. *Journal of Obstetric, Gynecologic, & Neonatal Nursing: Clinical Scholarship for the Care of Women, Childbearing Families, & Newborns* 38, 22–34. DOI: 10.1111/j.1552-6909.2008.00307.x.

Zeiders, K.H., Doane, L.D. and Adam, E.K. (2011) Reciprocal relations between objectively measured sleep patterns and diurnal cortisol rhythms in late adolescence. *Journal of Adolescent Health* 48, 566–571. DOI: 10.1016/j.jadohealth.2010.08.012.

4 A Transactional Analysis of Occupational Therapy in an Equine Environment for Youth with Autism Spectrum Disorder

B. Caitlin Peters[1]*, Christof Bentele[2] and Wendy Wood[3]
[1]Department of Animal Sciences, Colorado State University; [2]Department of Occupational Therapy, Colorado State University; [3]Departments of Animal Sciences and Occupational Therapy, Colorado State University

Abstract
Existing empirical investigations of occupational therapy services that incorporate horses for youth with autism spectrum disorder (ASD) are limited by a lack of standardized protocols. The current study sought to support manualization of one approach to occupational therapy in an equine environment for youth with ASD (OT[ee] HORSPLAY), by vividly describing the intervention from a transactional perspective. Direct, real-time videos plus field notes of OT[ee] HORSPLAY were collected and then subjected to qualitative content analysis and descriptive statistical analysis. Analytical methods followed Ricoeur's hermeneutic arc of naïve interpretation, explanation through structural analyses, and comprehensive description. Following this arc, we found many architectural spaces, sequences of occupational opportunities, and actions of practitioners and youth that fluidly shifted in relationship to one another. The results of this study will inform the development of intervention protocols that will be tested for feasibility, acceptability, and preliminary efficacy in future investigations of OT[ee] HORSPLAY.

Keywords: occupational therapy, autism spectrum disorder, equine-assisted services, transactional perspective

Introduction

Autism spectrum disorder (ASD) is a complex neurological disorder defined by limitations in social communication and restricted and repetitive behaviors (American Pyschiatric Association, 2013). Youth with ASD often demonstrate decreased engagement in social situations, including during psychosocial interventions, therefore creating barriers to effective intervention. A growing body of evidence supports that including animals in interventions for youth with ASD may increase participants' social engagement, leading to a variety of benefits (O'Haire, 2017). In particular, recent reviews have found that various equine-assisted services demonstrate promise for improving social interaction, social communication, and behavior in youth with ASD (McDaniel-Peters and Wood, 2017; Srinivasan *et al.*, 2018).

*Corresponding author: Caiti.Peters@colostate.edu

One such review identified three studies where occupational therapists incorporated horses into their services for youth with ASD (McDaniel-Peters and Wood, 2017). Findings demonstrated that occupational therapy that incorporates horses may improve outcomes such as social engagement (Llambias *et al.*, 2016), motor skills (Liddiard, 2009; Ajzenman *et al.*, 2013), and performance and participation in related functional activities such as handwriting and self-care (Liddiard, 2009; Ajzenman *et al.*, 2013). Despite these promising findings, studies remain in early stages of scientific development, limited by small sample sizes and a lack of standardized protocols, particularly intervention manuals (McDaniel-Peters and Wood, 2017).

To address the need for increasingly rigorous research to guide the inclusion of horses in occupational therapy services, our team has followed a phased approach developed by the National Institute of Mental Health to the scientific development and evaluation of psychosocial interventions for youth with ASD (Smith *et al.*, 2007). Fig. 4.1 illustrates the phases within Smith's recommended approach, and how our previous and current research is situated within those phases. We first conducted a phase one study, which included a single-case experimental trial of OT^{ee} HORSPLAY with 7 participants, and interviews with providers and parents. This phase one study demonstrated that occupational therapy in an equine environment for youth with ASD was acceptable to parents (Kalmbach *et al.*, 2020), improved participants' goal performance and social functioning, and decreased participants' irritability and hyperactivity (Peters *et al.*, 2020). Given these promising results, we are now entering into Phase 2 of scientific development, focused on protocol development and manualization.

Creating a manual is an essential step in the development and evaluation of complex interventions, as it standardizes procedures to facilitate future research, and makes the intervention more widely available for clinical use once the manual is disseminated (Smith *et al.*,

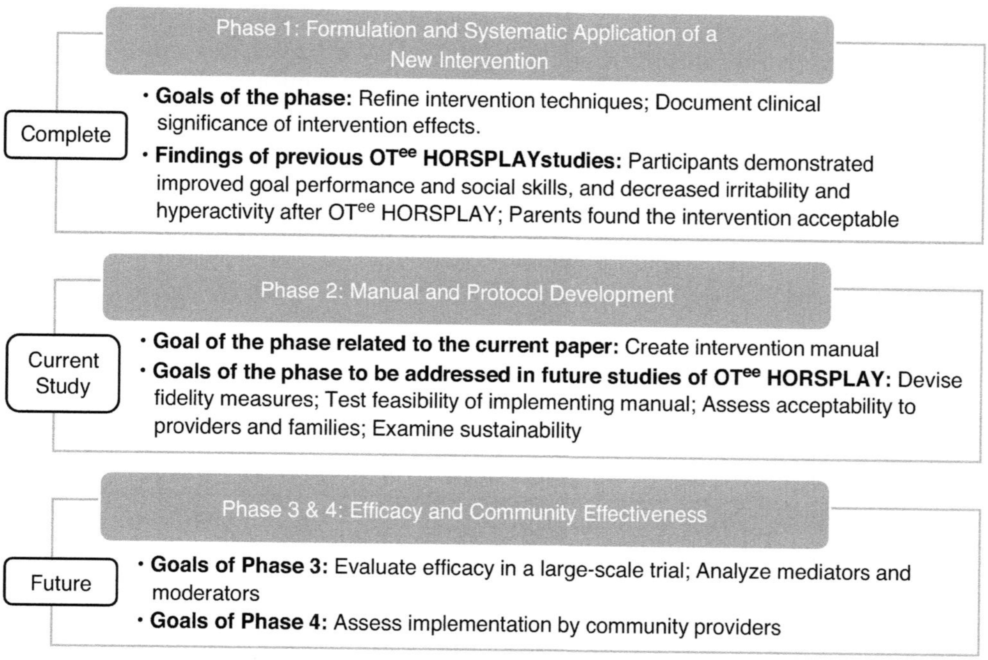

Fig. 4.1. OT^{ee} HORSPLAY Research Situated within Smith *et al.*'s (2007) Framework of Intervention Development and Evaluation for Individuals with Autism
Note: Depiction of how the ongoing program of research on OT^{ee} HORSPLAY is situated within Smith *et al.*'s (2007) recommended phases for intervention development and evaluation.

2007). During the process of manualization, our team named the intervention under investigation **O**ccupational **T**herapy in an **E**quine **E**nvironment: **H**arnessing **O**ccupation for Self-**R**egulation, **S**ocial Communication and **Play** in **Y**outh with Autism (OT^ee HORSPLAY). To guide manualization, our research team is drawing on several sources of data, one of which is the focus of the present paper: videotapes of OT^ee HORSPLAY sessions. The present paper reports on detailed analyses our team conducted of videotaped intervention sessions, in order to elucidate the content of the intervention and support manual development.

Applying a Transactional Perspective to the Detailed Description of OT^ee HORSPLAY

Occupational therapists recognize that their clients' occupations—the everyday things that they want to do, need to do, or are expected to do (Law *et al.*, 2014)—are unique to each individual client and the contexts in which they live their lives. Therefore, in accordance with best practices in occupational therapy for youth with ASD, OT^ee HORSPLAY is tailored to participants' individualized occupational performance goals, collaboratively determined by an occupational therapist, parent, and child during an initial evaluation (See Table 4.1). In determining how to best analyze videotaped intervention sessions, we sought a method that could account for the individualized, occupation-centered nature of OT^ee HORSPLAY. In this pursuit, we identified transactional perspectives of occupation developed in occupational science as an appropriate theoretical framework to guide an analysis that explicates the inherent complexities and dynamic nature of OT^ee HORSPLAY.

Over the past decade, the concept of transactionalism has reached a level of prominence in the occupational science literature, often regarded for its potential utility and advantageous perspective in studying occupation (Aldrich, 2008; Cutchin and Dickie, 2013; Wood, 2019). In defining transactionalism, John Dewey, an American pragmatist philosopher, described people as live creatures who live *through* the external environment in a constant codependent relationship, rather than understanding the environment as a container *in which* people live (Dewey and Bentley, 1949). When studying experience from a transactional perspective, all observable elements of situations are co-defining; to understand one element, one must understand how that element transacts with other elements. Thus, examining the content of OT^ee HORSPLAY through a transactional lens reveals a perspective of the intervention whereby all elements of the intervention are understood as interdependent and co-defining.

Guided by this understanding, we determined that three broad constituent units comprise individual sessions of OT^ee HORSPLAY. *Actors* are conscious beings that interact with their surroundings through actions.

Table 4.1. Participant Characteristics

Pseudonym	Gender	Age	Race	NVIQ	Occupational Performance Goal Area
Kayla	Female	9	Caucasian	123–137	Transitions
Ryan	Male	8	Caucasian	105–117	Reciprocal Conversation
Fisher	Male	6	Caucasian	99–112	Shoe Donning
Josh	Male	8	Native American	71–83	Transitions
Jorge	Male	11	Hispanic	43–55	Transitions
Maya	Female	13	Hispanic	103–116	Emotional Regulation
David	Male	13	Multi	71–83	Community Safety

Note: NVIQ = Nonverbal IQ as measured by the Leiter International Performance Scales, Third Edition (M = 100, SD = 15)

Architectural spaces refer to the physical environments designed by humans that are accessed during the intervention. Finally, *occupational opportunities* refer to how individual sessions of OT[ee] HORSPLAY avail, to the youth with ASD, opportunities to engage in various occupations, such as riding a horse or playing a game. These occupational opportunities are carefully selected and tailored by occupational therapists in order to address the participant's individualized goals. Hence, to grasp each session of the intervention in its entirety, we presumed that each constituent unit depended on its configuration with the other two.

Therefore, the aim of the present study was to illustrate and describe the constituent units of OT[ee] HORSPLAY and their dynamic transactions concretely and vividly, in order to support manualization of the intervention. To address this aim, we asked these four questions:

1. In what architectural spaces, and for how long, does OT[ee] HORSPLAY typically transpire?
2. What occupational opportunities comprise individual sessions of this intervention, and how are they sequenced throughout the sessions?
3. What do the youth with ASD and their occupational therapists do throughout the sessions of OT[ee] HORSPLAY?
4. How do the constituent units of the intervention, as analyzed in questions one through three, transact with one another?

Method

This paper stemmed from a larger pilot study of OT[ee] HORSPLAY, the design and methods of which are reported elsewhere (Peters *et al.*, 2020). For the present paper, we comprehensively analyzed videotaped data of individual sessions of OT[ee] HORSPLAY, which had previously been collected during the pilot study. Colorado State University's Institutional Review Board approved all study procedures, and the Institutional Animal Care and Use Committee designated the study as exempt from review. Occupational therapists and caregivers provided informed consent; youth with ASD provided verbal assent if able.

Participants

For the pilot study (Peters *et al.*, 2020), the first author distributed fliers to local organizations to recruit eight youth with ASD who met the following criteria: ages 6–13, diagnosed with ASD confirmed by meeting clinical cut-offs on the Social Communication Questionnaire and the Autism Diagnostic Observation Schedule, nonverbal IQ ≥ 55, and combined score ≥11 on the irritability and hyperactivity subscales of the Aberrant Behavior Checklist, Community. Participants were excluded if they had participated in equine-assisted services for 2 + hours in the previous six months, weighed more than 200 pounds, or did not meet physical, mental or emotional standards set forth by the Professional Association of Therapeutic Horsemanship, International (PATH, Intl). Eight youth enrolled into the study, but one dropped out due to anxiety about missing school. The remaining seven participants are described in Table 4.1.

Two occupational therapists who worked at the riding center provided the intervention; both had completed Hippotherapy Treatment Principles 1 & two courses by the American Hippotherapy Association (AHA) and were certified by PATH Intl. During the intervention there was a 1:1 ratio between youth and an occupational therapist. Youth with similar social communication abilities were paired into dyads and received OT[ee] HORSPLAY simultaneously, to allow for peer interaction. Volunteers trained in safety procedures served as horse leaders and side-walkers. When possible, we paired youth with the same occupational therapist, peer, horse and volunteers throughout the 10 weeks, although these changed on occasion due to absences.

Intervention

As previously reported in Peters *et al.* (2020), the intervention occurred at a PATH Intl Premiere Accredited Center that met all required safety standards (equine selection and training, facility set-up, helmet use, etc.). Prior to the first session, youth and parents participated in an occupational therapy evaluation that included collaborative goal setting; goals were related to social participation and self-regulation. The intervention consisted of 10 weekly sessions that each lasted 45–60 min and followed a general structure: greetings, activities with horses, debriefings with parents, goodbyes. In accordance with best practice in occupational therapy for youth with ASD, OT^{ee} HORSPLAY was highly individualized to participants' goals (Tomchek and Koenig, 2016). Occupational therapists designed intervention activities based on these critical elements: a) individualization of activities to address participant goals, b) facilitation of social interaction, c) positive reinforcement of communication, d) use of visual aids, and e) use of *hippotherapy* (AHA, 2021). The AHA defines hippotherapy as 'the purposeful manipulation of equine movement as a therapy tool to engage sensory, neuromotor, and cognitive systems to promote functional outcomes' and encourages therapists to integrate hippotherapy 'along with other therapy tools and/or strategies' (AHA, 2021). In the current study, equine movement was integrated into OT^{ee} HORSPLAY for these reasons: 1) motivating youth to participate in mounted intervention activities, 2) providing proprioceptive, vestibular, and tactile stimulation to help youth achieve an optimal state of arousal, and 3) as a positive reinforcement for communication.

Data collection

Because 8 participants were paired into dyads, and the intervention lasted 10 weeks, there were a total of 40 videotaped sessions, for a total of 38 hours of video data. As most sessions included two youth with ASD due to the design of the intervention, the focus of videotaping was on one participant for an entire session; the focus alternated between youth weekly. When a youth with ASD was not the direct subject of videotaping, the youth was referred to as the peer. Videotaping was conducted using a Sony Handycam and a tripod where possible, otherwise handheld filming was necessary to capture the dynamic changes and frequent movement across architectural spaces. Videotaping followed a predetermined schedule. We sampled 21 of the 40 sessions to be analyzed, choosing one videotaped session from each child's first, middle, and final sessions; overall, 19 hours and 15 min of video were sampled for analysis.

The second author also collected field notes during and following each day of data collection that included information not readily available from videotapes such as prevailing smells, textures, and sounds; nuances in temperature and weather; and direct quotes from actors. The field notes were also used to memo revelations inspired from witnessing the therapy in person.

Data analysis

The hermeneutic arc in Ricoeur's (1976) theory of interpretation is congruent with inquiry grounded in a transactional perspective of occupation. According to Ricoeur, hermeneutic interpretation transpires within a hermeneutic arc (Fig. 4.2). When following this arc, an observer successively moves back and forth among parts of a subject matter and its projected whole until comprehensive understanding of the subject matter is realized. Given its congruence with a transactional perspective, the hermeneutic arc guided our analysis.

The second author analyzed data following Ricoeur's hermeneutic arc. As next developed, entering the hermeneutic arc begins with a naïve interpretation, then an explanation, and finally moves into a comprehensive understanding.

Naïve interpretation

The *naïve interpretation* is a basic and initial understanding of the subject matter removed from the author's intent, forging a new relationship between the interpreter and the subject matter (Ricoeur, 1976). This first interpretation is essentially a guess that may or may not be validated through the explanation of the subject matter. After videotaping initial sessions of the therapy, but before conducting any further analysis, the second author formed a naïve interpretation by writing a simple description of OT^{ee} HORSPLAY. In doing so, he attempted to summarize field notes taken and the lived experience of being present in, but not a participant of the therapy. After reaching a naïve interpretation of the intervention as it was actually delivered,

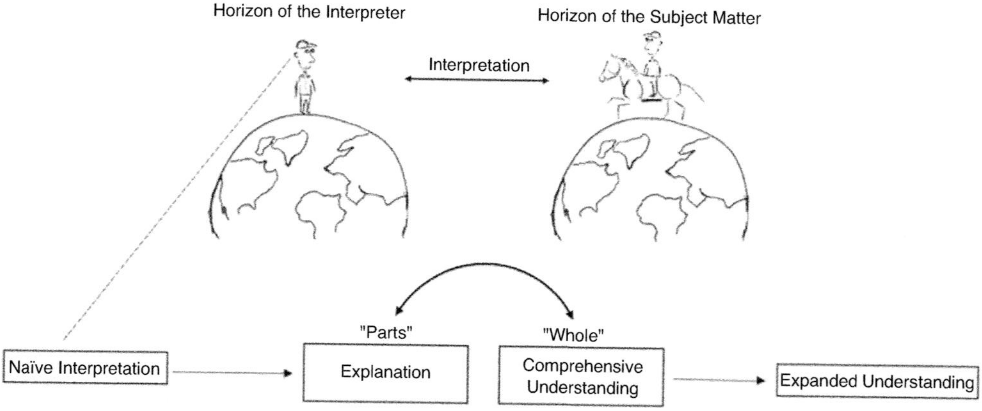

Fig. 4.2. Illustration of the Hermeneutic Arc
Note: The interpreter begins with a naïve understanding, then delves into the subject matter by moving back and forth successively between a part and the projected whole. This conversational process between the world of the interpreter and the world of the subject matter seeks to narrow the distance that exists between their two worlds.

the research team (all three authors) sought to explain this initial understanding through structural analysis.

Explanation through structural analysis

Related to *explanation*, or the structural analytical phase of the hermeneutic arc, Ricoeur (1976) recommended extracting and examining constituent units out of context before putting them back together in a unified sequence and whole. Our structural analysis involved examining each constituent unit of the intervention before integrating findings pertaining to them within a comprehensive understanding of OT[ee] HORSPLAY.

This examination was conducted using qualitative content analysis (QCA) of collected videotaped data. Because QCA can generate precise descriptions and descriptive statistics of salient information entrenched in data (Schreier, 2012), it is well-suited for explanation in the hermeneutic arc. In QCA, the coding scheme is deductively developed to answer research questions that were theoretically derived; our research questions formed the theoretical basis of the three broad constituent unit codes: architectural spaces, occupational opportunities, and actors and their actions (coding scheme available in Supplemental Table S4.1). QCA also permits inductive formation of codes from the data. For example, within the parent code architectural spaces, a child code 'sensory trail' was discovered during data analysis. The coding scheme is created during early stages of QCA and to ensure that multiple investigators are reliably applying it, agreement coefficients can be calculated. Applying QCA in this study, investigators developed a coding scheme and completed all coding and analyses of video data using the qualitative software program, NVivo for Windows, version 11.4.1.1064 (QSR International, 2017).

To develop the coding scheme, videotaped sessions were uploaded into NVivo on a weekly basis. Weekly also, the team watched and re-watched uploaded videotapes to develop, deliberate, refine, and reach agreement on all codes and their definitions. The evolving coding scheme was applied to the sampled videos weekly, forming more comprehensive descriptive results with each pass. Before the sampled data were exhausted, a point was reached in which no new codes were found, thus obtaining theoretical saturation. Analysis then continued until all sampled sessions were examined. A research assistant independently trained in the coding scheme achieved 96.84% agreement with coding by the research team, suggesting strong interrater reliability.

The final coding scheme included several inductively derived child codes within each constituent unit. Specifically, the constituent unit of *actors and their actions* included 15 child codes describing the youth's actions; 14 child codes describing occupational therapists' actions; seven child codes describing volunteers' actions; and one child code denoting a family member's presence. Especially detailed child codes described actions of occupational therapists given their prominence in delivering the intervention. All codes describing actions of adults pertained to their interactions with the youth. Because multiple actions often occurred concurrently, action child codes were not mutually exclusive. The constituent unit of *architectural spaces* included six mutually exclusive and exhaustive child codes denoting the spaces that were occupied during the intervention. The constituent unit of *occupational opportunities* included 16 mutually exclusive and exhaustive child codes denoting specific occupations offered to the youth both when horses were and were not present.

In the constituent units of *occupational opportunities* and *actors and actions*, the final coding scheme also included inductively developed modifiers. Specifically, modifiers to *occupational opportunities* elaborated on the opportunities. For example, two modifiers of the opportunity, 'riding horse—standing,' indicated the youth was also being offered an opportunity to play 'Simon Says' or 'Red Light, Green Light.' Hence modifiers denoted opportunities to participate in layered activities, which required more complex skills than the base opportunity alone, as in this example, simply to sit atop a standing horse. Modifiers of *actors and their actions* described actions more precisely. For example, the action code 'riding horse—standing' denoted the youth had accepted an opportunity to ride and was mounted on a standing horse. If applied, modifier codes of 'Simon Says' or 'Red Light Green Light' denoted that the youth was simultaneously playing one of these games.

After the coding scheme was finalized, all 19 hours and 15 mins of sampled video data were continuously coded. For each included videotaped session, investigators numbered the order in which specific occupational opportunities were presented; for example, a session opened with opportunities for social interaction followed by opportunities to transition to different activities and then to manage equipment. Lastly for the structural analysis, the investigators queried frequencies and durations of all codes, child codes, and modifiers in NVivo.

Comprehensive understanding

To form a *comprehensive understanding*, investigators synthesized findings pertaining to individual constituent units of the intervention into a comprehensive understanding and coherent narrative. Specifically, the second author determined sequences of occupational opportunities and corresponding architectural spaces during videotaped sessions. The investigator then visually aggregated these sequences in three figures (Figs 4.2, 4.3, and 4.4 as introduced under findings), which reflected the three main time segments of the intervention identified during the phase of naïve interpretation. To further explore dynamic relationships, matrix queries were run in NVivo to determine durations of overlap between salient codes and modifiers that had been computed during structural analyses of the explanation phase. One of these matrix analyses showed, for instance, that when mounted activities were offered (an occupational opportunity), the youth's actions reflected positive social interactions (a youth action code). Another analysis showed that the youth displayed more disengaged behaviors (a youth action code) when occupational therapists were instructing them or talking with family members (therapist action codes). Thus, to generate a comprehensive understanding of OT[ee] HORSPLAY, investigators iteratively moved back and forth across the hermeneutic arc. This iterative process allowed investigators to create a sequential, whole, and coherent narrative of the intervention, which encompassed its dynamic unfolding across individual sessions as next reported.

Results

Naïve interpretation

OT[ee] HORSPLAY as provided to youth with ASD was a highly social therapy program. This program involved a wide selection of

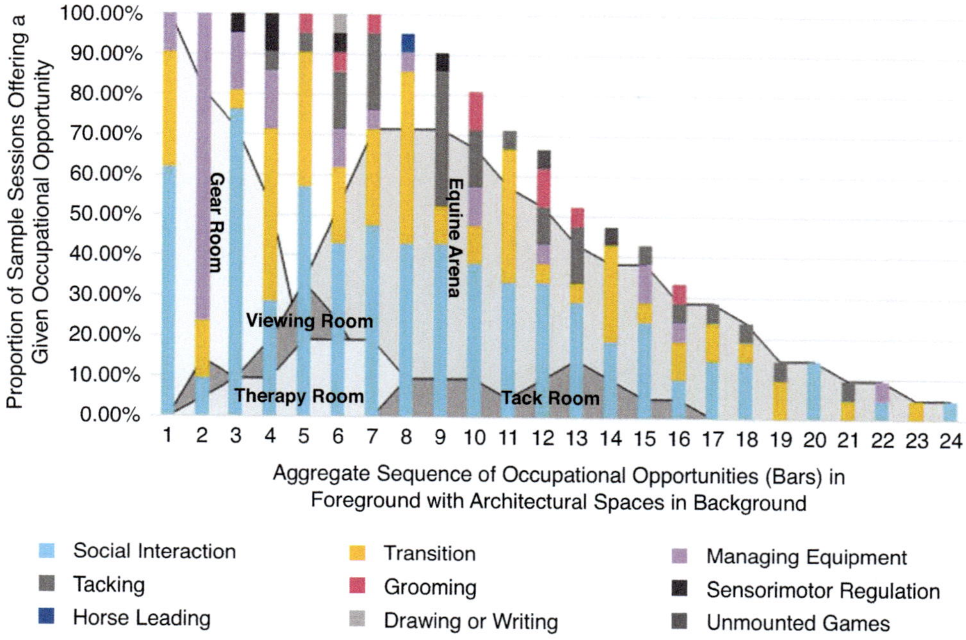

Fig. 4.3. Pre-Mounted Sequence of Occupational Opportunities Within Architectural Spaces
Note: This figure depicts aggregated sequences across all sampled sessions of architectural spaces and occupational opportunities throughout the pre-mounted segment. The background depicts proportions of sessions making use of different architectural spaces across the ordered sequence of occupational opportunities. The horizontal axis is numbered to reflect the sequence of offered occupational opportunities aggregated across all sessions. The vertical axis shows proportions of sampled sessions that corresponded with each numbered order of offered occupational opportunities. Drops in the stacked bars below 100% indicate that a proportion of the sampled sessions have completed this segment and transitioned into mounted segment. For example, by bar #18, nearly 80% of sampled sessions had transitioned into the mounted segment of the intervention.

occupational opportunities, physical spaces, and social environments that were ordered in a distinctive pattern. Broadly, this pattern took the shape of three distinct segments that occurred during every session: pre-mounted time, mounted time, and post-mounted time.

Explanation (Structural Analysis)

OT[ee] HORSPLAY consisted of ten sessions that averaged 55 min in duration. Detailed structural findings related to each constituent unit are next presented; these findings were informed by descriptive statistics (frequencies and durations) of all codes, child codes, and modifiers (Table 4.2 and Supplemental Tables S4.1–S4.3).

Architectural spaces

Six distinct architectural spaces were observed, as depicted in Supplemental Figure S4.1: the equine arena, the gear room, an outside 'sensory' trail, the grooming and tack room, the therapy room, and the viewing room. All seven participants were observed in the equine arena, gear room, sensory trail and viewing room; four participants were observed in the therapy room; and three participants were observed in the grooming and tack room. The majority of the intervention was offered in the equine arena, representing 61% of session time on average (Supplemental Table S4.2). The equine arena (area≈950 m^2) had an enclosed dirt floor room, direct access to the sensory trail, the grooming and tack room, and the viewing room, and contained mounting

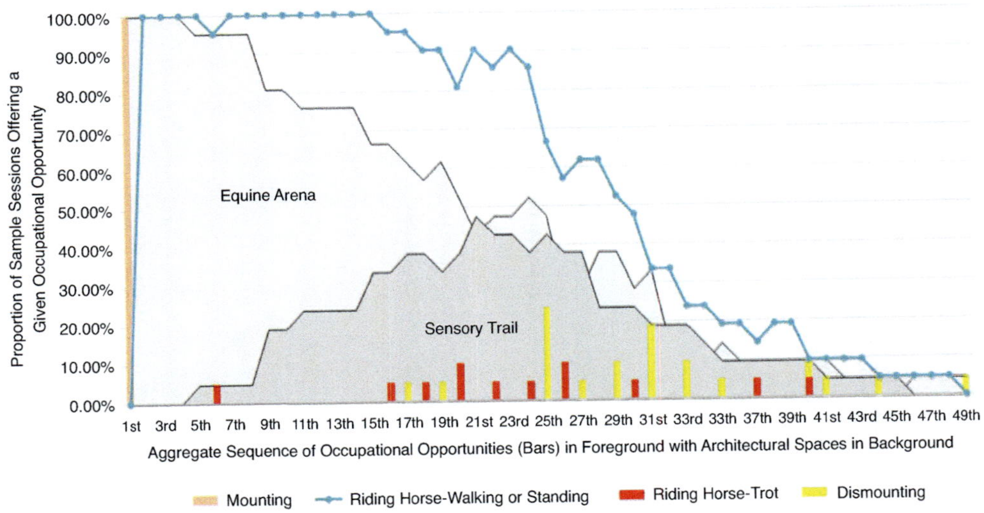

Fig. 4.4. Mounted Sequence of Occupational Opportunities Within Architectural Spaces
Note: This figure depicts aggregated sequences across all sampled sessions of architectural spaces and occupational opportunities throughout the mounted segment. The background depicts proportions of sessions making use of different architectural spaces across the ordered sequence of occupational opportunities, specifically the riding arena and the outdoor sensory trail. The horizontal axis is numbered to reflect the sequence of offered occupational opportunities aggregated across all sessions. Thus, mounting the horse is shown to have been the first offered opportunity in 100% of the sampled sessions. Because the prevailing opportunities involved being mounted on a standing horse or riding a walking horse, these opportunities are combined into a single blue line with dots that correspond with the order of presented opportunities. The large number of occupational opportunities, 49, reflect all of the possible modifiers of riding horse- standing, riding horse-walking, and riding horse-trotting. When the dotted blue line shows 100%, it represents the occupational opportunities from all sampled sessions in the mounted segment. Drops in the dotted blue line below 100% indicate that a proportion of the sampled sessions have completed this segment and transitioned into the post-mounted segment.

blocks, a white board, and various obstacles. About one fifth (19% on average) of session time occurred on the sensory trail outside. This looping dirt trail (distance≈730m) contained a whiteboard, garden, bridge, glockenspiel, a 'car wash' (a rectangle frame with hanging firehose), a gazebo, birdhouses, hills, a 'rubber tire dragon' (obstacle), a 30m 'race track' and a section of poles and platforms (obstacles). The temperature, wind, and ground precipitation of the sensory trail varied greatly across sessions. On average, 12% of session time took place in the gear room during pre- and post-mounted segments. The wood paneled gear room contained helmets, gait belts, and other safety equipment. To lesser extents, the intervention took place in the therapy, viewing, and grooming and tack rooms, or for 4%, 3%, and 1% of all session time on average respectively. The small therapy room had a fold-down therapy mat and whiteboard. The small viewing room had five chairs and three windows into the equine arena. The somewhat larger grooming and tack room contained saddles, reins, lead ropes, and six cross-ties for horses.

Occupational opportunities

Seventeen different occupational opportunities were observed (Supplemental Table S4.3), many of which were modified by 14 different modifiers (Supplemental Table 4.1). Every session offered the following occupational opportunities at least once: mounting a horse, being mounted on a horse that was standing, being mounted on a horse that was walking, dismounting, transitions, managing equipment, and social interaction. Modifiers described more specific activities, such as playing basketball while

Table 4.2. Actions of Youth with ASD and Occupational Therapists: Average Proportion of Time per Session Spent Performing Actions

Actions	% Total Session (mm:ss)	% Total Session During Pre-mounted	% Total Session During Mounted	% Total Session During Post-mounted
		Youth with ASD		
Social Interaction	92% (50:44)	30%	53%	10%
Riding Horse-Stand	28% (15:11)	0%	28%	0%
Riding Horse-Walk	25% (13:36)	0%	25%	0%
Verbalization	10% (05:42)	4%	5%	2%
Managing Equipment	5% (02:51)	3%	< 1%	2%
Disengaged Behaviors	5% (02:51)	2%	< 1%	2%
Transitioning	4% (02:06)	3%	< 1%	1%
Tacking	4% (01:59)	3%	< 1%	< 1%
Grooming	1% (00:42)	1%	0%	< 1%
Petting Horse	1% (00:38)	< 1%	< 1%	< 1%
Mounting	< 1% (00:13)	0%	< 1%	0%
Sensorimotor Regulation Tasks	< 1% (00:11)	< 1%	0%	< 1%
Dismounting	< 1% (00:10)	0%	< 1%	0%
Riding Horse-Trot	< 1% (00:08)	0%	< 1%	0%
Drawing or Writing	< 1% (00:05)	< 1%	0%	< 1%
Horse Leading	< 1% (00:02)	< 1%	0%	0%
		Occupational Therapists		
Instructing the Child	15% (08:18)	6%	8%	1%
Physical Facilitation	4% (02:27)	2%	3%	< 1%
Instructing Parents	3% (01:38)	< 1%	0%	3%
Eliciting Communication	3% (01:36)	< 1%	2%	< 1%
Facilitating Social Interaction	2% (01:19)	1%	< 1%	< 1%
Positive Reinforcement	2% (00:59)	< 1%	< 1%	< 1%
Physical Support	2% (00:58)	0%	2%	< 1%
Individualizing Equine Equipment to Child	2% (00:54)	< 1%	2%	< 1%
Instructing Volunteers	1% (00:34)	< 1%	< 1%	< 1%
Eliciting Choice-making	< 1% (00:22)	< 1%	< 1%	< 1%
Collaboration with other OT	< 1% (00:17)	< 1%	< 1%	< 1%
Singing	< 1% (00:16)	< 1%	< 1%	0%
Modeling	< 1% (00:15)	< 1%	< 1%	< 1%
Promoting Sensorimotor Regulation	< 1% (00:10)	< 1%	0%	< 1%

mounted on a horse, and varied widely across sessions. Durations of opportunities to participate in mounted occupations extended over half of all sessions, or 53% on average. Durations of opportunities to participate in unmounted occupations with the horse present extended over nearly one quarter of all sessions, or 24% on average. Durations of opportunities to participate in unmounted occupations when horses were not present took up just under one fifth of all sessions, or 18% on average.

Actors and their actions

Average percentages of session time in which different actors were present were as follows: youth with ASD (participant who was the focus of the videotape; 99.79%), occupational therapist (97.16%), volunteers (85%), horses (80.15%), peer (participant who was not the focus of the videotape; 78.98%), and family members (15.64%). Table 4.2 presents the most common actions of youth with ASD and occupational therapists. The youth with ASD were observed to spend the largest proportion of sessions, 92% on average, engaged in various positive or neutral social interactions. The youth with ASD engaged in shared communication with other actors while mounted and unmounted, clearly demonstrating face and body orientation to a social partner, self- initiated and therapist-prompted verbal contributions, and directed listening with the other actors. The nature of each social interaction was characterized by the architectural spaces and the recent or upcoming occupational opportunities—often helping to process or prepare for past and future activities. For example, before transitioning into the equine space where the youth would meet the horse and volunteers, social interaction was focused on helping the youth understand the social or physical expectations of the coming activities such as making a wide arc when walking behind the horse. Frequencies of actions varied across segments. For example, about four times on average during pre-mounted segments, the youth action code, 'disengaged behaviors,' was recorded to represent refusals to participate or unsafe or disengaged behaviors; these behaviors were rarely observed during the mounted segments.

The occupational therapists were observed providing instruction to the youth with ASD over the largest proportion of sessions (15% of session time). Other notable actions were performed frequently yet over smaller proportions of sessions, such as eliciting communication (average of 36.14 times per session, but only 2.92% of session time), verbal positive reinforcement (23.5 and 1.8%), facilitating social interaction (13.6 and 2.39%), physical facilitation (nine and 4.45%), instructing volunteers (6.14 and 3%) and eliciting choices (5.3% and 0.65%). At least two volunteers were present for each youth during mounted times; one led the horse and the other walked by the horse's side. Volunteers physically supported the youth during 18% of all sessions on average, by providing arm-over-thigh holds or ankle-holds to increase stability. Volunteers also periodically instructed and elicited communication from the youth.

Comprehensive understanding

A coherent storyline that provides a comprehensive understanding of OT[ee] HORSPLAY is next presented. Figs 4.2, 4.3 and 4.4 partly illustrate this storyline by presenting sequences of utilized occupational spaces in relationship to sequences of offered occupational opportunities, as aggregated on average across all sessions. We next chronicle the progression of dynamic interactions among architectural spaces, actors and their actions, and occupational opportunities across the pre-mounted, mounted, and post-mounted time segments of the intervention. The following storyline integrates the visual data presented in Figs 4.2, 4.3 and 4.4 with further analyses of video data and fieldnotes.

Pre-mounted time

On average, pre-mounted time lasted 18:25 mm:ss and was primarily devoted to preparing the youth for mounting and riding the horse (Fig. 4.3). Every session began in the gear room where occupational therapists greeted the youth with ASD, who would soon transition to donning safety equipment with some physical facilitation from the therapist or a volunteer. This transition from a family context into an unfamiliar social context often appeared to elicit, in some of the youth, disengaged behaviors. As a thoroughfare for the youth, therapists, and volunteers—with as many as 13 different people present at a given time—the gear room was potentially overstimulating. Of the total duration that the youth were disengaged, about half occurred during pre- mounted time. Although challenging for the youth at times, this segment was also rich with preparatory therapeutic activities between the youth and the therapist such as social interaction related to goal behaviors and rapport building.

Following a period of social interaction where the occupational therapist instructed the youth about the contents of the day's session, most sessions transitioned by the fifth occupational opportunity into the viewing and therapy rooms or equine arena. Family members typically accompanied the youth from the gear room into the viewing room, but rarely joined them in the equine arena, therapy room, or grooming and tack room during pre-mounted time. The arrival of the peer was inconsistent across sessions; sometimes the youth prepared together and other times they did individualized introductory activities and met in the equine arena for shared mounted time. Sessions that utilized the therapy room prior to mounted time did so for two purposes: to facilitate sensorimotor self-regulation tasks or as an opportunity to collaboratively plan session activities with the youth. When utilized, the grooming and tack room primarily functioned as a social space for familiarizing the youth with their horse and volunteers and affording them opportunities to choose which equine equipment they needed for eventually tacking and grooming horses in the equine arena. Every pre-mounted time concluded by transitioning into the equine arena, where the primary occupational opportunities included social interaction, tacking, grooming and managing equipment.

Thus, during pre-mounted time, the youth's relationship to architectural spaces and occupational opportunities evolved over the course of the intervention. The youth tacked horses in every session, but as they became more familiar with the space and its rules, the youth required less instruction and grooming time before mounting horses or playing unmounted games. During tacking or grooming, some youth talked about the equine environment (it has a cooler temperature...it makes me calmer'), how the horse made them feel ('her walk, [the horse's] it has a nice feel to it, side to side') and that they wished to continue working with horses in the future. Most sessions concluded pre-mounted time after 14 occupational opportunities.

Mounted time

Mounted time averaged 29:06 mm:ss. This segment began by the youth's mounting of horses, which the occupational therapists always physically facilitated. Once mounted, the youth rode while horses alternated between walking, standing, or infrequently, trotting. Compared to pre- and post-mounted time, the occupational therapists performed all actions at their highest frequencies and longer durations (except parental instruction and sensorimotor regulation promotion). Thus, mounted time elicited especially persistent and diverse skilled actions by the occupational therapists. Also compared to pre- and post- mounted time, the youth were rarely disengaged during mounted time.

Because mounted time superficially appeared repetitive given its same basic format across sessions, Fig. 4.4 highlights moments of variation. Indeed, mounted time was especially rich with activities that modified the occupational opportunities of horse riding. On average, the youth spent 04:20 mm:ss participating in 12 activities that modified their actions while atop a horse, including mounted basketball, follow the leader, Simon says, red-light/green-light and obstacle navigation among others. Specific activities offered by the occupational therapists and chosen by the youth often related to youth's goals. For example, one youth participated in a question cards activity to develop his reciprocal conversation skills. On average, each session's mounted occupations were modified by six or more activities. Volunteers mostly physically supported the youth, thus acting as extensions of the occupational therapists and allowing them flexibility in their proximity to the youth. The youth always dismounted on the side of the occupational therapists, who provided physical facilitation with periodic support from

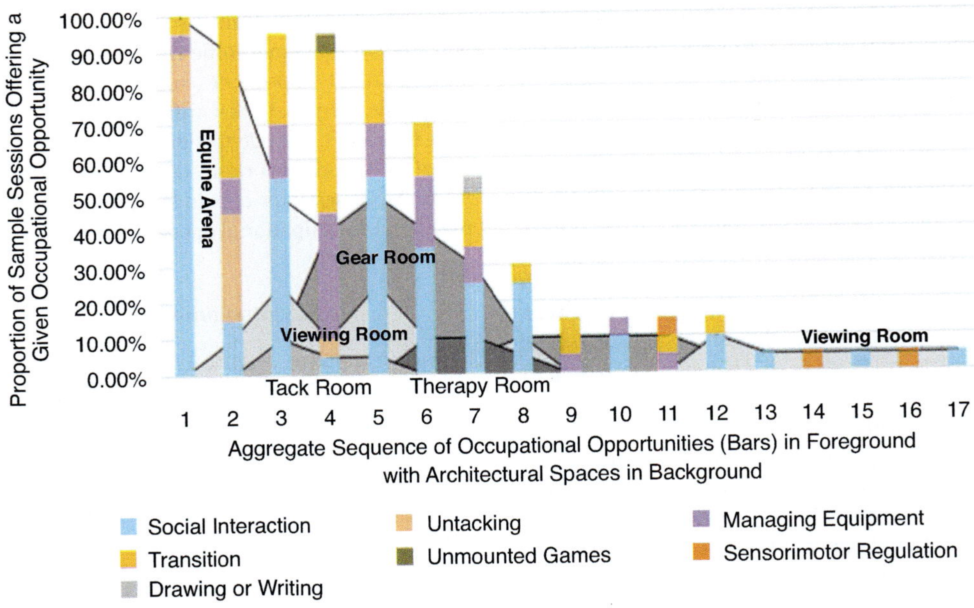

Fig. 4.5. Post-mounted Sequence of Occupational Opportunities Within Architectural Spaces
Note: This figure depicts aggregated sequences across all sampled sessions of architectural spaces and occupational opportunities throughout the post-mounted segment. The background depicts proportions of sessions making use of different architectural spaces across the ordered sequence of occupational opportunities. The horizontal axis is numbered to reflect the sequence of offered occupational opportunities aggregated across all sessions. The vertical axis shows proportions of sampled sessions that corresponded with each numbered order of offered occupational opportunities. For example, bar #1 shows that the first occupational opportunity was always offered in the equine arena, with nearly 75% of all sessions starting the post-mounted segment with opportunities for social interaction. Drops in the stacked bars below 100% indicate that a proportion of the sampled sessions have completed this segment, marking the end of the session.

volunteers. Most sessions concluded mounted time after 23 occupational opportunities.

Post-mounted time

Post-mounted time averaged 07:21 mm:ss and was primarily devoted to transitioning socially and physically from mounted time and to debriefing with family members (Fig. 4.5). Most sessions transitioned out of the equine arena by the fourth occupational opportunity and moved into the gear room by way of the viewing or grooming and tack rooms.

After dismounting, the occupational therapists facilitated the youth thanking horses and volunteers. The youth then typically untacked the horse and restored equipment to the original locations. In the final session, this moment provoked frequent reflections by the youth. For instance, the youth expressed feelings of personal growth ('I accomplished a lot'), the bond they developed with their horse ('I love Gavin' [horse]), context-relevant reflections ('I feel like a cowboy') and sadness that the experience was ending ('I'll miss you' [speaking to the horse]). As therapists debriefed sessions with parents, absences of occupational opportunities that might have directly engaged the youth were common, as were disengaged behaviors by the youth. The majority of sessions concluded this segment after eight occupational opportunities.

Discussion

The present study aimed to vividly describe a novel approach to incorporating horses into occupational therapy for youth with ASD, OT^{ee} HORSPLAY. A transactional perspective of occupation helped to generate findings that illuminated the fast-moving and dynamically occupation-centered nature and complexity of the intervention.

To advance the science of novel interventions for people with ASD, investigators must first identify and fully elucidate active elements that comprise how an intervention is actually implemented (Smith *et al.*, 2007; Craig *et al.*, 2013). Systematic mapping reviews of equine- assisted services for diverse populations, including youth with ASD, suggest that investigators typically neglect or pay just scant attention to this crucial early research step (McDaniel-Peters and Wood, 2017; Wood and Fields, 2019). For example, Shurtleff and Engsberg (2012) offered one of the more comprehensive depictions of hippotherapy, a treatment strategy focused on equine movement; their description included a list of gaits, gait changes, sitting positions, mounted activities, walking patterns, and duration, frequency and location of sessions. Llambias *et al.* (2016) also provided brief descriptions of on- and off-horse activities in their descriptions of equine-assisted occupational therapy. Both studies identified occupational therapists as providers. However, sparse descriptions of their respective interventions did not foreground occupation and its dynamic manifestations and variations, thus leaving much to the imagination of other interested occupational therapists and researchers.

In contrast, the present study's specific and vivid elaborations of OT^{ee} HORSPLAY offer realistic examples of how occupational therapists and youth with ASD can fluidly engage in shared occupations with horses and other people throughout fast-moving, intensive therapy sessions. Findings clearly elucidated these key interworking parts of the intervention: utilized architectural spaces, the progression of opportunities to engage in different occupations when horses were and were not present, and associated actions of the youth with ASD and the practitioners, and, to lesser extents, other people and the horses. Findings further distinguished how these interworking parts successively arose, endured, and faded, eventually coalescing in complete 45 to 60 min sessions that broadly unfolded in three main segments of time. This concrete and vivid illumination of the rich and dynamic flows of sessions is one source of data that can inform development and revision of the OT^{ee} HORSPLAY intervention manual, and can help other researchers and occupational therapists implement this intervention and tailor it to the needs of specific participants with ASD. This thorough description highlighted the dynamic nature of the intervention, whereby occupational therapists intentionally created unique occupational opportunities tailored to the goals of each youth, and created in-the-moment adaptations in response to the various actors and their actions. Thus, OT^{ee} HORSPLAY cannot be reduced to a simple list of activities, but is a highly tailored intervention approach that must be customizable to particular facilities, therapists, and youth. Creating an intervention manual that standardizes the approach, while also allowing for intensive individualization will no doubt be challenging; yet, standardization that allows for individualization is a hallmark characteristic of high-quality intervention manuals for youth with ASD (Smith *et al.*, 2007).

The results of this study also reveal potential mechanisms by which inclusion of horses in occupational therapy may affect outcomes. For example, the analysis revealed that youth were significantly less likely to disengage from activities while mounted on the horse. This finding lends support to the hypothesized mechanism that horses capture participants' attention and simultaneously support physiological regulation, therefore increasing participants' social engagement during therapy sessions (Peters *et al.*, 2019). Yet, the descriptive nature of this study cannot infer causal relationships between mounted occupations and social engagement. Future research can use inferential designs and validated behavioral coding tools, such as the Observation of Human-Animal Interaction in Research (Guérin *et al.* 2018), to confirm differences in social engagement while mounted.

Furthermore, this study revealed areas of the intervention that merit special attention by providers, or that can be improved. For example, frequent disengaged behaviors in the gear room were likely related to the commotion, large number of people in the small room, and prevalence of verbal instruction with the absence of activities to engage the participant. It may be helpful to include direction in the intervention manual for providers to be aware of their clients' sensory needs, and arrange the environment to best support their learning. For example, at this particular facility, that may mean limiting the number of people in the gear room, limiting the participants' time in that room, or providing activities for youth to engage in during the caregiver debrief. The fine-grain detail of each constituent unit revealed by the structural analysis, combined with the comprehensive understanding highlighting the interdependence of all constituent units, resulted in these detailed findings that have implications for how OTee HORSPLAY can be manualized and improved.

Limitations and future research

This study is limited by reliance on two data sources (videotapes and fieldnotes) to characterize one approach to integrating horses into occupational therapy for youth with ASD, OTee HORSPLAY. To fully understand the intervention, it will be important to elucidate the professional reasoning that guided how occupational therapists designed occupational opportunities for particular youth. Therefore, analyses of videotapes cannot alone inform an intervention manual; several other data sources such as interviews and document review of session notes will be necessary to fully capture the critical elements. Thus, although this paper advances understandings of OTee HORSPLAY for youth with ASD, it does not tell the full story of this intervention.

The current study does, however, fully elucidate the architectural spaces, actors, and occupational opportunities that comprised the OTee HORSPLAY intervention, and reveals the complex transactions among them. The process of manualization will require researchers to decide which elements of this description are critical and generalizable and therefore should be included in the manual, and which elements are necessarily specific to the particular facility, providers, and participants in this study. One challenge to manualization will be to standardize procedures in a manner that allows for individualization, which is critical to occupational therapy services for youth with ASD. Representing one small step forward in the rigorous development and evaluation of a novel intervention, the current study will inform the development and refinement of protocols that should be tested for feasibility, acceptability, and efficacy in future investigations of OTee HORSPLAY (Smith *et al.*, 2007).

Conclusion

This study produced a concrete and vivid description of a novel intervention incorporating horses into occupational therapy for youth with ASD, which can help to inform the development and refinement of an intervention manual. Such a thorough description that grapples with the complexity of an intervention is absent from the body of literature on equine-assisted services for youth with ASD, and therefore represents a pioneering step forward in the scientific advancement of this novel intervention. Our analytic method was effective in revealing the complex, dynamic, and fast-moving nature of OTee HORSPLAY, and also reveals potential challenges in manualizing the intervention. Ultimately, this study makes clear that OTee HORSPLAY cannot be reduced to a checklist of activities, but rather is a dynamically unfolding intervention responsive to emergent participant needs in the context of varied occupational opportunities.

Authors Note

We have no conflicts of interest to disclose. This research was funded by the Carl and Caroline Swanson Foundation. We are grateful for the families that participated in the study, and the staff and occupational therapists at Hearts and Horses Therapeutic Riding Center who made the study possible. We also thank Erin Kelly for her help in obtaining interrater reliability.

Supplementary Content

Refer to supplementary material on the CABI Digital Library: https://www.cabidigitallibrary.org/doi/book/10.1079/9781836991601.0000

References

Ajzenman, H.F., Standeven, J.W. and Shurtleff, T.L. (2013) Effect of hippotherapy on motor control, adaptive behaviors, and participation in children with autism spectrum disorder: A pilot study. *American Journal of Occupational Therapy* 67(6), 653–663.

Aldrich, R.M. (2008) From complexity theory to transactionalism: Moving occupational science forward in theorizing the complexities of behavior. *Journal of Occupational Science* 15(3), 147–156. DOI: 10.1080/14427591.2008.9686624.

American Hippotherapy Association (2021) *Welcome to the American Hippotherapy Association, Inc.* Available at: https://americanhippotherapyassociation.org/

American Pyschiatric Association (2013) *Diagnostic and statistical manual of mental disorders*, 5 ed. American Psychiatric Publishing.

Craig, P., Dieppe, P., Macintyre, S., Michie, S., Nazareth, I. *et al.* (2013) Developing and evaluating complex interventions: The new medical research council guidance. *Journal of Epidemiological Community Health* 66(12), 587–592. DOI: 10.1136/jech-2011.

Cutchin, M.P. and Dickie, V. (2013) Transactional perspectives on occupation: An introduction and rationale. In: Cutchin, M.P. and Dickie, V. (eds) *Transactional Perspectives on Occupation*. Springer, pp. 1–10.

Dewey, J. and Bentley, A.F. (1949) *Knowing and the known*. Beacon Press, Boston, MA.

Guérin, N.A., Gabriels, R.L., Germone, M.M., Schuck, S.E., Traynor, A. *et al.* (2018) Reliability and validity assessment of the observation of human-animal interaction for research (OHAIRE) behavior coding tool. *Frontiers in Veterinary Science* 5, 268.

Kalmbach, D., Wood, W. and Peters, B.C. (2020) Parental perspectives of occupational therapy in an equine environment for children with autism spectrum disorder. *Occupational Therapy In Health Care* 1–23. DOI: 10.1080/07380577.2020.1751903.

Law, M., Baptiste, S., Carswell, A., McColl, M.A., Polatajko, H.J. *et al.* (2014) *Canadian Occupational Performance Measure, Fifth Edition*. Canadian Association of Occupational Therapy Publications.

Liddiard, J. (2009) Hands up for hippotherapy: An investigation into the use of hippotherapy to promote handwriting legibility for children who have disabilities. *Scientific and Educational Journal of Therapeutic Riding* 19, 72–94.

Llambias, C., Magill-Evans, J., Smith, V. and Warren, S. (2016) Equine-assisted occupational therapy: Increasing engagement for children with autism spectrum disorder. *American Journal of Occupational Therapy* 70(6), 1–9. DOI: 10.5014/ajot.2016.020701.

McDaniel-Peters, B.C. and Wood, W. (2017) Autism and equine-assisted interventions: A systematic mapping review. *Journal of Autism and Developmental Disorders* 47(10), 3220–3242. DOI: 10.1007/s10803-017-3219-9.

O'Haire, M.E. (2017) Research on animal-assisted intervention and autism spectrum disorder, 2012–2015. *Applied Developmental Science* 21(3), 200–216.

Peters, B.C., Wood, W. and Hepburn, S. (2019) Theoretical development of equine-assisted activities and therapies for children with autism: A systematic mapping review. *Human- Animal Interaction Bulletin* 8(2), 1–35.

Peters, B.C., Wood, W., Hepburn, S. and Bundy, A. (2020) Pilot study: Occupational therapy in an equine environment for youth with autism. *OTJR: Occupation, Participation, and Health* 40(3), 190–202. DOI: 10.1177/1539449220912723.

QSR International (2017) NVivo for windows (Version 11.4.1.1064).

Ricoeur, P. (1976) *Interpretation theory: Discourse and the surplus of meaning*, 6th ed. The Texas Christian University Press, Fort Worth, TX.

Schreier, M. (2012) *Qualitative content analysis in practice*. SAGE, Thousand Oaks, CA.

Shurtleff, T. and Engsberg, J. (2012) Long-term effects of hippotherapy on one child with cerebral palsy: A research case study. *British Journal of Occupational Therapy* 75(8), 359–366. DOI: 10.4276/030802212x13433105374279.

Smith, T., Scahill, L., Dawson, G., Guthrie, D., Lord, C. *et al*. (2007) Designing research studies on psychosocial interventions in autism. *Journal of Autism and Developmental Disorders* 37(2), 354–366. DOI: 10.1007/s10803-006-0173-3.

Srinivasan, S.M., Cavagnino, D.T. and Bhat, A.N. (2018) Effects of equine therapy on individuals with autism spectrum disorder: A systematic review. *Review Journal of Autism and Developmental Disorders* 5(2), 156–175.

Tomchek, S. and Koenig, K. (2016) *Occupational therapy practice guidelines for individuals with autism spectrum disorder*. AOTA Press, The American Occupational Therapy Association, Incorporated, Bethesda, MD.

Wood, W. (2019) The 2018 Ruth Zemke lectureship in occupational science—The environment, time and occupation. *Journal of Occupational Science* 26(4), 456–469.

Wood, W.H. and Fields, B.E. (2019) Hippotherapy: A systematic mapping review of peer-reviewed research, 1980 to 2018. *Disability and Rehabilitation*. DOI: 10.1080/09638288.2019.1653997.

5 The Impact of an Equine Facilitated Learning Program on Youth with Autism Spectrum Disorder

Phyllis Erdman*, Darcy Miller and Sue Jacobson
Washington State University

Abstract

The purpose of this study was to assess the impact of an equine facilitated learning program on youth with Autism Spectrum Disorder (ASD) who were paired with typically developing peers. Even though there has been an increasing interest in the field of animal assisted interventions and the benefits for youth with ASD, to date there is no research on using equine facilitated, peer assisted learning programs, emphasizing ground work rather than riding, with youth with ASD. An exploratory, mixed methods case study design was used with 3 youth with ASD and 3 youth without disabilities. An equine program based on a natural horsemanship framework that emphasized groundwork was implemented over 10 weeks of after-school sessions. Data from behavioral observations, parent interviews, and the Social Responsiveness Scale indicate that the three youth with ASD improved their social awareness and social cognition, with two of the youth also improving their social motivation. Implications for future research and practice are discussed.

Keywords: equine facilitated learning, human animal interaction, autism spectrum disorder

Introduction

Youth with Autism Spectrum Disorder (ASD) are often challenged in the areas of social interactions and developing relationships (American Psychiatric Association, 2013). Problems with social communication, awareness, and cognition are typical characteristics of youth with ASD (National Research Council, 2001). These youth may demonstrate impaired communication skills, repetitive behaviors or speech, a lack of responsiveness, stereotyped movements, resistance to change, and unusual sensory responses (Individuals with Disabilities Education Improvement Act [IDEIA], 2004). Research has documented that children with ASD frequently engage with animals more easily than with human (Celani, 2002; Grandin and Johnson, 2005; Burrows *et al.*, 2008). Therefore, in this study we assessed the impact of an equine facilitated learning program on youth with ASD to determine if their interactions with horses contributed to improvement in social competence.

*Corresponding author: perdman@wsu.edu

Benefit of Animals with ASD

The benefits of therapy animals with children are multifaceted and vary in different contexts (Fine, 2010). For example, Burrows *et al.* (2008) identified several functions that canine therapy animals can have in the family system of a child with autism. The dog can act as a physical anchor for the child, serving as a protective safety measure; as a companion to facilitate learning new interactional skills; as a family respite to siblings and parents; and lastly, as a way to encourage an approachable presence by others to the child with autism. Others have suggested that animals, including horses, can serve as 'transactional objects' allowing a child to bond first to the animal and then to other people (George, 1988; Katcher, 2000; Fine, 2010).

When given a choice, individuals with ASD may also choose to interact with animals rather than humans (Celani, 2002). Redefer and Goodman (1989) found that the presence of a dog promoted more prosocial behaviors and decreased self-absorption in children who displayed autistic features, and the children showed fewer autistic behaviors. An even larger study was conducted by O'Haire *et al.* (2013) in which a child with ASD was grouped with 2 typically-developing peers and video-recorded during three free-play sessions with toys and three free-play sessions with guinea pigs. They found that children with ASD demonstrated more social approach behaviors (e.g. talking, making tactile contact) and received more social approaches from their peers in the play sessions with the guinea pigs than in the play sessions with toys. Children's preference for working with animals was further confirmed by Sams *et al.* (2006) in which they compared traditional occupational therapy techniques with therapy incorporating llamas with children with autism. The group receiving the llama techniques demonstrated greater use of language and greater improvement in social interactions than those receiving traditional therapy.

The Impact of Horses on Children with ASD

Research also suggests that horses can have an even stronger impact than other animals for children with ASD for several reasons. First, the size of horses creates a unique perception by children, who are usually awed by their size and power. Second, because they are prey animals they have highly effective communication systems based on body language and can easily pick up on human emotions and intentions, characteristics that children with ASD generally lack. Additionally, horses are herd animals and have strong social bonds that are based on cooperative living patterns, but they also have their own place in the herd and understand and accept the implications of their role in relation to others. Awareness of such subtle social nuances is also a characteristic lacking in children with ASD. Horses' inherent fear and need for safety, security, and clear communication demands that humans interacting with them model a calm and confident presence with a focus on the here-and-now. Such skills are generally lacking in children with ASD, so horses can serve as role models to help teach these skills (Burgon, 2011).

A recently published book and documentary by Rupert Isaacson (2009), entitled *The Horse Boy*, highlighted the healing role of horses in the life of the author's son, who has autism. Isaacson noted that although the horses didn't cure his son of autism, they opened up opportunities for his son to connect with the rest of the world in a way that nothing else had. Other researchers have documented the effectiveness of horses in therapeutic interventions, such as equine facilitated learning programs, on promoting social competence and social-emotional development in youth. Although some of these studies do not specifically target children with ASD, they do indicate success in showing improvement in interactional skills and social competence, which are skills all children need to function within society.

Burgon (2011) conducted a qualitative study with seven at-risk young people who participated in a therapeutic horsemanship program. The themes that emerged from the results of that study were related to self-confidence building, sense of mastery or self-efficacy, empathy, and opening positive opportunities. Results indicated participants developed empathy toward the horses and learned they could monitor their own behaviors and change them by modeling confident leadership with the horse. Trotter *et al.* (2008) compared equine assisted counseling (EAC) to a school-based counseling intervention. Their results indicated that EAC was effective for improving adaptive skills, such as leadership, adaptability, and social skills in children and adolescents. As reported by the participants and their parents, the program had a positive impact on the participants' internalizing behaviors particularly, which facilitates their abilities to internally cope with their problems, leading to fewer feelings of loneliness, nervousness, and anxiety. A recent study by found positive effects on youth who completed an 11-week equine facilitated learning program on aspects of social competence, such as improvement in self-awareness, self-management, personal responsibility, decision making, goal directed behavior, and relationship skills. Additionally, Pendry *et al.* (2014) found that the equine learning program resulted in participants' lower average stress hormones throughout the day, and particularly in the afternoon, suggesting that such an intervention might be a preventive measure associated with healthier development in adolescents.

Therapeutic Riding and ASD

Equine interventions that specifically include a riding element appear to benefit children with ASD across several domains of functioning. For example, Mason (2004) identified the role that sitting on a horse has on the vestibular system and other sensory processing systems of children with autism. In this study, it was found that riding increased the children's muscle tone, flexibility, and coordination, as well as their language skills, self-esteem, and social skills. The researchers also noted the potential impact on pro-social behaviors gained from interactions between the rider, the horse, and the instructor. Taylor *et al.* (2009) explored the effects of hippotherapy on the volition of three children (ages 4–6) with autism by using a single-subject A-B-B design. Volition was measured by the Pediatric Volitional Questionnaire, which defines volition as a means of assessing and explaining the motivation of children towards particular everyday activities. The improvement among the three children was inconsistent – one showed significant and consistent increase in volition; one did not show any improvement until the end; and one showed improvement after starting the program but then remained constant through the remainder.

Bass *et al.* (2009) evaluated the effects of therapeutic riding on autism by using an experimental design with 34 children who all met the criteria for DSM-IV-TR autism spectrum diagnosis. They found that the children in the experimental group improved in sensory integration and directed attention, demonstrated improved social motivation and sensory activity, and decreased inattention and distractibility. Ajzenman *et al.* (2013) conducted a pilot study looking at both structural and behavioral aspects of hippotherapy on six children (ages 5–12) with autism. They researched the impact of the horse's movement during hippotherapy activities on postural control, as well as its impact on adaptive behaviors and engagement in age-appropriate activities. Their results indicated a decrease in postural sway, and increases in overall adaptive behaviors (e.g. receptive communication and coping), participation in self-care, low-demand leisure, and social interactions. Holm *et al.* (2014) examined different doses of therapeutic riding with three boys, aged 6–8 years, with autism in a single subject design. The children participated in 1, 3, or 5 times/week of therapeutic riding to determine its impact on parent-identified behaviors typically associated with autism (e.g. tensing of muscles, pounding on surfaces, echolalia). Their findings indicated that the higher number of sessions was associated with more change, but also that the improvement was retained even after withdrawal of treatment (which for this study was defined as reverting back to 1 session/week). An important finding was that the behaviors generalized to home and community, as reported by the parents.

An interesting study by Kern *et al.* (2011) examined the effects of therapeutic riding on the severity of symptoms of autism but also looked at changes in sensory processing, quality of life, and parental treatment satisfaction. Twenty participants (aged 3–12 years) completed the entire 6 months program of riding. Even though this particular program emphasizes riding, they also built in relationship type activities with the horses (e.g. grooming, saddling, feeding them). Their results showed a significant decrease in severity of symptoms even after as little as 3 months, but more importantly, improvements were noted in quality of life by the parents over the entire 6 month course, suggesting that just enrolling in the program may have created a positive change in the participants and in their family's perception of quality of life.

Need for our study

We can conclude, then, that research supports that animals have a positive impact on children, including children with ASD. We can also conclude that horses might even have a more significant impact on children with ASD than other animals, but the majority of research in the last decade has focused on the value of therapeutic riding and/or hippotherapy with children with ASD, or on non-riding programs, such as equine facilitated learning programs on children in general, but not targeted for children with ASD. What has not been clearly documented is the impact that an equine facilitated learning program that emphasizes groundwork rather than riding, might have on children with ASD. Groundwork with horses, such as grooming and moving and leading the horse through obstacles, involves focusing, listening, communicating, and interacting with the horse and with other humans even more than riding. In most therapeutic riding programs, the riders usually have lead walkers who steer, or assist in steering, the horse for them. However, in groundwork activities, the child is solely responsible for the movement and behaviors of the horse so the need for effective interactive skills is particularly emphasized. Therefore, the purpose of our study was to document the impact that an equine facilitated learning program that emphasized groundwork had on youth with ASD who were paired with typically developing peers. The groundwork activities included grooming the horse, learning to read and understand the horse's movements (e.g. ears moving, tail twitching), desensitizing the horse to touch and unfamiliar sounds, walking them with lead ropes through an obstacle course, and learning to move the horse forward and backward with a finger touch or by wiggling the lead line. We paired them with typically developing peers because of the positive impact such a relationship could provide to the children with ASD. Interactions with nondisabled peers can have a positive impact on the development of social interaction and communication skills of children with ASD and have been shown to improve social communication, interactions, and awareness (Carter *et al.*, 2001; Carter *et al.*, 2011; Koegel *et al.*, 2012). Research on interventions that involve nondisabled youth and youth with ASD, has also shown that these peer mediated/assisted interventions can improve the social interaction skills, and expand the friendship networks and social connections of youth with ASD (Carter *et al.*, 2011). Social and relational skills are integral components of social competence, which is viewed as a larger construct that encompasses a diverse set of skills and abilities, related to perceptions, social awareness, social cognition, and emotional regulation (Bierman and Welsh, 1997; Dirks *et al.*, 2007). Possessing social competence is seen as critical for success in school and community for all children (Gresham, 1986; Semrud-Clikeman, 2007).

Definitions

In order to provide a common understanding of the field of equine interventions, we provide the following definitions, which are provided by the Professional Association of Therapeutic Horsemanship (PATH), International. PATH International is the professional association that accredits therapeutic riding programs, including therapeutic learning programs and hippotherapy, among other types of equine programs.

Therapeutic riding, as defined by PATH, International is:

> *... an equine-assisted activity for the purpose of contributing positively to the cognitive, physical, emotional and social well-being of individuals with special needs. (Learn about EAAT, 2014)*

Although many therapeutic riding programs focus primarily on riding, they also include varying degrees of ground work with horses, such as grooming, leading and walking with a horse, or causing a horse to move within a round pen.

Equine Facilitated Learning (EFL), as defined by PATH, International is:

> *... an educational approach to equine-assisted activities. EFL content is developed and organized by credentialed practitioners with the primary intent to facilitate personal growth and development of life skills through equine interactions. (Learn about EAAT, 2014)*

These programs usually provide a combination of riding and groundwork activities.

On the other hand, The American Hipppotherapy Association, Inc. defines hippotherapy, in which the focus is on riding as:

> *... a physical, occupational or speech therapy treatment strategy that utilizes equine movement. The word hippotherapy derives from the Greek word hippos, meaning horse. The term hippotherapy refers to the use of the movement of the horse as a treatment strategy by physical therapists, occupational therapists and speech/language pathologists to address impairments, functional limitations and disabilities in patients with neuromotor and sensory dysfunction. This treatment strategy is used as part of an integrated treatment program to achieve functional goals. (Learn about EAAT, 2014)*

Method

This exploratory, mixed methods case study (Merriam, 1998) was designed to examine the impact of an equine facilitated, peer-assisted learning program on youth with ASD, with attention to their social competence. We chose a case study design because our question was exploratory. There is no research that documents the interaction, involving ground work activities, between children with ASD and a horse. We also chose recorded observations so we could track any changes that occurred during the sessions. The participants' social skills and interactions with the horses, and with others, were examined by observing them at the beginning, middle, and end of the program. The parents' perspectives on the youths' social responsiveness were explored using both a standardized measure and qualitative interviews.

Participants

Four students diagnosed with ASD and four peers without disabilities were recruited through the local public middle and high schools to participate. One participant was dropped from the study due to incomplete data. Two counselors from the local middle and high schools were asked to nominate students with ASD using a set of criteria that included youth who: a) were between the ages of 11 to 15 years old; b) had the physical ability to participate in an equine assisted program, c) had the ability to follow verbal/physical instructions, and d) could verbally communicate. The counselors were also asked to nominate peers without disabilities who: a) were between the ages of 11–15 years old; b) had the temperament and patience to work with youth with ASD; c) expressed comfort in working with horses; and d) had the ability to physically complete the program. The four typically developing peers were not from the same school as the participants with ASD. We intentionally selected peers who were a grade or two older than the participants with ASD and had no previous history with them, assuming this would replicate a 'real life' situation where they had to initiate and practice interpersonal skills with strangers.

The participants with ASD had the diagnosis of moderate ASD and met the DSM-5 criteria for Autism Spectrum Disorders (American Psychiatric Association, 2013) as well as the criteria in the Individuals with Disabilities Education Improvement Act (2004). The three participants with ASD were males with a mean age of 12–6 (12-1, 14-0, 11-4) and the 3 peers without disabilities were females with a mean age of 13–8 (12-1, 14-2, 15-2). The gender of each group (all males for the ASD group, and all females for the typically developing peers) was based on nominations and availability and not intentional. The study was approved by the university's Institutional Review Board and the Institutional Animal Care and Use committee. All parents consented to have their children participate, and to be interviewed, and all participants assented.

Materials and procedure

In this mixed methods case study, both qualitative and quantitative data were collected. Behavioral observations were captured using digital recordings and behaviors were coded for frequency. Parents completed the Social Responsiveness Scale (SRS) (Constantino and Gruber, 2012) and were interviewed before and after the program implementation.

Behavioral observations

Research assistants assigned to each participant/peer pair observed them at the beginning of the program (week 2), in the middle of the program (week 6), and at the end of the program (week 9). Each participant/peer pair and their horse handler were video-recorded for an hour using a small, hand-held digital camera, as the youth completed the tasks outlined for each session (e.g. grooming the horse, leading the horse). At the completion of the program, two research assistants viewed each of the one-hour videos and coded the participants' with ASD social and behavioral interactions. The observation protocol was built on and aligned with the subscales of the SRS resulting in the following categories: social awareness, social cognition, social communication, social motivation, and restricted interests/repetitive behavior. Two additional categories of 'humor' and 'confidence' were added. All of the categories of interest were operationally defined using the criteria from the SRS as well as the research literature on social responsiveness and social competency (Semrud-Clikeman, 2007). Table 5.1 provides descriptive examples of the behaviors/social indicators that were coded for each category. One of the researchers and two graduate students completed a norming session to agree on what behaviors might be rated for each category. After viewing 2 hours of video and reaching inter-rater reliability of 0.85 (number of agreements/number of agreements + disagreements), the research assistants independently viewed all of the participants' videos. A second researcher or research assistant viewed the videos and coded behavior on all participants. Data recording occurred at the end of each 10 min segment during each 60 min video, resulting in six data sets for each participant with ASD for each hour of video. At the end of each 10 min segment, the observer provided a rating of what was observed for each category of behavior using the Likert-type scale of: 1 = Not observed, 2 = Observed some indication, 3 = Observed several occurrences, and 4 = Observed consistently throughout segment. Observers viewed the 10 min segment as many times as needed so that the data recorded would be as accurate as possible. Inter-rater reliability was tested by the researcher using observational data collected from 1 hour of video taken from the beginning, 1 hour in the middle, and 1 hour at the end of program, resulting in agreements of 0.85, 0.88, 0.87 (number of agreements/number of agreements + disagreements).

Social responsiveness

The Social Responsiveness Scale (Constantino and Gruber, 2012) was completed by each parent of the ASD group before and at the end of the program. The SRS contains 65 items that provide an indication of the level of functioning across five domains: social awareness, social cognition, social communication, social motivation, and restrictive/repetitive behaviors. The internal consistency of the SRS is high (Alpha 0.95) (Constantino, 2002; Constantino and Gruber, 2012).

The SRS items are responded to using a Likert scale with the following options: 1-Not true, 2-Sometimes true, 3-Often true, and 4-Almost always true. An example of one of the prompts is 'Has difficulty making friends, even when trying his or her best.' Some prompts, such as, 'Is able to communicate his or her feelings to others' are reversed scored. The subscales scores include raw and standardized scores (t scores) with the total SRS standardized score ranging from 90 to 30 (mean of 50; standard deviation of 10) indicating social responsiveness

Table 5.1. Categories Observed and Examples of Social/Behavioral Indicators

Observation Categories	Examples of Social/Behavioral Indicators Coded
Social Awareness	Behaviors or communication indicating ability to pick up on social cues; using any sense (kinesthetic, listening, vision) to pick up on social cues
Social Cognition	Behaviors or communication indicating ability to interpret social cues once they are picked up
Social Communication	Behaviors or communication that show expressive social interactions; these may include some or all of the following: expressing emotion appropriately, imitating appropriate actions/words, expressing empathy (comfort, words, etc.) toward others or the horse, relating to others, handling mood changes appropriately, demonstrating a give and take conversation, making eye contact, politeness, staying with an activity (not wandering aimlessly), laughing appropriately, answering questions clearly and appropriately, talking in normal tones, respecting personal space, showing flexibility, and/or being emotionally present (not distant)
Social Motivation	Behaviors and communication that show the extent to which the student is motivated to engage in social-interpersonal behavior; is not anxious, inhibited
Restricted Interest/Repetitive Behavior	Behaviors and communication that are bizarre, show perseveration, inflexibility, unusual sensory interest (mouthing, spinning, or strange interactions with items), inflexibility when stressed, touching others in odd ways, talking about the same thing over and over, doesn't like changes in routine
Confidence	Behaviors or communication that indicate a sense of empowerment or strength; verbal or behavioral indications that the student is proud of himself, physical use of the body to show confidence (arm up, arms up, thumbs up, standing up straight)
Humor	Behavior or communication that indicates sense of humor; verbal or nonverbal behavior indications that the student told a joke or made an appropriate funny comment; laughing appropriately at someone's joke; smiling at a joke or funny statement

Note: Social awareness, cognition, communication, motivation, and restricted interest/repetitive behavior descriptions adapted from SRS manual (Constantino and Gruber, 2012).

functioning levels (59 and below: within normal limits; 60–65: mild deficiencies; 66–75: moderate deficiencies; 76–90: severe deficiencies). The higher the SRS standardized scores are, the more impairment is indicated.

Parent interviews

Researchers conducted pre- and post-program interviews with parents of the ASD group. Interviews were recorded for analysis. The pre-interviews were semi-structured and included the following questions:

- What are your goals for your child for the program in terms of developing social interaction skills? In relation to peers/adults/community/sibling and/or family members?
- What do think are your child's strengths/weaknesses?
- Talk about how your child takes/or doesn't take initiative, particularly in regards to social interactions?
- What tends to happen when he/she is challenged by a particular task/ or when he/she get frustrated?

- In what areas does your child demonstrate independence, and what areas is he/she more dependent?
- Do you have any concerns about safety regarding any of your child's behaviors?
- Are there any other specific behaviors, personality traits, or characteristics that you think we should know about your child?

Post-program interviews were also semi-structured using the same questions, but more focused on any changes that the parents observed. We started the interview by asking the parents to think about how their child is doing now, after completing the program. We then asked the following questions:

- Do you think your child attained any of the goals you identified for him/her for the program in terms of developing social interaction skills? In relation to peers/adults/community/sibling and/or family members?
- Talk about any changes you see now in your child's strengths/weaknesses.
- Do you notice any changes now in how your child takes/or doesn't take initiative, particularly in regards to social interactions?
- Do you notice any changes now when your child is challenged by a particular task/ or when he/she get frustrated?
- Talk about any changes you see now in how your child demonstrates independence, and what areas is he/she more dependent?
- Do you have any concerns now about safety regarding any of your child's behaviors?
- Are there any other specific behaviors, personality traits, or characteristics that you think we should know about your child now?

Procedure

Each of the students with ASD was paired with a peer without disabilities, a horse handler, and a horse. Pairings of students and horse assignments were made by the equine instructor in consultation with one of the researchers. Criteria for pairing the students was based on best fit using information received from applications completed by the parents. The assignment of horses and horse handlers was made by the equine instructor based on the personality of the horse and the personality and experience of the horse handlers. All of the horses were owned by the local university and were experienced therapeutic riding horses that are part of a PATH, International Premier Accredited therapeutic riding program at the university. Horse handlers were volunteer undergraduate or graduate students with horse experience and had previous experience with the established therapeutic riding program. The person who provided the equine lessons was a PATH International certified therapeutic riding instructor and equine specialist in mental health/learning.

The 10-week program consisted of weekly after-school sessions that lasted about 75 min. The participants were bussed to the arena from their school, and parents picked them up at the end of the session. Although horse experience was not required for the program, some of the participants had some riding experience, usually with a relative's or friend's horse. None of the participants had any experience with groundwork, which was the focus of our program. We based our program on a natural horsemanship approach that emphasized primarily groundwork activities with horses, such as grooming, leading horses through obstacle courses, and learning how to move horses toward/away from the person and around obstacles through gentle touch. Natural horsemanship is based on concepts of trust and relationship building with horses before engaging in any kind of activities. Since horses are prey animals and naturally fearful of unfamiliar surroundings and people, it is important to first establish an environment of trust and safety. Once this relationship is established between the horse and the human, the horse is more receptive to learning, creating a mutually positive environment. The curriculum centered on weekly themes that were grounded in establishing trust and safety. Themes included identifying and showing respect, using nonverbal communication and body language, differentiating between assertive and aggressive behavior, building teamwork, establishing and recognizing boundaries, and gaining confidence with challenges. Each lesson started with an overview of safety rules, a description of the equine lesson and a demonstration, and followed with grooming their horse in pairs and practicing the equine lessons in pairs. Riding was not really the focus of the program, but since most children enjoy riding, it was an optional activity that we provided toward the end of three of the lessons. All children chose to ride, and while one participant of the pair rode, the other participants engaged in brief word games or a small group activity with one of the staff on site. Parents completed the Social Responsiveness Scale and were interviewed at the beginning of the program and again at the end.

Results

Quantitative data

Results from the quantitative measures included observed frequencies from videotaped sessions, which were coded and rated by observers, and parents' responses on the Social Responsiveness Scale (pre and post). A detailed analysis of these results follows.

Observed frequency of social responsiveness, humor, and confidence

The ratings assigned by each observer to each of the six, 10 min segments over the hour of observation were averaged, resulting in one rating for that category (e.g. social communication) for that hour. The two observers' ratings for each category and hour were then averaged resulting in the participants receiving one rating for each category for the beginning, middle, and end of program for each category (Table 5.2).

Participant 1 increased his frequency of observed behaviors in the categories of social awareness, social cognition, social motivation, and confidence. His social communication behaviors were consistently observed throughout the program. This participant began the program with consistently observed behaviors related to humor, with a slight decrease at the end of the program. Over the program, Participant 1 exhibited low or no frequency of restricted interests and repetitive behaviors.

Participant 2 showed increases in the observed frequency of his behaviors in all subscale categories except restricted interests and repetitive behaviors, neither of which were observed during the program. He showed the highest increase in behaviors in the category of confidence (2.6, 3.7, 3.9), and similar strong increases in the categories of social communication and social motivation. Participant 3 exhibited more frequent behaviors related to social awareness, social cognition, social communication, humor, and confidence over the program. His behaviors related to social motivation were observed consistently at the beginning and end of the program, with a drop in frequency during the middle of the program. Behaviors related to restricted interests and repetitive behaviors were observed less frequently at the end of the program.

Social responsiveness

Total score results and treatment subscale results were obtained from the administration of the SRS with parents pre- and post-program (Table 5.3). The scores indicate severity of ASD symptoms from the parents' perspectives. A positive outcome would be a reduction in the value of the scores from the pre- to post-program. Although the total score results for Participant 1 showed a slight increase in severity of ASD symptoms (72 to 74), his treatment subscale results indicate a reduction in severity of symptoms in the categories of social awareness and social cognition. Participants 2 and 3 both reduced the symptoms of ASD (88 to 82, 90 to 84, total scores respectively) and improved in all subscale categories except for a slight increase in symptoms related to social communication for Participant 3.

Qualitative data

Qualitative data analysis included identifying themes from parent interviews (pre and post). The researchers recorded the interviews and then coded them for themes (Rubin and Rubin,

Table 5.2. Frequency Ratings[a] for Social Responsiveness Categories of Behavior across the Beginning, Middle, and End of Program

Participants	Social Awareness			Social Cognition			Social Communication			Social Motivation			Restricted Interests and Repetitive Behaviors			Humor			Confidence		
	B[b]	M	E	B	M	E	B	M	E	B	M	E	B	M	E	B	M	E	B	M	E
Participant 1	3.6	3.9	4.0	3.6	3.9	4.0	4	4	4	2	3.4	3.5	1	1.8	1	4	4	3.6	2.6	2.8	3.5
Participant 2	3.4	3.6	3.8	3.2	3.8	3.8	2.7	3.2	3.7	2.9	3.4	3.9	1	1	1	3.1	3.5	3.6	2.6	3.7	3.9
Participant 3	3	3.25	4.0	2.5	3.25	3.0	3.0	3.5	3.3	4.0	3.2	4.0	2	2	1.3	1.5	2.25	2.3	1.5	2.5	3.3

[a]Not observed – 1; Observed some indication – 2; Observed several occurrences – 3; Observed consistently – 4
[b]Beginning of Program – B; Middle of Program – M; End of Program – E

Table 5.3. Total Scores and Treatment Subscale Results[a] from the Social Responsiveness Scale Pre-/Post-Program

Treatment Subscale Results

	Total Score Results	Social Awareness	Social Congnition	Social Communication	Social Motivation	Restricted Interests/ Repetitive Behaviors
	Pre/Post	*Pre/Post*	*Pre/Post*	*Pre/Post*	*Pre/Post*	*Pre/Post*
Participant 1	72/74	67/64	63/61	74/75	71/73	69/76
Participant 2	88/82	73/70	87/74	83/81	77/71	96/91
Participant 3	90/84	76/73	92/83	86/86	75/71	94/82

[a]t-Scores Interpretation: Severe: >=76; Moderate: 66-75; Mild: 60-65; Normal: <=59

1995). A summary of each parent's comments is included in (Table 5.4).

Parents' perspectives

Themes that emerged pre-program indicated parents' concerns with their children's social skill deficits. These included: lack of engagement in social interactions; frustration/low coping skills when confronted with a challenge; inability to express emotions appropriately; and lack of awareness of others and environment. During the post interviews, parents consistently revealed themes that indicated a change in the children's skills. These included: an increased willingness to ask for help; increased interactions with peers; and showing less frustration when challenged. The parents of Participant 1 also noted that he had established empathy with animals, he was not getting into as much trouble at school, and there was a dramatic increase in asking for help. At the start of the program, Participant 1's inability to ask for help was very frustrating to these parents. They noted that he would never ask for help with homework, and then would get very frustrated when it was wrong. The parents of Participant 2 expressed surprise that they now observed their child expressing confidence when he took charge of a situation and would problem solve – something they had not previously seen in him. The parent of Participant 3, who was the most severely withdrawn of all three participants, and had the most difficulty in respecting social boundaries, reported that he was more aware of boundaries and consequences of his action, and learned that 'you have to ask first' before touching someone.

Discussion

The purpose of this study was to assess the impact of an equine facilitated learning program on youth with ASD using qualitative and quantitative measures. The qualitative and quantitative results indicate that the participants improved in many areas of social competence. From the parents' perspectives (from interviews and the Social Responsiveness Scale) and the observations conducted, there were increases in the participants' social awareness, social cognition, social communication, social motivation, humor, and confidence.

The horse handlers, who did not code videos or collect data, often anecdotally commented to the researchers that the participants seemed to improve their conversational skills (with them and the peer participants). They also commented that they gained confidence grooming, handling, and leading the horses. An interesting observation was that they commented specifically about how the participants with ASD became more comfortable understanding humorous comments (e.g. appropriate jokes in context) offered by themselves or others. This was particularly important because appropriately engaging with peers in social interactions and grasping subtle nuances of casual joking, was one of the main concerns expressed by the parents of the ASD group. Such social skills are typically a difficult task for children with ASD and reinforces their social isolation. Hence, observing this kind of casual exchange was a positive indicator of increased comfort with others.

Table 5.4. Summary of Parent Interviews

Participant Parents	Pre-program Themes	Post-program Themes
Participant 1 parents	Not open to social interactions	Dramatic increase in asking for help/clarification appropriately
	Lacks social skills	
	Low coping skills when challenged	Very talkative when picked up after program/open to sharing – not talkative when picked up after school
	If asks for help but not assisted right away he might have a tantrum	
	Not always aware of environment/internally focused	Having an easier time at school, and not getting into trouble as much
		Able to establish an empathic relationship with animal which was great
Participant 2 parents	Doesn't engage in social interactions	Got to do something he felt good about, and it raised his confidence/self-worth
	Gets frustrated when peers do not like him and they get frustrated with him	Interacted with peers much more
	Doesn't like the unfamiliar	Ask for help appropriately
	Is not socially aware	Expressed confidence when he took charge of a situation and solved the problem – we were surprised
	Breaks down if challenged	
	Can't express his frustrations	
Participant 3 parents	Doesn't know physical, social, cultural boundaries	More aware of boundaries as a result of working on boundaries with horses
	Highly influenced by peers	Talks now about safety issues
	Hard to communicate appropriately	Pays attention to consequences of behavior
	Can't handle frustrations – gets angry	Before program he would have a tantrum first and then ask for help; now he asks for help first.
	Wants friends, but doesn't have many	

Participant 2 began the program exhibiting shyness and very few non-verbal or verbal indications that he was enjoying himself. As the program continued, he began to joke with and ask questions of Participant 1, his assigned peer, and his horse handler. He made more eye contact and at the end of the program strode confidently with the horse across the arena, with his arm held high clutching the lead rope, completing all of the required tasks. He had a big smile on his face as he returned to the starting position.

Of note was the decline in 'humor' of Participant 1 at the end of the program. Participant 1 consistently demonstrated humor-related behaviors but demonstrated less humor-related behaviors at the end of the program. In reviewing the video from that session, we attribute this possibly to the seriousness and intensity of the content of that session. For that session participants were required to lead their horse through several activities requiring them to work as partners with their peers and demonstrate that they had learned the proper techniques and strategies. Hence, they were required to demonstrate horsemanship skills, clear verbal and non-verbal communication, leadership skills, and teamwork. It appeared that Participant 1 took this lesson very seriously and was intent on completing all the activities correctly, adding to his more serious demeanor during this particular session.

The typically developing peers who were assigned to each participant with ASD were older than the ASD participants and had no history of knowing them previously. We believe this was probably a positive factor because social interactions with a stranger in an unfamiliar environment was probably a more challenging task for the ASD group than if they had known their peer partners.

Limitations

An obvious limitation of this study is the case study format, and lack of control group, which prevents us from concluding a cause and effect relationship between the positive outcomes on the variables of interest and the intervention. However, the multiple sources of data reinforce consistent patterns of improvement. There is no way of knowing whether these improvements in behaviors will be long term or will transfer to other contexts. A longitudinal follow up, with interviews from parents and others in different contexts (e.g. teachers) would help determine if the skills the participants learned could be generalized to other situations. However, the comments from parents in the post interviews did suggest that this was happening, at least for the short term. We cannot attribute improvements directly to the equine intervention. The lack of a comparison treatment group has been noted as an inherent limitation in much of the human-animal interaction research (Marino, 2012). The special time together, the consistency and predictability in how the lessons were delivered, the gender of the participants, or potential bias in parental reports could have also impacted the results. However, the power of horses with children with ASD is very well documented, and even though the documentation is often anecdotal it is hard to deny that there is something special in that relationship.

Implications for Future Research and Practice

This study contributes to and confirms existing research; but more importantly, creates additional questions for future researchers. It supports other research that indicates children with ASD can benefit from equine interventions, but also provides a unique contribution by focusing on groundwork rather than riding and pairing of youth with ASD with typically developing peers. Future research could build on this study by adding a comparison group where children with ASD are paired with typically developing peers in a different type of program without horses or are compared to a program that included dogs or different types of animals. This would not only tease out whether animals contribute a unique quality to the intervention, but more specifically, whether there is something particularly unique about horses. Additionally, a 6 month or longer follow-up would help determine the long term impact of the intervention. Also, a mid-point interview with the parents might be helpful to determine at what point they started to identify changes. Most equine programs tend to be 8–10 weeks, but there is no research that compares lengths of program. Perhaps one or two sessions would be equally effective as ten.

Diversity in gender of participants should also be explored. We found that most of the children in our area identified with ASD tended to be male. A future study that included mixed gender for both the children with ASD, as well as the typically developing peers, would also be helpful.

Acknowledgements

Special thanks to Heather Burchell, Ph.D. Candidate in Special Education, Washington State University, for assistance with video coding.

References

Ajzenman, H., Standeven, J. and Shurtleff, T. (2013) Effect of hippotherapy on motor control, adaptive behaviors, and participation in children with autism spectrum disorder: A pilot study. *The American Journal of Occupational Therapy* 67(6), 653–663.

American Psychiatric Association (2013) *Diagnostic and statistical manual of mental disorders*, 5th ed. Author, Washington, DC.

Bass, M., Duchowny, C. and Llabre, M. (2009) The effect of therapeutic horseback riding on social functioning in children with autism. *Journal of Autism and Developmental Disorders* 39(9), 1261–1267.

Bierman, K. and Welsh, J. (1997) Social relationship deficits. In: Mash, E.J. and Terdal, L.G. (eds) *Assessment of Childhood Disorders*. Guilford Press, New York, pp. 328–365.

Boyatzis, R. (1998) *Transforming qualitative information: Thematic analysis and code development*. SAGE Publications, Thousand Oaks, London, & New Delhi.

Burgon, H. (2011) Queen of the world: Experiences of 'at-risk' young people participating in equine-assisted learning/therapy. *Journal of Social Work and Practice* 25(2), 165–183.

Burrows, K., Adams, C. and Spiers, J. (2008) Sentinels of safety: Service dogs ensure and enhance freedom and well-being for families with autistic children. *Qualitative Health Research* 18(12), 1642–1949.

Carter, E., Hughes, C., Copeland, S. and Breen, C. (2001) Differences between high school students who do and do not volunteer to participate in a peer interaction program. *Research and Practice for Persons with Severe Disabilities* 26, 229–239.

Carter, E., Moss, C., Hoffman, A., Chung, Y. and Sisco, L. (2011) Efficacy and social validity of peer support arrangements for adolescents with disabilities. *Exceptional Children* 78, 107–125.

Celani, G. (2002) Human beings, animals and inanimate objects? What do people with autism like? *Autism* 6, 93–102.

Constantino, J. (2002) *The social responsiveness scale*. Western Psychological Services, Los Angeles.

Constantino, J. and Gruber, C. (2012) *Social responsiveness scale*, 2nd ed. Western Psychological Services, Torrance, CA.

Creswell, J. and Plano-Clark, V. (2011) *Designing and conducting mixed methods research*, 2nd ed. SAGE Publications, Thousand Oaks, CA.

Dirks, M., Treat, T. and Weersing, V. (2007) Integrating theoretical, measurement, and intervention models of youth social competence. *Clinical Psychology Review* 27, 327–347. DOI: 10.101016/j.cpr.2006.11.002.

Fine, A. (ed.) (2010) *Handbook on animal assisted therapies: Theoretical foundations and guidelines for practice*. Elsevier, Pomona, CA.

George, H. (1988) Child therapy and animals. In: Schaefer, C.E. (ed.) *Innovative Interventions in Child and Adolescent Therapy*. John Wiley, New York, pp. 400–418.

Grandin, T. and Johnson, T. (2005) *Animals in translation*. Houghton Mifflin Harcourt, New York.

Gresham, F. (1986) Conceptual issues in the assessment of social competence in children. In: Strain, P., Guralinick, M. and Walker, H.M. (eds) *Children's Social Behavior*. Academic Press, London, pp. 143–180.

Holm, M., Baird, J., Kim, Y., Rajora, K., D'Silva, D. *et al.* (2014) Therapeutic horseback riding outcomes of parent-identified goals for children with autism spectrum disorder: An ABA multiple case design examining dosing and generalization to the home and community. *Journal of Autism and Developmental Disorders* 44, 937–947.

Individuals with Disabilities Education Improvement Act of 2004 (IDEIA) (2004) Pub.L. No. 108-446, 118 Stat. 2647, 31 U.S.C.

Isaacson, R. (2009) *The horse boy*. Little Brown Company, New York, NY.

Katcher, A. (2000) The future of education and research on the human-animal bond and animal-assisted therapy. Part B: Animal-assisted therapy and the study of human-animal relationships: Discipline or bondage? Context or transitional object? In: Fine, A. (ed.) *Handbook on Animal-Assisted Therapy: Theoretical Foundations for Guidelines and Practice*. Academic Press, San Diego, pp. 461–473.

Kern, J., Fletcher, C., Garver, C., Mehta, J., Grannemann, B. *et al*. (2011) Prospective trial of equine-assisted activities in autism spectrum disorder. *Alternative Therapies* 17(3), 14–23.

Koegel, R., Fredeen, R., Kim, S., Danial, J., Rubinstein, D. *et al*. (2012) Using perseverative interests to improve interactions between adolescents with autism and their typical peers in school settings. *Journal of Positive Behavior Interventions* 14, 133–141.

Learn about EAAT (2014). Available at: http://www.pathintl.org/resources-education/resources/eaat (accessed 28 September 2014).

Marino, L. (2012) Construct validity of animal assisted therapy and activities: How important is the animal in AAT? *Anthrozoös* 25, 139–151.

Mason, M. (2004) Effects of therapeutic riding in children with autism. Unpublished dissertation, Capella University, Minneapolis, MN.

Merriam, S. (1998) *Qualitative research and case study applications in education*. Jossey-Bass, San Francisco.

National Research Council (2001) *Educating children with autism*. National Academic Press, Washington, DC.

O'Haire, M., McKenzie, S., Beck, A. and Slaughter, V. (2013) Social behaviors increase in children with autism in the presence of animals compared to toys. *PLOS ONE* 8(2), 1–10.

Pendry, P., Smith, A. and Roeter, S. (2014) Randomized trial examines effects of equine facilitated learning on adolescents' basal cortisol levels. *Human-Animal Interaction Bulletin* 2(1), 80–95.

Redefer, L. and Goodman, J. (1989) Brief report: Pet facilitated therapy with autistic children. *Journal of Autism and Developmental Disorders* 19(3), 461–467.

Rubin, H. and Rubin, I. (1995) *Qualitative interviewing: The art of hearing data*. Sage, Thousand Oaks, CA.

Sams, M., Fortney, E. and Willenbring, S. (2006) Occupational therapy incorporating animals for children with autism: A pilot investigation. *American Occupational Therapy Association* 60, 268–274.

Semrud-Clikeman, M. (2007) *Social competence in children*. Springer, New York, NY.

Taylor, R., Kielhofner, G., Smith, C., Butler, S., Cahill, S. *et al*. (2009) Volitional change in children with autism: A single-case design study of the impact of hippotherapy on motivation. *Occupational Therapy in Mental Health* 25, 192–200.

Trotter, K., Chandler, C., Goodwin-Bond, D. and Casey, J. (2008) A comparative study of the efficacy of group equine assisted counseling with at-risk children and adolescents. *Journal of Creativity in Mental Health* 3(3), 254–284.

6 Impact of Adaptive Devices on Horses, Assisting Persons, and Riders with Cerebral Palsy in a Therapeutic Riding Program

Lana Kaiser[1], Mathew J. Edick[6], Katherine Smith Buckler[1], Camie Heleski[2], Sherman Gorbis[3], Bonnie DePue[6], Liz Lewis[2] and LeeAnn J. Kaiser[5]

[1]Michigan State University Department of Medicine, College of Human Medicine; [2]Michigan State University Department of Animal Science, College of Natural Science; [3]Michigan State University College of Osteopathic Medicine; [4]Michigan State University Department of Large Animal Clinical Sciences; [5]Michigan State University College of Veterinary Medicine; [6]CHUM Therapeutic Riding

Abstract

Hypothesis: The goal of this study was to evaluate the impact of three different adaptive devices used to facilitate therapeutic horseback riding for riders with cerebral palsy (CP), the horses, and the assisting persons. The study hypothesis was that there is no difference in clinical benefit to the rider; stress or physical saddle pressure to the horse; or the level of exertion required from assisting persons when using any of the three adaptive devices. Population: Individuals with varying disability due to cerebral palsy who participated in a therapeutic horseback riding program at a Professional Association of Therapeutic Horsemanship (PATH) Premier Accredited Center.

Intervention: Therapeutic horseback riding sessions using one of three adaptive devices. Comparators: Two commonly used adaptive devices (English saddle and bareback pad with backrider), and one recently developed, less-used device (Independence saddle). Outcomes: Clinical benefit for the individual with cerebral palsy was evaluated using an osteopathic musculoskeletal exam score. Equine stress was evaluated using a horse stress behavior ethogram. Physical pressure under the saddle of the horse was evaluated using a Pliance saddle pad under each adaptive device. Required level of exertion by assisting persons was measured using an assistant exertion score.

Conclusion: Therapeutic riding using an English saddle, a backriding pad with backrider, or an Independence saddle, by individuals with various degrees of disability, does not cause excessive stress or result in excessive pressure to the horse. Using one assistive device over another also does not appear to change the likelihood of clinical benefit to the rider. However, exertion scores for assisting persons were lower when the Independence saddle was used compared to the English saddle or backriding pad. Adaptive devices that allow an individual with CP to improve function, ride more independently, provide some respite for the side walker and do not cause undue stress or frustration to the horse would be a great addition to the therapeutic regimen. Based on the results of this study, the Independence saddle is a viable alternative to the English saddle or the backriding pad for riders that would otherwise need extensive assisting person support in a therapeutic riding program.

Keywords: cerebral palsy, therapeutic riding, equine behavior

Introduction

Therapeutic riding (TR) is an equine-assisted activity that aims to improve the mental and physical well-being of participants through the act of riding a horse. Equine assisted activities (EAA) have been used as an intervention for individuals with a broad range of physical and psychological issues (Griffith, 1992; Engle, 1997; Huang *et al.*, 2006; All *et al.*, 1999), and anecdotal evidence and case studies have heralded the positive effects of EAA for decades (Mackinnon *et al.*, 1995; Engle, 1997). Many studies have focused on individuals with specific physical handicaps, particularly cerebral palsy (CP) and multiple sclerosis (Bertoti, 1988; MacKay-Lyons *et al.*, 1988) Early studies of EAA have been criticized for a lack of scientific rigor, small numbers of participants, lack of non-riding controls, and not including the most severely affected individuals (Sterba, 2007). However, with recent scientific interest in the interaction of humans and animals and the therapeutic role of animals, research teams are evaluating EAA for a variety of physical and psychological conditions in a more rigorous fashion. Despite the limitations, evidence suggests that TR and hippotherapy are efficacious, and may be medically indicated as therapy for gross motor rehabilitation for individuals with cerebral palsy (Mackinnon *et al.*, 1995; Engle, 1997; Benda *et al.*, 2003; Snider *et al.*, 2007; Park *et al.*, 2014; Laiou *et al.*, 2017).

The term cerebral palsy is used to describe a group of permanent movement and posture disorders that limit activity and are attributed to non-progressive insult to the fetal or infant brain (Rosenbaum *et al.*, 2002). The Gross Motor Function Classification System is a method of grading the severity of impairment of people with cerebral palsy (https://www.canchild.ca/en/resources/42-gross-motor-function-classification-system-expandedrevised-gmfcs-e-r; Palisano *et al.*, 1997; Palisano *et al.*, 2000). Grades I and II include individuals who are able to walk without assistance, whereas those in Grades III–V have increasingly limited self-mobility. More severely affected individuals may lack strength in their core musculature and consequently have poor control of their trunk movements.

To obtain physical benefits of TR and to ride safely, riders with CP and other physical disabilities may require adaptive devices and several assisting persons. One adaptive device, the Independence Saddle™ (IS), was developed to allow a person with 'severe physical challenges to sit safely and independently astride a horse, while the traditional physical dependency on side walkers is dramatically decreased' (https://www.independentstrides.com/). Additionally, the IS is a balanced saddle and may remove a source of stress and discomfort for the horse, since weight balance issues are a primary trigger for horse stress responses in this rider population (Foster and Wallach, 2015). Conversely, because of the increased size, bulk and mass of the IS when compared to traditional adaptive devices, the IS may be more stressful and aversive to the horse than other adaptive devices. The effect of the IS on the horse has not been investigated.

The goal of this study was to evaluate the impact of three different adaptive devices used to facilitate therapeutic horseback riding on the rider (an individual with cerebral palsy), the horse (a therapeutic riding horse), and the assisting persons. The study hypothesis was that there is no difference in clinical benefit to the rider, stress behaviors or physical saddle pressure to the horse, or the level of exertion required from assisting persons when using any of the three adaptive devices.

Methods

Riders

Ten individuals with CP, currently riding weekly, participated in this study (Table 6.1) at a Professional Association of Therapeutic Horsemanship International (PATH Intl. formerly NARHA) (http://www.pathintl.org) Premier Accredited Center. Inclusion criteria were: a diagnosis of CP; originally referred by a physician to the therapeutic riding program; currently participating in TR at CHUM; no ongoing medical conditions at the time of enrollment that might limit ability to participate (i.e. urinary tract infection, decubitus ulcers, chronic respiratory disease); and parents/guardians willing to be involved in extra rides for saddle pressure measurements that would require additional travel to CHUM. Individuals with CP meeting these criteria that

Table 6.1. General characteristics of riders.

ID	Age (yr)	Gender	Weight (kg) / Height (in)	Motor Classification / GMFCS	Topographical Classification	Able to Ambulate	Verbal	Independent (life)	Years Riding
5	18	M	55/54	Spastic / III	Paraplegic	With Lofstrand crutches	Yes	Yes	15
6	21	M	54/54	Spastic / V	Quadriplegic	No, wheelchair	No	No	17
7	2	F	12/37	Spastic / V	Quadriplegic	No	No	No	2
8	9	F	21/48	Spastic / V	Quadriplegic	No	No	No	5
9	20	F	59/66	Spastic / II	Hemiplegic	Yes, with orthotics	Yes	Yes	10
10	16	M	60/66	Spastic / II	Quadriplegic	Yes, with orthotics	Yes	Yes	16
11	8	F	27/60	Athetoid / V	Quadriplegic	Power chair ambulatory	Yes	No	10
12	28	M	34/48	Spastic / V	Quadriplegic	No	No	No	15
13	13	F	28/48	Spastic / V	Quadriplegic	Power chair ambulatory	Yes	No	6
14	24	M	47/66	Spastic / V	Quadriplegic	Power chair ambulatory	Yes	No	15

Characteristics of the individuals with cerebral palsy who participated in the study. All riders were already involved in therapeutic riding and rode weekly. A physician had referred riders to the program. The goal of referral was to improve functional mobility. Age, height, weight, and years riding were determined at the beginning of the study. Motor and topographical classification, ability to ambulate, verbal category (yes/no) and ability to perform activities of daily living (independent) were determined by medical care provider. GMFCS = Gross Motor Function Classification System for cerebral palsy, with I = minimal impairment in mobility and V = severe impairment in mobility (Palisano *et al.*, 1997). Participants were currently riding in a therapeutic riding program one time per week, and one of the indications for the sessions was to improve core strength. Orthotics = custom external braces (http://www.cerebralpalsy.org/information/mobility/orthotics). Power chair = wheel chair operated by the individual with cerebral palsy (or care giver) using a joystick.

had not been participating in the TR program at CHUM for at least 1 year were excluded from the study. The goal of the original physician referral to CHUM for TR was to improve functional mobility, including stamina, strength, balance, coordination, and core strength. Because these participants were riders in an existing program, they were able to continue riding on their usual schedule, i.e. the evaluation was scheduled around the riders' schedule to minimize differences in the study intervention and typical TR sessions. Most riders had spastic paresis and were severely disabled (Table 6.1). While some participants are capable of performing select functions of independent living, none of the participants in the study were capable of independent living.

The family (or caregiver) burden in the current study is notable. In order to participate, 6 of 10 riders needed a special van with a lift for transportation. Families traveled at least 70 min, and up to >5 hours (round trip), not including entry and exit in to the vehicle, to participate. For example, one rider lived in a group home, so to participate, a parent drove from home to the group home, to CHUM, and then back, with travel time exceeding 3 hours per visit. Another rider had 8 additional siblings with significant special needs in the family. If a caretaker was not available, all 9 individuals had to be loaded into a vehicle and taken to CHUM for one rider to participate.

Horses

Six horses participated in this study (Table 6.2). This study did not involve any activities other than those that the horses routinely performed in their role in equine assisted activities and therapies. In compliance with PATH guidelines, horses were not ridden more than 3 hours per 12-hour day and not ridden more than 4 days per week. The size, gait, temperament, height, and body width determine if a horse is suited for an adaptive device. For example, a bareback pad (BP) with backrider is more appropriate on a horse of 1100 pounds than a 900-pound horse. To match the rider and device, PATH Certified Instructors follow PATH safety guideline and use their knowledge of horsemanship to make decisions regarding horserider-adaptive devices combinations. In the current study, the PATH certified instructor made all rider-horse decisions in the absence if input or intentional influence by riders, parents/guardians, or researchers.

The study closely mimicked the conditions in a working therapeutic riding program. Of the six horses in the study, four horses were ridden in the English saddle (ES); three horses in BP; and three horses with the IS. One of the six horses in the study was ridden with all three adaptive devices (ES, BP, IS); two were ridden with two devices (IS and ES); and three with one adaptive device (BP = 2; ES = 1; Table 6.2).

The project was approved by Michigan State University Institutional Review Board (IRB#03–047) and explained to the individual and his/her parent/guardian. Informed written consent was obtained from all parents/guardians and assent from riders. The project was exempt from filing an Institutional Animal Care and Use Committee Animal Use Form. The exemption was approved because the horses were kept under generally accepted agricultural management practices and the research involved only observation of the horses doing their usual job.

Program

Riding and evaluation took place at CHUM (Children and Horses United in Movement) Therapeutic Riding, Dansville, Michigan (Kaiser *et al.*, 2006a). CHUM is a Premier Accredited Center by Professional Association of Therapeutic Horsemanship International and is run as a 501C3. The facility is 35 acres, and includes a barn with horse stalls and hay storage, pasture, an indoor heated arena (66x 128 ft.), outdoor riding ring (80x 160 ft.), and trails. Horses have 24/7 access to pasture and run in stalls and hay, and are fed concentrate indoors twice daily. Twelve horses are housed at the facility, and participate in various therapeutic riding and lesson riding activities as well as in hippotherapy. Six horses participated in this study. Routine veterinary care, hoof trimming, dental care, chiropractic manipulation, acupuncture, and massage were provided as needed.

Table 6.2. Characteristics of horses

Horse	Age, yrs	Breed	Years in program	Equine activities					
				Therapeutic Riding (TR)	Hippotherapy (HPOT)	Sport / recreation	4H	Trails / camping	Parades / demos
1	12	Spanish mustang	4	X (ES, IS)	X	X	X	X	X
2	4	QH	1	X (ES, IS)		X	X	X	X
3	7	Draft X QH	2	X (BP)	X	X	X	X	
4	12	Saddlebred	10	X (ES, BP, IS)	X	X	X	X	X
5	4	Mustang X QH	2	X (ES)		X	X	X	
6	7	QH	5	X (BP)	X	X	X	X	X

ES - English saddle; BP - bareback pad; IS - Independence Saddle.

Evaluation of the rider, horse behavior (except behavior at pasture), and side walker exertion was done in the indoor arena. Part of the horse management protocol is that all riders mount from a 4x 8 platform. The platform is accessed by a built-in ramp and can accommodate caregivers who carry riders, wheel chairs, riders with crutches, and independent riders. The horse stands facing the arena while the rider mounts from the platform.

Rider evaluation

Osteopathic musculoskeletal examination (OME): We used the OME of the Hospitalized Patient Form to clinically evaluate the rider before and after riding. The OME is the standard to determine areas requiring osteopathic manipulation, to assess the effectiveness of manipulation, and evaluates symmetry/asymmetry, range of motion restriction, texture of the skin, and tone of the skeletal muscles in various regions of the spine, pelvis and the rib cage (Sleszynski *et al.*, 2004). The OME was completed by the same osteopathic physician (SG), and a clinical comparison made between the before and after riding OME. Categories of the assessment were clinically improved, no change, or clinically worse. The frequency of riders experiencing clinical improvement, no change, or clinical worsening was compared using the Pearson's Chi Squares test except in cases where any category had less than 5 observations. In this situation, Fisher's Exact test was used to make the comparison.

Riding protocols

Two different riding protocols were used in this study: a therapeutic riding lesson riding protocol and a pressure pad protocol.

Therapeutic riding lesson riding protocol

For behavioral assessment of the horses, assistant exertion, and osteopathic medicine evaluation, riding was scheduled for 1 hour, one time per week. To reduce the burden of participation on the rider and their caregivers, riding lessons that were a part of this study were scheduled on the day of the week and at the time of day the rider typically participated in a riding lesson. Equine behavior, side walker exertion, and OME data were obtained for the three different adaptive devices: ES, BP, and the IS. Each rider rode one time on each device. On occasion, due to the riders' schedule, two riders rode on different horses using different devices during the same lesson time. When this occurred two observers were available to assess behavior and videotape each horse-rider combination. The therapeutic riding lesson protocol was scheduled for weekly rides over a 12-week period. Because individuals with cerebral palsy are at risk for developing infections and other medical complications, make-up rides for this protocol were scheduled at the caregivers' convenience. All rides were completed within an eight-month time frame.

The therapeutic riding lesson was approximately 1 hour in length with approximately 30 min of functional riding activity. Riders with severe disability require additional time and staff to mount the horse. Riders mounted the horse with stand-by assist (n=3); two person standing pivot (n=2); and total lift (n=5). Riding activities included, with the horse at a walk, weaving through cones, circling barrels, walking over poles on the ground, halt walk transitions, and upper extremity play. Activities were carried out moving in both a clockwise and counterclockwise direction.

Pressure pad protocol

Because of the complexity associated with saddle pressure measurements involving riders with physical disabilities, core strength, and balance issues, and because the equipment was used during the week at the university, pressure measurements were scheduled separately from the therapeutic riding lesson protocol and took place on weekends. Extra time was given to accommodate the needs of each rider. To measure saddle pressure, the rider had to be able to ride without any assistance for 4 minutes. Saddle pressure was measured for one rider, on one horse, with one device at a time. Pressure recordings were made for each rider on each device. One rider-device combination was recorded per day and there was a minimum of 7 days between pressure recording sessions.

Riders with limited core strength and serious balance issues were unable to complete the entire saddle pressure protocol for some devices. Due to issues of rider fatigue, imbalance and limited core strength, as well as repeated technical issues with the Pliance system, data were not recorded for all riders on all devices. Ultimately pressure data were obtained from eight riders on the ES, five on the BP; and four on the IS.

When technical difficulties occurred, make up rides were scheduled at the parent/guardians' convenience; however, health issues of the riders, financial constraints, and caregiver illness/death made it impossible to schedule make up rides within a reasonable period.

For the pressure pad protocol, each device was evaluated using the same pattern of riding, consisting of a series of straights and turns moving in both clockwise and counterclockwise directions at the walk.

Adaptive devices

Therapeutic riding utilizes a number of adaptive devices and practices to enable individuals with physical disabilities to ride.

Traditional adaptive devices include the conventional English style dressage or hunt seat saddle and the backriding pad. In this study the BP was only used with a backrider.

English saddle

The ES provides minimal medio-lateral (ML) or anterior-posterior (AP) support for the rider. Riding independently in this saddle requires sufficient core strength, balance, and the ability to adduct and stretch the legs over the sides of the horse and into the stirrups. With this saddle, individuals with insufficient core strength, and those unable to ride independently, require side walkers. Side walkers must have sufficient strength and height to provide both ML and AP support to the rider, while walking along the side of the horse (Fig. 6.1). We have previously shown that riders with cerebral palsy have significantly greater anterior-posterior and medio-lateral range of motion in the ES when compared to able-bodied riders (Clayton *et al.*, 2011). The ES is positioned on the horse in the cranial and mid thoracic spine. Use of the ES for individuals with extremely poor balance or uncontrolled athetoid movements, raises concern for the welfare of the horse. Sudden and unexpected movements of the rider may cause a startle response, or result in abnormal pressure, pain, or imbalance to the horse. The ES used in this study weighed 5.6 kg (with pad), was 22 in wide (from bottom of stirrup flap to bottom of stirrup flap), 17 in long (front to back), with a maximum height in the rear of 3.5 in.

Backriding pad

Backriding involves the use of the backriding pad (BP; also referred to as a bareback pad) with the rider sitting in front of a trained individual (the backrider) who must have sufficient strength and balance to provide both ML and AP support for the rider (Fig. 6.1). The horse is fitted with a saddle pad and backriding pad. The backrider generally supplies support for the rider, and the side walkers generally provide little or no physical support for the rider. The BP is very useful for young or extremely small individuals, where the total weight of both the rider and the backrider is not greater than one larger adult. However, using the BP may cause welfare issues for the horse not only because the horse may be carrying more weight, but also because of how the weight is distributed. With the BP, most of the weight is carried farther back over the horse's caudal thoracic and lumbar spine, which is less well supported than the cranial mid thoracic spine. Furthermore, the different movements of the two riders during backriding could be stressful to the horse. Backriding is not appropriate for full sized adults, regardless of the level of disability. It is typically recommended that a horse not carry more that 20–25% of its body weight (Powell *et al.*, 2008); thus a 1000-pound horse would be expected to safely carry less than 250 pounds of rider(s) and equipment. The estimated weight of two full sized adults is between 370 and 400 pounds (https://www.cdc.gov/nchs/fastats/body-measurements.htm) and would require a large horse, weighing 1500 pounds or more. The backriding pad and blankets used in this study weighed 8.2 kg, were 15 in wide (from

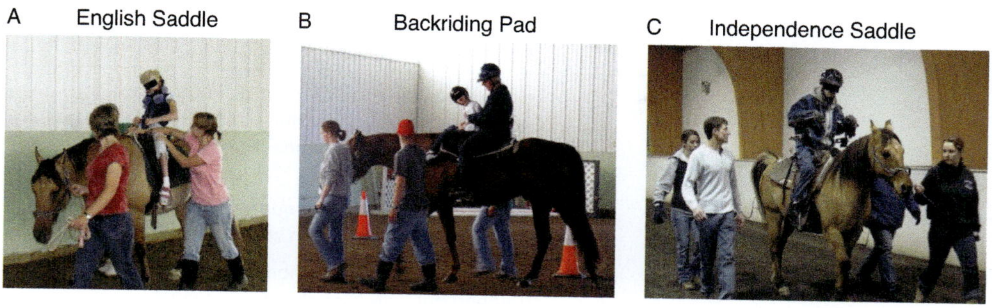

Fig. 6.1. *Adaptive Devices In Use.* A. Rider and assisting persons using the English saddle (ES). B. Rider, backrider, and assisting persons using the backriding pad (BP). C. Rider and assisting persons using the Independence saddle (IS). Note that with the ES the side walkers have both hands on the rider to maintain balance and safety (A). With the BP and IS (B and C) side walkers are not touching the rider. The ES weighed 5.6 kg (with pad), was 22 in wide (from bottom of stirrup flap to bottom of stirrup flap), 17 in long (front to back), with a maximum height in the rear of 3.5 in. The BP and blankets weighed 8.2 kg, was 15 in wide, 17 in long, with a maximum height in the rear of 1 in. The total weight of the BP and back rider was 97.6 kg. The IS weighed 30.1 kg, was 26 in wide, 22 in long, with a maximum height in the rear of 18 in.

bottom of stirrup flap to bottom of stirrup flap), 17 in long (front to back), with a maximum height in the rear of 1 in. The total weight of the backriding pad and backrider (instructor) was 97.6 kg.

The independence saddle

This is a recently developed adaptive device with adjustable and removable headrest, backrest, forearm supports, handgrips, quick-release safety stirrups, and a safety release system (Fig. 6.1). With the IS, severely disabled individuals are able to sit independently in the saddle. The parts of this saddle that support the rider are fully adjustable and can be configured or removed in accordance with the individual rider's needs. When using the IS, the horse is led and side walkers may be required to assist in repositioning individuals with poor core strength or athetoid upper body movements. Although the IS may provide a disabled individual the ability to ride with minimal assistance, there are potential equine welfare concerns with this device. The IS is longer, heavier, more rigid, and more upright than conventional English or Western saddles, potentially resulting in increased total weight or abnormally distributed weight and pressure. The IS may also affect the horse's balance. Neither the physical nor the psychological effects on the horse have been investigated previously.

The IS used in this study weighed 30.1 kg (saddle = 16.4; arms and back = 9.5; wool pad = 4.2 kg; the head rest was not used). The IS was 26 in wide (from bottom of stirrup flap to bottom of stirrup flap), 22 in long (front to back), with a maximum height in the rear of 18 in (without the head rest).

Evaluation of equine behaviour

Certain equine behaviors are associated with stress in the horse, and observation of these behaviors can be a useful way to assess if horses perceive various activities as stressful (Kaiser *et al.*, 2006a; Heleski *et al.*, 2009). We have developed a tool (equine ethogram, Table 6.3) for evaluating the behavior of the horse under different conditions, including therapeutic riding (Kaiser *et al.*, 2006a). We use the word 'stress' to include behaviors indicating conflict, frustration, or irritation.

Sessions were both live assessed and video recorded. Video recording was used for later assessment and to calculate inter observer reliability. Two observers were trained to assess the behaviors based on the ethogram. Due to the

Table 6.3. Horse behavior ethogram

Behavior	Descriptions
Head toss	horse moves head quickly out of 'neutral' position (i.e. where it is positioned the majority of the ride); oscillation; will contain some movement in medial-lateral plane
Head raised	head held higher than the normal carriage with nose extended upward and with slight extension of the neck
Head down	head held lower than the normal carriage, neck may be stretched out with nose pushed forward
Head shake	movement in vertical plane dorsally or ventrally; might be only nose
Head turn	lateral cervical flexion not in line with rest of horse's body; horse may appear to be making a threat display toward one of handlers
Head tilt	nasal midline not perpendicular to ground; deviation of angle of nasal midline to left or right
Ears pinned back	ears pressed caudally against the poll area of the neck
Tail lash/Tail swish	dorsa-ventral movement of caudal vertebrae; movement of tail beyond that of simple rhythmic swaying of the tail; and/or circular (or medial-lateral) movement of caudal vertebrae

**Other behaviors representing conflict/frustration/irritation such as frequent defecation, baring of teeth, kicking, striking, rearing or bucking were never observed in this population. We did not record chomping of the bit because most horses were not ridden with bits.

Ethogram of stress (conflict/frustration/irritation) behaviors observed in horses while being ridden in a therapeutic riding program (Kaiser *et al.* 2006a). The ethogram represents behaviors previous identified in a therapeutic riding and research setting. We did not observe the following behaviors and they were not included in the ethogram: arched neck threat; ears laid back /threat; snapping; bite threat or bite; strike threat or strike; kick threat or kick. Frequent defecation, baring of teeth, kicking, striking, rearing or bucking were never observed in this population. Most horses were not ridden with a bit, so bit chomping was not included in the ethogram.

nature of the adaptive devices, blinded behavioral assessment was not possible.

Fifty-five therapeutic riding sessions were observed to assess stress behaviors of the horses ridden under three different conditions (ES, n=21; BP, n=14; IS, n=20). Behavior was assessed from all riders during the therapeutic riding lesson protocol with the following observation protocol: behaviors during mounting and dismounting were observed (average of 3 min data), along with the following time points: 0–1 min of the riding session, 5–6 min, 10–11 min, 15–16 min, 20–21 min, and 25–26 min, thus approximately 9 min of behavioral data were collected for each riding session. Behavior was also assessed during the saddle pad pressure protocol during mounting and dismounting and each time saddle pressure was recorded. Saddle pressure was recorded three times when the horse was at the walk going straight. The total observation data for horse behavior was 495 min of data. Due to the constraints of conducting this research in an ongoing therapeutic riding program, we did not have the identical number of trials per rider, per horse, or per condition. To determine that horses were capable of exhibiting stress behaviors, behavior was additionally assessed on the same horses used in this study during four traditional lessons (using the therapeutic riding lesson protocol) and over a four-hour period (using the assisting persons protocol; 1 minute of observation every 5 minutes) when the same horses were at pasture.

To determine the number of individuals required to detect statistically significant differences, we performed power analysis for equine behavior and side walker exertion. Using equine behavior as an example, we used a model that included the effect of treatment (saddle), but also the effect of rider, since this factor is an important source of variation in our study. This requires data from each rider in all treatments. Power analysis showed that with 30 samples (i.e. 10 riders who each ride in the three different adaptive devices) we have a power near 0.99 (99% probability to find differences between treatments). With only 3 samples in each treatment (9 samples total) we have a power of 0.86. (Power analysis was conducted by the MSU College of Agriculture and Natural Resources Statistical Consulting Center). Similar numbers were obtained when statistical power was determined using preliminary data for side walker exertion.

Measurement of saddle pressure

Force and pressure measurement

A thin elastic electronic pressure mat (Pliance Saddle System, Novel GmbH, Germany) was used to measure the pressure distribution between the adaptive device and the horse, as previously described (de Cocq *et al.*, 2009; Clayton *et al.*, 2011). The mat consists of two distinct sides measuring 512 mm x 570 mm (each of which has 128 sensors arranged in 8 columns and 16 rows), that measure pressure applied perpendicular to their surface.

The pressure mat was calibrated in a pneumatic calibration frame prior to the start of data collection. The pressure mat was placed in a similar location on each horse and the sensors on the mat were zeroed to match the contour of the horse's back. The appropriate adaptive device was then placed on top of the mat. The girth was tightened gradually to the preferred length. The rider was then assisted, as needed, to mount the horse from a high platform, and any necessary adjustments to the tack were made.

Saddle pressure was measured for one rider, on one horse, with one device at a time. For riders with physical disabilities, set up time for each pressure measurement could take up to 15 min. For each rider, three 10 s pressure recordings were made at a sampling frequency of 60 Hz as the horse walked in a straight line. If a rider became obviously unbalanced or was assisted, pressure recordings during that period were discarded. Three trials per rider per device were obtained and analyzed. Pressure recordings were made for each rider on each device. One rider-device combination was recorded per day and there was a minimum of 7 days between pressure recording sessions.

Each trial was analyzed using proprietary (Novel Electronics) and custom software (Matlab, Mathworks). The force on each sensor was calculated by multiplying the pressure (kPa) by the sensor's area (9.375 cm^2). To analyze the results, the area of the saddle pad was subdivided into left and right halves, and each half further subdivided into front, middle, and back thirds.

This allowed an evaluation of weight distribution on different areas of the horse's back. Because of the number of variables (horse, device, riders), and the low number of riders per device with useful data, we were unable to make meaningful statistical comparisons between adaptive devices in regards to saddle pad pressure.

Assistant exertion

Therapeutic riding for individuals with physical disabilities can require a great deal of physical effort on the part of the side walkers, most of whom are volunteers. The backrider may also require a great deal of physical strength to hold the rider in the appropriate position. Thus, adaptive devices that would not only improve the quality of the ride for the rider, but also decrease the physical work of the volunteers, assistants and staff could be useful. An exertion scoring scale (Table 6.4) was developed for side walkers and the backrider. Exertion was rated on a scale of 0 to 3; with 0 representing no assistance given to the rider and 3 representing high degree of assistance, i.e. the side walker must work to keep rider in the saddle. Total assistance score was acquired by combining exertion scores of all volunteers involved; this included one or two side walkers and the backrider

Table 6.4. Assistant exertion scale

Score	Description
0	Assistance not needed; side walker not needed; walked next to rider as a safety precaution, but no assistance given
1	Minimal assistance; side walkers arms generally at sides, with brief assist
2	Medium assistance; side walkers or back rider provides light support, or supports with only one hand
3	High support; side walker or back rider must work to keep rider in saddle, both hands used

For the safety of the horse and rider, and to enable individuals with handicaps to obtain the physical and psychological benefit of the motion of the horse, side walkers walk along the side of the horse when being ridden by riders that need assistance. Side walkers provide anterior-posterior and medial-lateral support for riders with poor core strength, balance, and coordination. Side walkers were present at all times when the riders rode with the three adaptive devices. The scale ranges from a minimum of 0 (no support needed or provided) to a maximum of 3 (support of the rider required by side walkers throughout the entire ride).

when applicable. Exertion data were assessed live as well as from video recordings. Exertion was assessed for three different conditions for approximately one minute every five minutes during therapeutic riding lesson protocol (ES, n=7; BP, n=4; IS, n=7).

Results

Outcome 1. Clinical benefit for the individual with CP

Pre and post OME evaluations were done for all participants, for all rides, on all devices (n=30). Without considering the particular adaptive device used, almost two-thirds (19/30, 63%) of riding lessons resulted in clinical improvement after a riding lesson compared to before a riding lesson. Six individuals were clinically worse and there was no change in five when pre/post OME evaluations were compared. Looking specifically at OME data from each individual with CP after using the device chosen by the therapist, almost two-thirds (6/10, 60%) showed clinical improvement after riding in their usual adaptive device chosen by the PATH Certified instructor, two showed no change, and two were clinically worse (Table 6.5). There was no difference in the frequency of riders in each of these categories when using either of the other two devices not chosen by the instructor (13/20 with clinical improvement, 4/20 with clinical decline, and 3/20 with no change; p>0.9). Furthermore, there was no difference in the frequency of riders who were clinically improved, the same, or clinically better when comparing IS to ES, IS to BP, and ES to BP (data not shown).

Table 6.5. OME clinical evaluation

ID	ES	BP	IS
5	-	+	+
6	+	-	+
7	-	+	+
8	nc	+	-
9	+	+	+
10	+	+	-
11	-	+	nc
12	+	+	+
13	nc	nc	+
14	+	+	nc

Comparison of OME results after riding to results obtained before riding for the three different adaptive devices; ES = English saddle; BP = backriding pad (with backrider); IS = Independence saddle. Clinical evaluation assessed as improved (+), no change (nc) or clinically worse (-). Bold red indicates the adaptive device chosen by the PATH certified instructor that the rider usually used when riding. There was no difference in the frequency of patients who did or did not show clinical improvement whether comparing use of one device to another or comparing those who were using the device recommended by the therapist, versus any other device (data not shown).

Outcome 2. Stress and pressure on the horse

Stress behaviors were evaluated from 55 TR riding sessions; a total of 495 min of data. An average of 3.2 stress behaviors per session were observed in sessions using the IS (n=20), compared to 1.7 stress behaviors per session using the ES (n=21) and 3.6 stress behaviors per session when using BP (n=20) demonstrating no difference in stress behaviors between the three devices. For comparison, the frequency of stress behaviors demonstrated by these horses in a pasture setting was 5.2 stress behaviors per session. However, stress behaviors were more frequent during traditional lessons (12.4 stress behaviors per session) than any of the other conditions ($P<0.02$).

Kilopascal is the unit of pressure that is measured by the Pliance pressure mat; force (newtons) is calculated by multiplying pressure by area; and in turn the weight of the subject and device is used when calculating mass normalized force (N/kg). Average rider weight (kg) for each device was: 35.2 ± 16.8 (ES); 28.3 ± 16.0 (BP); and 41.8 ± 12.3 (IS). Device weight (kg) was 5.6 (ES); 89.4 (BP); and 30.1 (IS). Note the BP weight included the weight of the device and the backrider.

Fig. 6.2 shows three-dimensional pressure scans at the point of maximal total force for the ES (left), BP (center), and IS (right) for the same rider (Rider # 11). The front of the saddle is at the top of the scans. Overall, the maximal total force with any of the adaptive devices used in this study is not excessive, though use of the BP shows areas of higher pressure near the withers underneath the girthing system. This area of higher pressure was consistently seen with the BP, but not with other adaptive devices. No physical trauma or lesions were observed on the horse with any of the devices.

Table 6.6 shows average force and pressure variables for the three adaptive devices. Backriding showed higher pressures overall, some of which is related to the girthing system.

Outcome 3. Required level of exertion by assisting persons

Assistant exertion was assessed for the ES (n=7); BP (n=4); and IS (n=7) for approximately 1 minute every 5 minutes during therapeutic riding lesson protocol. The average exertion scores while using the IS (1.2 +/- 0.28) were significantly lower ($p<0.0025$) when compared to ES (2.1 +/- 0.25) or BP (3.7 +/- 0.64).

Discussion

Adaptive devices that allow an individual with CP to improve function, ride more independently, provide some respite for the assisting

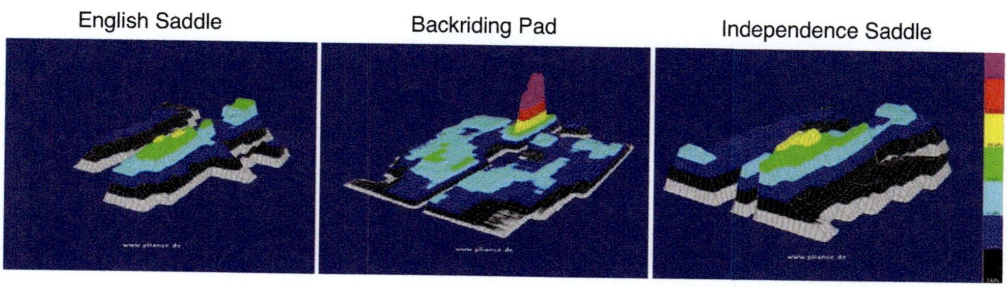

Fig. 6.2. *Representative three-dimensional pressure scans at the point of maximal total force for the English saddle (ES, left), backriding pad (BP, center), and Independence saddle (IS, right) for the same rider.* The front of the saddle is at the top of the scans. Pressure exerted on each sensor is color-coded based on the pressure scale (right), with black indicating the lowest and bright pink being the highest pressure measured. Scans are a good representative of the pressure for each device and illustrate the differences in the magnitude and distribution of pressure between devices. The high pressure on the BP scan represents the girth area of the backriding pad. This area of high pressure is also observed with the backriding pad without a rider (data not shown).

Table 6.6. Force and pressure variables for the English saddle (ES), back riding (BP), and Independence saddle (IS)

	English saddle	Backriding	Independence saddle
Total force (N)	400 ± 133.72	974 ± 83.49	723 ± 170.25
Mass normalized force (N/kg)	10.3 ± 2.3	7.8 ± 0.5	10.1 ± 1.5
Maximal pressure on any sensor (kPa)	28.75	63.75	28.00
Average number of sensors measuring pressure >40 kPa	0 ± 0	148 ± 65	0 ± 0
Average pressure (kPa)	4.65 ± 0.75	5.41 ± 0.20	4.60 ± 0.37
Left front (kPa)	4.54 ± 0.68	6.18 ± 0.33	4.71 ± 1.15
Right front (kPa)	4.27 ± 0.60	4.42 ± 0.25	4.72 ± 0.20
Left middle (kPa)	4.47 ± 0.77	5.74 ± 0.52	4.15 ± 0.44
Right middle (kPa)	4.45 ± 0.70	4.37 ± 0.87	4.25 ± 0.53
Left rear (kPa)	4.59 ± 1.74	6.08 ± 0.66	4.87 ± 0.38
Right rear (kPa)	4.94 ± 1.64	5.29 ± 0.77	5.02 ± 0.13

Mean values (±SD) for force and pressure variables measured under the ES (*n*=8 riders), BP (*n*=5), and IS (*n*=4). The table shows force in Newtons and pressure in kilopascals. Kilopascal is a unit of pressure measurement that has widely replaced pounds per square inch. For comparison the total force for a rider using a dressage saddle was 843 Newtons and a treeless saddle was 794 Newtons (Belock *et al.*, 2012). The force is also normalized for the weight of the rider and assisting devices (Mass normalized force (N/kg). The maximum pressure on any sensor refers to the highest pressure reading obtained during recording. For comparison the maximal pressure on any sensor was 20.7 kPa (Belock *et al.*, 2012). The pressure for the right and left sides and front, middle, and back were also recorded. The high pressure in the left front in back riding corresponds to the area where the pad was girthed. The average number of sensors measuring >20, 40, and 60 KPa was evaluated, and back riding had more sensors measuring >40 KPa of pressure.

persons, and not cause undue stress or frustration to the horse would be a great addition to the therapeutic regimen. Families with children with CP often lead complicated lives (http://www.cerebralpalsy.org/about-cerebral-palsy/definition). They have to manage more doctor appointments, more therapy sessions, and more physical accommodations/equipment than most parents with typically developing children. Participating in a study that puts additional demands on the family is often a hard sell if the study is science for the sake of science. However, the fact that families were willing to participate in the current study and take on the added demands as a result suggests that they believe in TR and that it has value as part of their child's therapy. An often-overlooked threat to therapeutic riding programs like CHUM offering therapeutic riding to children with CP is the availability of assisting persons, who are generally volunteers or family members. Without these team members, the capacity of a program to serve children with CP is dramatically reduced. Considering that clinical outcomes for the rider and stress behaviors for the horse were similar using the Independence Saddle compared to either of the other devices, the Independence Saddle should be considered a viable option that may incentivize the assisting person(s) with a lighter workload.

Cerebral palsy is a chronic incurable condition that requires ongoing life-long treatment and therapy that is uniquely tailored to the individual (http://www.cerebralpalsy.org/about-cerebralpalsy/definition). The goal of the treatment plan is to utilize different therapies to maintain and promote mobility and function. Based on the OME, clinical improvement was observed after the majority of riding lessons (63%). No single device had a higher likelihood of improving clinical outcomes than any other in this study. While these results suggest that the PATH certified therapist could make horse assignments solely of the basis of safety of the rider and temperament of the horse, it is recognized that this conclusion is based on a single modality for assessing clinical outcome. The OME is a recognized clinical outcome that is appropriate for assessing improvement after TR. In this study one doctor of osteopathic medicine (SG) an osteopathic manipulation specialist assessed each rider before and after each ride on each device. However, additional clinical outcome measures, and patient/family reported outcomes measures should be considered for future work.

This study found no difference in the average number of stress behaviors, suggesting that when compared to other adaptive devices, the IS does not result in more stress behaviors in the horse. Further, these data support our earlier work showing that therapeutic riding, at least at this center, with this set of horses, is not perceived by the horses as being mentally stressful (Fig. 6.3). Horses did not perceive one adaptive device as more aversive than another. In addition, horses in this study did not show significantly more stress behaviors than when they were ridden in traditional lessons.

We have previously shown that horses in a therapeutic riding program show low levels of stress behaviors, but these levels increase when the same horses are ridden in a horse show (Kaiser *et al.*, 2006a). Taken together these studies suggest that therapeutic riding with multiple adaptations and with individuals with various degrees of disability is not necessarily stressful to the horse when equine temperament, physical ability, and rider's needs are correctly matched. It should be noted that stress behaviors in all situations were low.

Our results are in agreement with another study that found no difference in salivary cortisol (used to assess stress) in therapeutic riding horses ridden in a hunt seat saddle (a specific English saddle) by children at a riding camp or by children with emotional issues, compared to horses at rest (McKinney *et al.*, 2015). Using our equine ethogram (Kaiser *et al.*, 2006a) to assess behavior, they also measured salivary

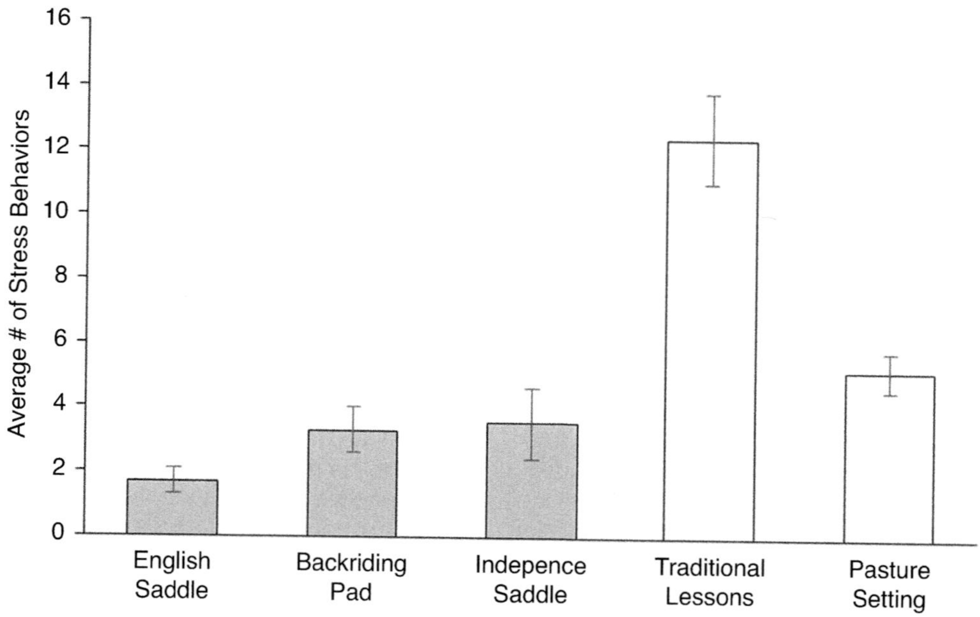

Fig. 6.3. *Stress (conflict/frustration/irritation) behaviors exhibited by horses when ridden with three adaptive devices and comparing with traditional lessons and pasture setting.* Stress behaviors were evaluated from 55 riding sessions as well as traditional lessons being ridden by the horses in this study. In addition stress behaviors were evaluated with the horses in a pasture setting. Stress behaviors were more prevalent during traditional lessons than any of the other conditions (*P*=0.02). Error bars represent standard error of the mean (SEM).

cortisol three times per week over 6 weeks. To eliminate circadian changes in cortisol, samples were obtained at the same time, after mounting (3 PM), after riding (3:30 PM), and after 1 hour of riding (4 PM). Similar to our study, horses exhibited few stress behaviors. There was no significant difference in delta cortisol between the groups. Salivary cortisol was not higher in therapeutic riding, when compared to recreational riding or rest, and the hypothesis was not supported.

Whether or not therapeutic riding with disabled riders resulted in increased stress to the horse when compared to recreational riding by healthy individuals with no previous riding experience has also been investigated (Fazio, 2013). To assess equine stress the hypothalamic pituitary axis (HPA) was monitored by measuring beta-endorphin, ACTH, and cortisol. The results suggest that the HPA in therapeutic riding horses is less responsive to disabled riders than recreational riders. One interpretation of these results is that the horses have the ability to evaluate the needs of the rider and perform their job. It could also be argued that the disabled riders made fewer demands on the horse to perform.

Another study compared the responses to acute, novel stimuli (restraint by covering the head for an hour and to a startle stimulus, shaking a garland) of therapeutic riding horses and jumping horses (Minero, 2006). They measured heart rate (HR; as an assessment of stress), heart rate variability (HRV; as an assessment of autonomic nervous system activity), and equine behavior. Basal HR of therapeutic riding horses was greater than jumping horses, likely due to the better physical condition of the jumping horses. Both groups of horses reacted to novel stimuli, and therapeutic riding horses did not react less than jumping horses. The authors suggest that this should be considered in therapeutic riding programs. Another interpretation is that TR horses should be exposed to as many different situations and stimuli as possible so that they are comfortable in unusual and unexpected situations. This type of exposure would result in decreased risk of injury to disabled individuals in a TR program. Heart rate variability has been used to assess acute and chronic stress. One group of investigators used Holter monitors to measure HRV in nine horses (15.9 +/- 7.7 years) actively working in an equine-assisted activities setting (Gehrke, 2011). Recordings were taken with horses in a two-acre pasture. When compared to previous studies of two-year old Thoroughbreds, diurnal and nocturnal high frequency (as an assessment of parasympathetic nervous system activity) and low frequency (as an assessment of sympathetic nervous system activity) powers were not different in the horses in this study. Although it has been suggest that comparison of HRV parameters between studies be done with caution (Stucke *et al.*, 2015), due to technical inconsistencies, the difference may be due to age, breed, experience, environment, or nutrition.

A study published in 2017 evaluated equine behavior, based on our ethogram (Kaiser *et al.*, 2006a), and physiologic measurements of stress (cortisol, ACTH, glucose) in horses working with veterans with post-traumatic stress disorder (PTSD) (Johnson *et al.*, 2017). The comparison group, at a different time point, was experienced riders. Although physiologic and behavioral patterns varied, these authors also concluded that TR was not stressful to the horse in this study.

Peer reviewed studies using a variety of methods to assess stress in horses involved in therapeutic riding suggest that the activity is not inherently aversive or stressful to the horse. One could argue that horses in a TR program have coped with their condition by developing learned helplessness (Hall *et al.*, 2008). To rule out learned helplessness, we observed the horses in our study in a pasture setting with conspecifics (other horses) and in traditional lessons. In these situations, the horses in our therapeutic riding study expressed typical horse behaviors, including bite threats, kick threats, ears pinned back, and tail lashing. These behaviors were not observed in the TR setting. Similarly, in our previous study, horses at a horse show exhibited typical horse behaviors that were not seen in the TR setting (Kaiser *et al.* 2006a). The evidence of normal horse behaviors at pasture or in traditional lessons suggests these horses were capable of excellent calm behavior when 'doing their job,' and were not displaying learned helplessness.

Areas of high pressure underneath adaptive devices may be uncomfortable or painful for the horse and may lead to stress behaviors.

Direct pressure measurements beneath each device were taken to evaluate the magnitude of pressure being experienced by horses for each device. The highest pressures were noted when using the BR, and the pressure measurements for the ES and IS devices are of similar magnitude. None of the pressure measurements are alarming when compared to able-bodied riders on a conventional or treeless saddle (Belock *et al.*, 2012).

Various aspects of horse behavior, temperament and personality have been studied (Houpt *et al.*, 2000; McDonnell, 2003; Waring, 2003). However, equine behavior while being ridden has infrequently been examined (Odberg, 1987; Weeks and Beck, 1996; Rivera *et al.*, 2002) and stress behaviors of horses while being ridden in a therapeutic riding setting have received little attention (Kaiser *et al.*, 2006b; McKinney *et al.*, 2015; Johnson *et al.*, 2017; De Santis *et al.*, 2017).

A study conducted in 1999 concluded: ‘it is very difficult to objectively determine the suitability of horses for therapeutic riding programs regarding their temperament and reactivity, probably because other traits are considered important’ (Anderson *et al.*, 1999). Interestingly, since that time, the number of programs offering therapeutic riding and other equine associated activities have increased, yet there is no consistent method assessing the welfare of the horse in a therapeutic setting (De Santis *et al.*, 2017). Clearly, each center is different, and it is important for the instructors to understand not only the needs of the client, but also the needs of the horse.

The decision of which adaptive device to use is based on the needs of the rider. However, when several adaptive devices may equally benefit the rider, the amount of exertion required from volunteer assistants should be considered. When riders with limited core strength sit in

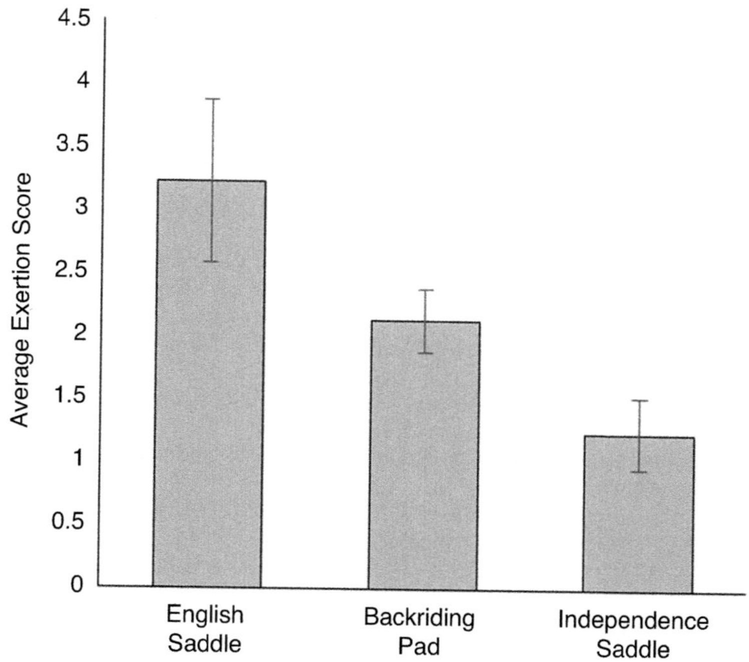

Fig. 6.4. *Average exertion score for assistants during use of the English saddle (ES), back riding (BP), and the Independence saddle (IS).* Exertion was assessed in the three different conditions for approximately 1 minute every 5 minutes during therapeutic riding lesson protocol (ES, *n*=7; BP, *n*=4; IS, *n*=7). Exertion scores while using the IS were significantly lower when compared to ES or BP. The effect of device type on overall exertion scores was significant (*P*=0.0025). Error bars represent standard error of the mean (SEM).

an English saddle, they often need considerable physical support from side walkers, one on each side of the horse. When riders are in need of a backrider, the backrider puts forth considerable exertion and may need an additional side walker. This study shows that assistant exertion is less with the IS than other adaptive devices tested (Fig. 6.4).

With the changing societal values regarding the role of animals, it would behoove programs to objectively assess and monitor the welfare of the horses in their equine assisted activities and therapies programs. Further, the Animal Welfare Act (Public Law 89–544, 1966), which 'requires that minimum standards of care and treatment be provided for certain animals bred for commercial sale, used in research, transported commercially, or exhibited to the public', and the Professional Animal Auditor Animal Certification Program (https://animalauditor.org/index.php), which 'promotes the humane treatment of animals through education and certification of animal auditors,' could serve as templates to develop scientific criteria to assess the welfare of horses involved in therapeutic riding. Audits designed to promote continuous improvement in therapeutic riding and hippotherapy programs would likely improve the welfare of the horse.

Acknowledgements

This work was funded by a FACT grant from Michigan State University, East Lansing, MI.

References

All, A.C., Loving, G.L. and Crane, L.L. (1999) Animals, horseback riding, and implications for rehabilitation therapy. *Journal of Rehabilitation* 65, 49–57.

Anderson, M.K., Friend, T.H., Evans, J.W. and Bushong, D.M. (1999) Behavioral assessment of horses in therapeutic riding programs. *Applied Animal Behaviour Science* 63, 11–24.

Animal Welfare Act, Public Law 89-544 (1966) United States Code, Title 7, Sections 2131 - 1256. Available at: https://www.nal.usda.gov/awic/animal-welfare-act

Belock, B., Kaiser, L.J., Lavagnino, M. and Clayton, H.M. (2012) Comparison of pressure distribution under a conventional saddle and a treeless saddle at sitting trot. *Veterinary Journal* 193, 8791.

Benda, W., McGibbon, N.H. and Grant, K.L. (2003) Improvements in muscle symmetry in children with cerebral palsy after equine-assisted therapy (hippotherapy). *Journal of Alternative and Complementary Medicine* 9, 817–825.

Bertoti, D.B. (1988) Effect of therapeutic horseback riding on posture in children with cerebral palsy. *Physical Therapy* 68, 1505–1512.

Center for Disease Control National Center for Health Statistics (2016). Available at: http://www.cdc.gov/nchs/fastats/body-measurements.htm

Clayton, H.M., Kaiser, L.J., Pue, B. and Kaiser, L. (2011) Center-of-pressure movements during equine-assisted activities. *American Journal of Occupational Therapy* 65, 211–216.

de Cocq, P., Clayton, H.M., Terada, K., Muller, M. and Leeuwen, J.L. (2009) Usability of normal force distribution measurements to evaluate asymmetrical loading of the back of the horse and different rider positions on a standing horse. *Veterinary Journal* 181, 266–273.

de Santis, M., Contalbrigo, L., Borgi, M., Cirulli, F., Luzi, F. *et al.* (2017) *Equine Assisted Interventions (EAIs): Methodological Considerations for Stress Assessment in Horses* 4.

Engle, B.T. (1997) *Rehabilition with the Help of a Horse: A Collection of Studies*. Barbara Engle Therapy Services, Durango, CO.

Fazio, E. (2013) Hypothalamic-pituitary-adrenal axis responses of horses to therapeutic riding program: Effects of different riders. *Physiology & Behavior* 118, 138–2013.

Foster, R. and Wallach, K. (2015) Equine stress-related behaviors in therapeutic riding classes. *In International Society for Equitation Science 2015 Annual Meeting*.

Gehrke, E.K. (2011) Heart rate variability in horses engaged in Equine-Assisted activities. *Journal of Equine Veterinary Science* 31, 78–2011.

Griffith, J.C. (1992) Chronicle of therapeutic horseback riding in the United States, resources and references. *Clinical Kinesiology, Journal of the American Kinesiotherapy Association* 46, 2–7.

Gross Motor Function Classification System - Expanded & Revised (GMFCS - E&R) (2017). Available at: https://www.canchild.ca/en/resources/42gross-function-classification-system-expanded-revised-gmfcs-e-r

Hall, C., Goodwin, D., Heleski, C., Randle, H. and Waran, N. (2008) Is there evidence of learned helplessness in horses? *Journal of Animal Welfare Science* 11, 249–266.

Heleski, C.R., McGreevy, P.D., Kaiser, L.J., Lavagnino, M., Tans, E. *et al.* (2009) Effects on behaviour and rein tension on horses ridden with or without martingales and rein inserts. *Veterinary Journal* 181, 56–62.

Houpt, K.A., Eggleston, A., Kunkle, K. and Houpt, T.R. (2000) Effect of water restriction on equine behaviour and physiology. *Equine Veterinary Journal* 32, 341–344.

Huang, X., Lin, J. and Demner-Fushman, D. (2006) Evaluation of PICO as a knowledge representation for clinical questions. *American Medical Informatics Association Annual Symposium* 359–363.

Independent Strides; Building Personal Dignity (2016). Available at: http://www.independentstrides.com/saddle.htm

Johnson, R.A., Johnson, P.J., Megarani, D.V., Patel, S.D., Yaglom, H.D. *et al.* (2017) Horses working in therapeutic riding programs: Cortisol, adrenocorticotropic hormone, glucose, and behavior stress indicators. *Journal of Equine Veterinary Science* 57, 77–85.

Kaiser, L., Heleski, C.R., Siegford, J. and Smith, K.A. (2006a) Stress-related behaviors among horses used in a therapeutic riding program. *Journal of the American Veterinary Medical Association* 228, 39–45.

Kaiser, L., Smith, K.A., Heleski, C.R. and Spence, L.J. (2006b) Effects of a therapeutic riding program on at-risk and special education children. *Journal of the American Veterinary Medical Association* 228, 46–52.

Laiou, A., Christakou, A. and Kaminiotis, V. (2017) Impacts of hippotherapy on children with cerebral palsy from parents' perspective: A qualitative research. *International Journal of Physiotherapy* 2, 947–957.

MacKay-Lyons, M., Conway, C. and Roberts, W. (1988) Effects of therapeutic riding on patients with multiple sclerosis: A preliminary trial. *Physiotherapy Canada* 40, 104–109.

Mackinnon, J.R., Noh, S., Laliberte, D., Allan, D.E. and Lariviere, J. (1995) Therapeutic horseback riding. *Physical & Occupational Therapy in Pediatrics* 15, 1–15.

McDonnell, S. (2003) *A Practical Field Guide to Horse Behavior: The Equid Ethogram*, 1 ed. Eclipse Press, Lexington, KY.

McKinney, C., Mueller, M.K. and Frank, N. (2015) Effects of therapeutic riding on measures of stress in horses. *Journal of Equine Veterinary Science* 35, 922–928.

Minero, M. (2006) A note on reaction to novel stimulus and restraint by therapeutic riding horses. *Applied Animal Behaviour Science* 97, 335–2006.

My Child at CerebralPalsy.org (2017). Available at: http://www.cerebralpalsy.org/about-cerebral-palsy/definition

Odberg, F.O. (1987) Chronic stress in riding horses. *Equine Veterinary Journal* 19, 268–269.

Palisano, R., Rosenbaum, P., Walter, S., Russell, D., Wood, E. *et al.* (1997) Development and reliability of a system to classify gross motor function in children with cerebral palsy. *Developmental Medicine and Child Neurology* 39, 214–223.

Palisano, R.J., Hanna, S.E., Rosenbaum, P.L., Russell, D.J., Walter, S.D. *et al.* (2000) Validation of a model of gross motor function for children with cerebral palsy. *Physical Therapy* 80, 974–985.

Park, E.S., Rha, D.W., Shin, J.S., Kim, S. and Jung, S. (2014) Effects of hippotherapy on gross motor function and functional performance of children with cerebral palsy. *Yonsei Medical Journal* 55, 1736–1742.

Powell, D.M., Bennett-Wimbush, K., Peeples, A. and Duthie, M. (2008) Evaluation of indicators of weight-carrying ability of light riding horses. *Journal of Equine Veterinary Science* 28, 28–33.

Professional Animal Auditor Certification Organization (2016). Available at: https://animalauditor.org/index.php

Professional Association of Therapeutic Horsemanship International (PATH Intl.) (2016). Available at: http://www.pathintl.org

Rivera, E., Benjamin, S., Nielsen, B., Shelle, J. and Zanella, A.J. (2002) Behavioral and physiological responses of horses to initial training: The comparison between pastured versus stalled horses. *Applied Animal Behaviour Science* 78, 235–252.

Rosenbaum, P.L., Walter, S.D., Hanna, S.E., Palisano, R.J., Russell, D.J. *et al.* (2002) Prognosis for gross motor function in cerebral palsy: Creation of motor development curves. *Journal of American Medical Association* 288, 1357–1363.

Sleszynski, S.L., Glonek, T. and Kuchera, W.A. (2004) Outpatient osteopathic single organ system Musculoskeletal exam form series: Validation of the outpatient osteopathic SOS Musculoskeletal exam form, a new standardized medical record. *Journal of the American Osteopathic Association* 104, 423–438.

Snider, L., Korner-Bitensky, N., Kammann, C., Warner, S. and Saleh, M. (2007) Horseback riding as therapy for children with cerebral palsy: Is there evidence of its effectiveness? *Physical & Occupational Therapy in Pediatrics* 27, 5–23.

Sterba, J.A. (2007) Does horseback riding therapy or therapist-directed hippotherapy rehabilitate children with cerebral palsy? *Developmental Medicine and Child Neurology* 49, 68–73.

Stucke, D., GroBe Ruse, M. and Lebelt, D. (2015) Measuring heart rate variability in horses to investigate the autonomic nervous system activity - Pros and cons of different methods. *Applied Animal Behaviour Science* 166, 1–10.

Waring, G.H. (2003) *Horse Behavior*, 2nd ed. Noyes Publications, Norwich, NY.

Weeks, J. and Beck, A.M. (1996) Equine agitation behaviors. *Equine Practice* 18, 23–24.

Part 2: Adolescents/Young Adults

7 Integration of Equine Assisted Services and Social-Emotional Learning: A Preliminary Evaluation of a Life Skills Training Program for at-Risk Youth

Ashleigh Zaker[1*], Lee Dudley[2], Mary Slouka[2] and Lorann Stallones[1,3]
[1]Department of Psychology, 1876 Campus Delivery, Colorado State University, Fort Collins, USA; [2]The Equine Partnership Program, 40495 County Road 13 Elizabeth, Colorado, USA; [3]One Health Institute, Colorado State University, Fort Collins, USA

Abstract

This study was a preliminary evaluation of a life skills training program conducted by the Equine Partnership Program (EPP) integrating Equine Assisted Services (EAS) with social emotional learning (SEL) competencies. SEL programs aim to foster character development, curb maladaptive behaviors, and enhance social skills in at-risk youth by promoting positive relationships and emotional regulation. Many youth still require more targeted or alternative approaches to effectively nurture these skills. EAS has the potential to enhance SEL through its experiential learning approach and use of horses as a bridge, which is believed to be effective because of the horse's innate ability to mirror and respond to human emotion. The purpose of this study was to assess the impact of EAS on the social–emotional development of at-risk youth as measured by the Depression, Anxiety, Pain, and Stress (DAPS) visual analog tool. It was hypothesized that participants would demonstrate measurable improvements in emotional regulation post intervention. Data were collected before and after the EAS experience from 33 participants. Participants were asked to rate their levels of DAPS ranging from a high of 10 to a low of 1. Using t-test comparisons, statistically significant decreases in scores were found in depression ($M = -0.67$, $p = 0.02$), and anxiety ($M = -1.18$, $p < 0.001$). Emotional pain and stress also showed decreases, though not statistically significant. This intervention shows promise to serve as an effective modality in addressing resilience and pro-social skills necessary for improved outcomes in behavioral and emotional development, thereby improving future academic, career, and overall life success, including improved social relationships.

Keywords: equine assisted services (EAS), social emotional learning (SEL), experiential learning, emotion regulation, depression, anxiety, at-risk youth, resilience, pro-social, skills, adolescent mental health

Introduction

According to the National Alliance on Mental Illness (National Alliance of Mental Illness (NAMI), 2021), 7.7 million youth in the USA, between the ages of 6–17, have experienced a mental health disorder. Of these, 50.6% received treatment with an average delay

*Corresponding author: ashleigh.zaker@colostate.edu

between onset of symptoms and treatment is 11 years (National Alliance of Mental Illness (NAMI), 2021). Mental health disorders, such as depression and anxiety, commonly occur together among youth (Centers for Disease Control and Prevention (CDC), 2021). Among children with depression, 73.8% also have a diagnosis of anxiety, and 47.2% display behavioral problems with an increasing number of youths diagnosed with mental health disorders over time (Centers for Disease Control and Prevention (CDC), 2021). These statistics highlight the importance of identifying interventions which mitigate risks and foster protective factors for today's youth.

Social emotional learning

Social-emotional learning (SEL) is a promising approach, focusing on equipping individuals with the necessary skills and attitudes to navigate emotions and promote overall well-being (Collaborative for Academic, Social and Emotional Learning (CASEL), 2021). Youth can learn to set and attain goals, have empathy for self and others, form positive relationships, and conscientiously make responsible choices (Domitrovich *et al.*, 2017). The goal of SEL programs is to encourage individual character growth, reduce maladaptive behaviors, and build social skills through promoting opportunities that encourage positive relationships and strong social support (Taylor *et al.*, 2017; Moreno-Gómez and Cejudo, 2019).

There is an increasing body of research that supports the long- term effectiveness of SEL interventions, not only in addressing specific behavioral issues but also in prevention (Taylor *et al.*, 2017; Moreno-Gómez and Cejudo, 2019). By developing emotional skills and improving prosocial behaviors, youth are better equipped to navigate a variety of life challenges. These skills are utilized as protective barriers to development of behavioral issues that arise during times of adversity and stress. Research has shown that advanced skills, rather than attitudes, predict long-term benefits of SEL (Taylor *et al.*, 2017). Furthermore, intrapersonal and interpersonal competencies enhance academic performance and are impactful to future career and life success.

SEL in classroom settings

Research suggests that school-based interventions are crucial in addressing mental health issues among youth, with SEL playing a pivotal role in K-12 education (Durlak *et al.*, 2011) SEL programs focusing on the development of emotional intelligence, and coping skills have been shown to improve mental health outcomes, particularly in reducing anxiety and depression symptoms, and serve as early intervention tools for youth with emerging mental health concerns (Taylor *et al.*, 2017). SEL has been shown to improve social and emotional skills, enhance academic achievement, and reduce negative behaviors such as bullying (Song and Kim, 2022).

Despite the integration of SEL into school curricula, many youth require more targeted or alternative approaches to effectively nurture these skills (Lawson *et al.*, 2019). This need is particularly pronounced among youth classified as 'at-risk,' typically defined as those exposed to factors that negatively impact their educational and life outcomes, such as poverty, family instability, or limited access to educational resources (Oldeide *et al.*, 2020). To address this need, resilience training programs have gained popularity, complementing SEL by teaching youth positive response techniques and stress management strategies (Pinto *et al.*, 2021).

Building resilience

Resilience is commonly defined as the ability to maintain a positive mindset in the face of challenging circumstances (Van Katwyk and Seko, 2018; Mahdiani and Ungar, 2021). It plays an important role as both an outcome and a facilitator for SEL by recognizing the intricate interplay of individual circumstances, neurobiological development, and environmental factors in mental health (Durlak *et al.*, 2011; Taylor *et al.*, 2017; Van Katwyk and Seko, 2018; Mahdiani and Ungar, 2021). Understanding the complexity of these factors, including genetics and brain development, is an important aspect of child development. A holistic approach offers a comprehensive framework to support the mental health and social- emotional development of youth.

Emotional regulation

For many youths with mental health or behavioral diagnoses, improving emotional regulation may be beneficial (Ehrenreich-May *et al.*, 2015; Suveg *et al.*, 2015). Emotional regulation equips youth with the ability to assess and modulate their emotional responses to challenges, which is essential for resilient behavior (Ehrenreich-May *et al.*, 2015; Daros *et al.*, 2021). Emotional regulation involves evaluating emotional experiences and aligning them with self-protective goals, functioning as part of a broader self-regulatory system encompassing both internal and external responses. Challenges with emotional regulation are associated with symptoms found in various psychological disorders, including depression and anxiety (Daros *et al.*, 2021). By managing emotions, individuals can better cope with adversity, maintain a positive outlook, and recover more quickly from setbacks, thus improving their mental health.

Importance of active engagement

Studies indicate that active participation in stressful situations is vital for enhancing resilience among youth (Tugade and Fredrickson, 2007; Xie, 2013). By confronting and navigating stressful situations in real time, youth can develop and enhance coping skills, foster resilience, and improve their ability to regulate emotions (Ehrenreich-May *et al.*, 2015). This process leads to improve long- term outcomes in behavioral and emotional development. Youth with mental illness or behavioral diagnoses often face difficulties in identifying and comprehending emotions, which can result in errors in interpreting others' emotional expressions (Suveg *et al.*, 2015). Establishing healthy emotional regulation and developing positive coping skills is important for youth to effectively self- regulate and work toward personal and relational goals.

Equine assisted services

Youth must have opportunities to practice SEL skills in safe environments that encourage growth and promote recognition of skills in real-life situations (Collaborative for Academic, Social and Emotional Learning (CASEL), 2021). The need for real- life experiences highlights the role of community partners and sponsored activities. Literature supports that task-focused approaches to developing SEL skills and building resilience have both immediate and long-lasting results (Jones, 2017). Equine assisted services (EAS) serve as an example of a task-focused approach, highlighting problem solving, cognitive reconstruction, and a method of skill development.

EAS is an emerging field that offers a unique learning experience through hands-on interactions with horses and is a term that incorporates a wide variety of programs including, but not limited to, equine assisted therapies (EAT) and equine assisted learning (EAL) (PATH International, 2021). EAS interventions have resulted in reductions in stress, depression, anxiety, and increases in self-esteem, responsibility, concentration, and motivation (Harvey *et al.*, 2020). EAS may help youth increase their capacity to recover from relational and emotional stress, improve their social interactions, and increase their positive behavioral measures. Activities in EAS widely vary. The basic techniques of EAS include grooming, tending to the physical needs of the horse such as providing food and water, therapeutic riding, and spending time around the animal (Jones, 2017).

Special nature of horses as co-facilitators

EAS have the potential to be more effective than traditional learning interventions in addressing problem-solving, trust, focus, communication, self-esteem, and relatedness (Mueller and McCullough, 2017). This heightened impact can be attributed to the innate sensitive nature of horses (Jones, 2017). Unlike other animals, horses may respond to human behaviour much the same way another person would because the horse is a prey animal and therefore are attuned to their environments to ensure survival. In contrast, dogs and cats are predators and do not react in the same way to threatening situations. As a result, horses may display certain vulnerability toward their handler that is not always seen in other domesticated animals (Jones, 2017).

Influence of the EAS setting

The locations of treatment facilities may contribute to the effectiveness of EAS (Mueller and McCullough, 2017). EAS are often experienced in farm settings and almost always have outdoor access. Travelling to a natural setting can offer a sense of anonymity and freedom and allow individuals to experience the psychological and physiological benefits of nature (Jones, 2017). In the barn, youth can connect emotionally and physically because they feel protected and can let down their internal defenses (Mueller and McCullough, 2017).

Methods

Drawing upon the existing literature on SEL and the potential benefits of EAS, it is evident that creating safe environments for youth to practice SEL skills and fostering resilience through task- focused approaches can have significant positive outcomes (Staudt and Cherry, 2017; Jones, 2017; Harvey *et al.*, 2020; Collaborative for Academic, Social and Emotional Learning (CASEL), 2021). The goal of this study was to evaluate the effectiveness of a school-based life skills training program conducted by the Equine Partnership Program (EPP) by assessing the impact of EAS on the social-emotional development of at-risk youth. Outcomes were measured by the DAPS, a visual analogue self-assessment tool that measures depression, anxiety, emotional pain, and stress. It was hypothesized that participants would demonstrate measurable improvements in emotional regulation post-intervention as reflected in decreased DAPS scores.

Participants

Participants for this study were students in a transition program located in Denver that offers support for at-risk youth between the ages of 18–21 years who are seeking support for post-secondary education, employment, and independent living. This program is part of the school district's goal to help youth focus on life options beyond school and to support special education with community- based education programs focusing on life skills, career, study skills, and adult literacy. Youth in this program represent a range of disabilities and possess an Individualized Education Plan (IEP). All participants chosen to participate in the life skills training program were selected by a team of school administrators including teachers and school social workers.

Program overview

EPP is a non-profit organization based in Elizabeth, Colorado dedicated to providing services to facilitate community development and growth by partnering with inner-city schools in Denver and surrounding areas. Their life skills training program aligns with the Equine Assisted Growth and Learning Association (Equine Assisted Growth and Learning Association (EAGALA), 2018) model, incorporating horses as part of a therapeutic team. The team included a mental health professional, who facilitated the therapeutic journey and discussions, and an equine specialist, who ensured the safety of all participants and interpreted the horses' behaviors. The goal of the program that was evaluated was to build SEL skills where participants learned to analyse their environment, to understand and improve relationships, and to identify solutions in partnership with professional facilitators.

The EPP life skills training program incorporated specific activities designed to promote SEL skill development. Participants engaged in structured problem-solving sessions, working collaboratively to address real-life challenges. These sessions encouraged the application of SEL skills such as teamwork, communication, and conflict resolution. Cognitive restructuring was integrated into the program through targeted sessions aimed at identifying and reframing negative thought patterns. Participants were guided through exercises that helped them recognize and challenge unproductive thought processes and foster emotional regulation and resilience.

Participants in the study chose which horse they worked with throughout the program, allowing for the development of unique bonds based on personal affinity and comfort. The same horses were consistently available during the program, ensuring that participants could work with their preferred horse each week, thereby enhancing continuity and depth of the human-horse relationship. Horses, central to the therapeutic process, mirrored and responded to human emotions, facilitating insights through non-verbal communication, and establishing respect.

For a comprehensive overview of the program's structure, including semesters, session times, and participant engagement, see Table 7.1. The progression and objectives of weekly activities are outlined in Table 7.2.

Measurement

Quantitative outcomes were measured using the DAPS tool, a visual analogue self-assessment instrument measuring depression, anxiety, emotional pain, and stress. This research utilized pre-existing data for which all necessary ethical considerations were retrospectively acknowledged. The study was subsequently reviewed by the Colorado State University IRB under Protocol number 21–10518 hour. The EPP staff selected the DAPS form due to its applied nature, accessibility, and ease of understanding. Given its simplicity and intuitive design, the instrument was considered suitable for the at-risk youth participants, aiding in the collection of self-reported data on their emotional and psychological states before and after each session.

Data collection

Participants were given a DAPS form and asked to gauge feelings of depression, anxiety, emotional pain, and stress on a visual analogue ranging from a high of ten to a low of one. Data were collected before and after each session. Team leaders explained the meaning of each category and remained available to assist, as necessary. Surveys were collected by a designated staff member and kept anonymous. Data on participants' prior experience with the program, familiarity with the equine facility, or previous horse experience was not systematically collected. Any personal information was de-identified. The data were coded with a participant number, date, gender, and session time, am for morning and pm for afternoon. Due to challenges with inconsistent participant attendance, we established a minimum participation threshold of 5 weeks (non-consecutively) to ensure that the data included in our analysis reflected meaningful engagement with the program. This resulted in a sample of 33.

Results

To test the effectiveness of the life-skill training program in improving self-assessed DAPS scores a dependent samples t-test was conducted for depression, anxiety, emotional pain, and stress. The results from the pre-test for depression ($M = 2.21$, $SD = 1.87$) and post-test ($M = 1.55$, $SD = 1.00$) indicated a statistically significant decrease in depression scores, $t(32) = -2.22$,

Table 7.1. Overview of program structure by semester and session.

Semester	Session time	Duration	Key activities	Focus areas
Spring 2019 (March–May)	Morning (9 am–11:30 am)	2.5 hour. 9 weeks	Grooming, groundwork, therapeutic riding	Interpersonal competence, conflict resolution, self-control, teamwork, improved communication
	Afternoon (12 pm–3 pm)	2.5 hour. 9 weeks		
Fall 2019 (September–November)	Morning (9 am–11:30 am) 2.5 hour.	9 weeks	Grooming, groundwork, therapeutic riding	Interpersonal competence, conflict resolution, self-control, teamwork, improved communication
	Afternoon (12 pm–3 pm)	2.5 hour. 9 weeks		

Table 7.2. Weekly program activities and therapeutic goals.

Week	Key activities	Focus area	Detailed activity description
Week 1	Building rapport	Introduction	Interactive discussions on respect, program outline, and expectation setting. Participants were introduced to each other and the horses, including learning about the horses' backgrounds. This session focused on building initial connections and understanding the stories of each horse to foster empathy and rapport.
Week 2	Haltering; leading	Responsibility	Emphasis on elements of group interaction and self-awareness. The session concentrated on haltering techniques and safety protocols. Participants began learning how to lead horses, emphasizing the importance of clear communication and trust.
Week 3	Grooming	Decision making	Focused on prioritizing choices to enhance personal and others' safety. Grooming activities were introduced, allowing participants to engage in caretaking practices while reinforcing safety lessons. This session aimed to deepen the bond between participants and horses.
Week 4	Ground reining	Dependability	Participants engaged in ground reining exercises, initially practicing on each other with blindfolds to simulate the experience from a horse's perspective. Then moving on to horse partners, focusing on non-verbal commands and trust.
Week 5	Navigating obstacles	Communication	Building on previous ground reining skills, participants guided horses around obstacles, enhancing their communication skills and understanding of how to work collaboratively with the horses. Covered verbal and nonverbal communication focusing on the importance of clear messaging.
Week 6	Therapeutic painting	Acceptance	A creative session where participants painted on horses with safe paints, expressing themselves artistically and sharing the stories behind their artwork, facilitating emotional expression and connection. Highlighted inclusive behaviors, listening as allies, and confronting self-perception challenges.
Week 7	Riding	Teamwork	Participants were introduced to riding. They learned skills such as mounting and basic commands. This phase was pivotal for nurturing an understanding of non-verbal cues and fortifying the bond of trust with their equine partners, aligning with the program's emphasis on communication and empathy development
Week 8	Videography	Organization	A special session where participants practiced their skills for a video project, capturing their journey and the skills learned throughout the program. Focused on skills for managing setbacks and navigating life's challenges.
Week 9	Culmination	Celebration	More riding activities and a reflective session to mark the program's conclusion, focusing on the achievements of the participants and their experiences with the horses.

p=0.016. The mean of differences between depression scores was −0.67. The results from the pretest for anxiety (M = 3.00, SD = 1.80) and post-test (M = 1.82, SD = 1.10) indicated a statistically significant decrease in anxiety scores, $t(32) = -4.44, p<0.001$. The mean of differences between anxiety scores was −1.18. The results from the pre-test for emotional pain (M = 2.79, SD = 2.45) and post-test (M = 2.09, SD = 1.55) were in the hypothesized direction but did not indicate a statistically significant decrease in emotional pain scores, $t(32) = -1.50$, p=0.07. The mean of differences between emotional pain scores was −0.69. The results from the pre-test for stress (M = 2.70, SD = 2.88) and post-test (M = 2.00, SD = 1.66) were in the hypothesized direction, but did not indicate a statistically significant decrease in stress scores, $t(32) = -1.57$, p=0.06. The mean of differences between stress scores was −0.69.

To correct for the increased chance of Type I errors arising from these multiple comparisons, the Bonferroni correction was applied. The p-value threshold was adjusted by dividing the standard significance level ($p<0.05$) by the number of tests conducted. As a result, with four tests in our analysis, the significance level for each individual test was recalibrated to $p<0.0125$. This calibration ensured a more conservative criterion for determining statistical significance and effectively controlled for the familywise error rate. The correction affirmed the significance of improvements in anxiety, underscoring this area as particularly responsive to the intervention. The Bonferroni-adjusted p-values for each comparison were as follows: depression, p=0.064; anxiety, p=0.004; emotional pain, p=0.28; and stress, p=0.24.

Discussion

The aim of this study was to evaluate the effectiveness of an equine assisted life-skill training program in improving self-assessed DAPS scores among participants. We conducted a dependent samples t-test to determine if there were statistically significant changes in DAPS scores before and after the program. The results of the study showed that there were statistically significant improvements in depression and anxiety scores and indicated that interventions targeted towards enhancing depression and anxiety could be advantageous for individuals experiencing such difficulties.

It is important to consider the lack of significant improvements in emotional pain and stress scores. While in the hypothesized direction, the modest changes observed could reflect the nuanced nature of these psychological constructs, which may require more time or targeted interventions to shift. Additionally, the sample size may have influenced the ability to detect changes in emotional pain and stress scores.

The 9-week duration may not have provided enough time for significant changes to occur in emotional pain and stress. Emotional growth is a process that often unfolds over extended periods, suggesting that longer-term interventions might provide a more suitable opportunity for participants to deeply engage with and apply new emotional regulation skills (Durlak *et al.*, 2011; Taylor *et al.*, 2017).

The observed improvements in DAPS scores among participants after engaging in the EAS program underscore the potential of such interventions to foster emotional and psychological well- being. These findings provide support for use of EAS interventions among at-risk youth with mental health or behavioral diagnoses, and those who struggle with emotional regulation and difficulties with social interactions. Although our study did not directly measure resilience, the observed improvements in DAPS scores are indicative of enhanced emotional and psychological well-being, which are critical components of resilience and SEL development.

Our findings are consistent with other research on at-risk youth (Taylor *et al.*, 2017; Harvey *et al.*, 2020; Haig and Skinner, 2022) suggesting that engaging with horses in structured activities allows youth to navigate challenges, thereby building a foundation for resilience (Gabriels *et al.*, 2015). Resilience equips at-risk youth with the tools necessary for enduring adversities. Thus, EAS emerges as an intervention that can be incorporated as part of a holistic intervention program targeting development of resilience.

Limitations

Our study utilized pre-existing data and did not specifically track the individual horse engagement, which could influence participants' experiences. The autonomy given to participants to work with the same or different horses could potentially impact their perception of the experience, contributing to the therapeutic benefits observed. Future research could explore this aspect in more detail to understand the dynamics between horse and participant interactions.

Since the DAPS assessment form was used by the EPP life skills training program and was not originally intended for research, there are inherent limitations. The use of well validated scales would have provided a more robust evaluation of outcomes collection. Without a control group, caution must be used in interpreting reported results as other factors could have contributed to the observed outcomes. One factor is the potential for regression to the mean, where initial measurements at the extremes tend to shift toward the average in later assessments. Strengthening research design with more structured data collection methods, validated instruments and including a control group could help assess the impact of the intervention.

The use of self-report measures may not have fully captured the nuanced changes in emotional regulation and stress management due to potential inaccuracies in self-perception, recall, and the challenge of articulating complex emotional states. Additionally, the current analysis, constrained by the use of pre-existing data, lacked specific inclusion criteria regarding cognitive abilities. Given that youth in this program possessed IEPs, reflecting a range of disabilities, it is possible that participants may not have had the ability to accurately report information. To ensure a more comprehensive evaluation of the program, it would be beneficial for future research to include data collected from observations by team leaders, parents, and teachers.

A final consideration is the existence of this program as part of an overall schedule of the transition program. The EPP life skills training program was one component of a comprehensive educational curriculum. The contribution of classes to the observed outcomes cannot be separated from those specific to the EPP life skills training program. Future studies would benefit from identifying unique contributions of each component of the curriculum to the overall developmental outcomes.

Future research

Recommendations for future research of this program include more robust assessments to improve the quality of data outcomes. Studies conducted by Gabriels *et al.* (2015) and by Peters *et al.* (2022) offer valuable insights toward specific measures to assess the youth-horse relationship as an active mechanism of change. They utilized the Social Responsiveness Scale (SRS) to measure social impairments and the Aberrant Behaviour Checklist-Community (ABC-C) to evaluate problem behaviors in understanding therapeutic outcomes. These measures may provide a comprehensive understanding of how equine assisted interventions facilitate improvements in social functioning and emotional well-being.

Future studies could explore the benefits of longer programs, allowing participants more time to engage with, reflect on, and incorporate new skills for regulating emotions. Such extended involvement might lead to deeper insights and more substantial changes in how participants handle emotional pain and stress, suggesting a potential path for enhancing the efficacy of interventions aimed at emotional well-being and SEL development.

Conclusion

Incorporating EAS into interventions for at-risk youth has shown promise in fostering resilience and emotional well-being (Haig and Skinner, 2022). EAS encompasses a range of activities, including equine therapy and equine assisted learning, focusing on groundwork and interaction with horses. These activities aim to improve social, emotional, and psychological outcomes among youth by promoting experiential learning through direct interaction with horses. Involvement in EAS can help at-risk youth develop coping strategies, enhance self-esteem, and improve overall mental health (Harvey *et al.*, 2020).

Participants in this study engaged in structured problem-solving sessions, where they collaboratively tackled real-life challenges, promoting essential skills such as teamwork, communication, and conflict resolution. Skill-building activities provided participants with opportunities to practice SEL competencies in a hands-on, real-life context. These activities included equine assisted problem-solving scenarios, where participants navigated challenges alongside their equine partners, as well as communication exercises that required effective emotional expression and interaction with the horses.

The findings of this study offer valuable insights, contributing to the existing literature on the effectiveness of EAS in promoting SEL skill development among youth, and highlight the potential for EAS as a supportive educational tool. Specifically, our findings demonstrated statistically significant improvements in participant scores related to depression and anxiety. These results underscore the potential advantages of interventions aimed at enhancing these specific areas of emotional well-being.

The hands-on nature of EAS activities, centered around the interaction with horses, offers at-risk youth a unique and impactful environment to practice and apply SEL skills. This hands-on, experiential approach to SEL provided a clear visual representation of maladaptive relational and coping skills, allowing for their replacement with healthier alternatives. The horse served not only as a catalyst for engagement but also as a mirror reflecting the emotional states and behaviors of the participants. This dynamic interaction facilitated a deeper understanding and mastery of SEL competencies in a tangible and immediate context. Considering the outcomes of this preliminary evaluation, EAS programs have the potential to offer a comprehensive approach to SEL that addresses the specific needs of this vulnerable population.

Acknowledgements

We wish to extend our gratitude to Mr Neil Yetz for his invaluable contributions. His role primarily involved providing critical insights and acting as a consultative resource during the data analysis process.

Funding

The research project did not receive any financial support.

Conflict of Interest

The authors declare no conflicts of interest to disclose.

Author Contributions

The authors' contributions to this study are as follows:

- Ashleigh Zaker: Responsible for the study's conception and design, data collection and analysis, interpretation of results, and drafting of the manuscript.
- Lee Dudley: Director of the Equine Partnership Program, and collaborator in the development of the Equine Assisted Social Emotional Learning program.

- Mary Slouka: Collaborator in the development of the Equine Assisted Social Emotional Learning program.
- Lorann Stallones: Academic advisor who offered guidance, performed editing, provided final approval, and contributed to the scholarly process as the principal investigator.

Ethics Statement

This research has received all necessary ethical approvals. It was reviewed by the Colorado State University IRB (FWA0000647) and determined to be Not Human Subjects Research (NHSR) Protocol number: 21-10518H.

References

Centers for Disease Control and Prevention (CDC) (2021) Data and Statistics on Children's Mental Health. Available at: https://www.cdc.gov/childrensmentalhealth/data.html

Collaborative for Academic, Social and Emotional Learning (CASEL) (2021) Social and Emotional Learning. Available at: https://casel.org/overview-sel

Daros, A.R., Haefner, S.A., Asadi, S., Kazi, S., Rodak, T, *et al.* (2021) A metaanalysis of emotional regulation outcomes in psychological interventions for youth with depression and anxiety. *Nature Human Behaviour* 5, 1443–1457. Available at: https://www.nature.com/articles/s41562-021-01191-9

Domitrovich, C.E., Durlak, J.A., Staley, K.C. and Weissberg, R.P. (2017) Social-emotional competence: An essential factor for promoting positive adjustment and reducing risk in school children. *Child Development* 88(2), 408–416. DOI: 10.1111/cdev.12739.

Durlak, J.A., Weissberg, R.P., Dymnicki, A.B., Taylor, R.D. and Schellinger, K.B. (2011) The impact of enhancing students' social and emotional learning: a meta-analysis of schoolbased universal interventions. *Child Development* 82(1), 405–432. DOI: 10.1111/j.1467-8624.2010.01564.x.

Ehrenreich-May, J., Kennedy, S.M. and Remmes, C.S. (2015) Emotion regulation interventions and childhood depression. American Psychological Association. DOI: 10.1037/14641-008.

Equine Assisted Growth and Learning Association (EAGALA) (2018) The EAGALA Model Standards. Available at: https://www.eagala.org/certification

Gabriels, R.L., Pan, Z., Dechant, B., Agnew, J.A., Brim, N. *et al.* (2015) Randomized controlled trial of therapeutic horseback riding in children and adolescents with autism spectrum disorder. *Journal of the American Academy of Child & Adolescent Psychiatry* 54(7), 541–549. DOI: 10.1016%2Fj.jaac.2015.04.007.

Haig, L. and Skinner, K. (2022) Use of equine-assisted services to improve outcomes among at-risk and indigenous youth: a scoping review. *Frontiers in Public Health* 10. DOI: 10.3389/fpubh.2022.730644.

Harvey, C., Jedlicka, H. and Martinez, S. (2020) A program evaluation: equine-assisted psychotherapy outcomes for children and adolescents. *Child and Adolescent Social Work Journal* 37(6), 665–675.

Jones, C.J.V. (2017) Differentiating the Effects of Animal-Assisted Verses Equine Facilitated Psychotherapy on Positive Psychological Outcomes Through a Stress Induction Context. Available at: https://digitalcommons.georgiasouthern.edu/etd/1701

Lawson, G.M., McKenzie, M.E., Becker, K.D., Selby, L. and Hoover, S.A. (2019) The core components of evidence-based social emotional learning programs. *Prevention Science* 20(4), 457–467. DOI: 10.1007/s10826-016-0648-6.

Mahdiani, H. and Ungar, M. (2021) The dark side of resilience. *Adversity and Resilience Science* 2, 147–155. DOI: 10.1007/s42844-021-00031-z.

Moreno-Gómez, A.-J. and Cejudo, J. (2019) Effectiveness of a mindfulness-based social – emotional learning program on psychosocial adjustment and neuropsychological maturity in kindergarten children. *Mindfulness* 10(1), 111–121. DOI: 10.1007/s12671-018-0956-6.

Mueller, M.K. and McCullough, L. (2017) Effects of equine-facilitated psychotherapy on posttraumatic stress symptoms in youth. *Journal of Child and Family Studies* 26(4), 1164–1172. DOI: 10.1007/s10826-016-0648-6.

National Alliance of Mental Illness (NAMI) (2021) Mental Health by the Numbers. Available at: https://www.nami.org/mhstats

Oldeide, O., Holsen, I. and Fosse, E. (2020) Youth perspective on outreach service: a safety net for at-risk youth in a municipality. *Children and Youth Services Review* 116, 105234. DOI: 10.1016/j.childyouth.2020.105234.

PATH International (2021) Professional Association of Therapeutic Horsemanship International Standards for Certification & Accreditation. Available at: https://pathintl.org/resources/standards-manual

Peters, B.C., Pan, Z., Christensen, H. and Gabriels, R.L. (2022) Self- regulation mediates therapeutic horseback riding social functioning outcomes in youth with autism spectrum disorder. *Frontiers in Pediatrics* 10, 884054. DOI: 10.3389/fped.2022.884054.

Pinto, T.M., Laurence, P.G., Macedo, C.R. and Macedo, E.C. (2021) Resilience programs for children and adolescents: a systematic review and meta-analysis. *Frontiers in Psychology* 12. Available at: https://www.ncbi.nlm.nih.gov/pmc/articles/PMC8645691

Song, Y.-M. and Kim, S. (2022) Effects of a social and emotional competence enhancement program for adolescents who bully: a quasi-experimental design. *International Journal of Environmental Research and Public Health* 19(12), 7339. DOI: 10.3390/ijerph19127339.

Staudt, M. and Cherry, D. (2017) Equine facilitated therapy and trauma: current knowledge, future needs. *Advances in Social Work* 18(1), 403–414. DOI: 10.18060/21292.

Suveg, C., Davis, M. and Jones, A. (2015) Emotion regulation interventions for youth with anxiety disorders. In: *Self-Regulated Learning Interventions with At-Risk Youth: Enhancing Adaptability, Performance, and Wellbeing*. American Psychological Association, Washington, DC, pp. 137–156. DOI: 10.1037/14641-007.

Taylor, R.D., Oberle, E., Durlak, J.A. and Weissberg, R.P. (2017) Promoting positive youth development through school-based social and emotional learning interventions: a metaanalysis of follow-up effects. *Child Development* 88(4), 1156–1171. DOI: 10.1111/cdev.12864.

Tugade, M.M. and Fredrickson, B.L. (2007) Regulation of positive emotions: emotion regulation strategies that promote resilience. *Journal of Happiness Studies* 8, 311–333. DOI: 10.1007/s10902-006.

Van Katwyk, T. and Seko, Y. (2018) Resilience beyond risk: youth re-defining resilience through collective art-making. *Child and Adolescent Social Work Journal* 36(6), 609–619. DOI: 10.1007/s10560-018-0590-0.

Xie, H. (2013) Strengths-based approach for mental health recovery. *Iranian Journal of Psychiatry and Behavioral Sciences* 7(2), 5–10.

8 Equine-Assisted Activities for Adolescents: Ethogram-Based Behavioral Analysis of Persistence During Horse-Related Tasks and Communication Patterns with the Horse

Hilde Hauge[1], Ingela L. Kvalem[2], Ingeborg Pedersen[3] and Bjarne O. Braastad[3]*

[1]Norwegian University of Life Sciences, Ås; [2]University of Oslo; [3]Norwegian University of Life Sciences, Ås

---

Abstract

Earlier studies have indicated that learning to handle a horse through tasks and activities can lead to a feeling of mastery which may have an impact on self-efficacy. The aim of this study was to examine how adolescents conducted horse-related tasks presented to them in an intervention in a farm environment, and whether there was a change during the intervention in persistence on tasks with the horse. Furthermore, we wanted to examine the behavior of the adolescents towards the horse and the response from the horse. Each participant was given an intervention once a week for approximately 16 weeks consisting of tasks with the horse, riding, grooming, and stable work. The sample presented in this study consisted of 29 participants who were successfully video-recorded in the beginning and at the end of the intervention. Petting the horse was the most frequent way of initiating contact with the horse, and the distributions of contact behaviors were the same at both time spots. The response of the horse was mainly neutral or positive. When participants did not succeed at their first attempt when trying to solve a horse-related task or an exercise during riding, their subsequent behavior was recorded as either trying again or not trying again. Early in the intervention, these two options were chosen with about the same frequency, while at the end of the intervention trying again was chosen significantly more often than not trying again. This was operationalized as an increase in persistence when having difficulties in solving tasks with the horse. The increased persistence late in the intervention in retrying tasks may indicate that the adolescents developed a feeling of mastery, which is an important factor in development of self-efficacy.

Keywords: adolescents, horses, intervention, mastery of task, equine-assisted activities

---

*Corresponding author: Bjarne.braastad@umb.no

Adolescent's Persistence During Equine-Assisted Activities and the Relation to Task-Specific Mastery

Interactions with horses may contribute to development of social skills and self-efficacy (Traeen and Wang, 2006; Trotter *et al.*, 2008; Keino *et al.*, 2009; Forsberg and Tebelius, 2011; Burgon, 2011). To understand why activities with horses may have a positive impact it is important to study the interaction process between the participant and the horse. In this study we wanted to analyze the experience of mastery with a typical group of adolescents through activities with horses. Methods of video analysis used in animal behavior research allowed us to investigate the interaction between the adolescents and the horses, and investigate development of mastery of tasks related to the horses during the intervention.

The horse in activities and therapy

Equine-assisted therapy (EAT) is a treatment of physical or mental disabilities with the horse as a co-therapist (Fine, 2010). In treatment of physical disabilities the horse works as a therapist to increase balance, stability, and flexibility in the patient's musculature (Bertoti, 1988; Davis *et al.*, 2009; Herrero *et al.*, 2010). Equine-assisted therapy might also enhance rehabilitation from physical illnesses (Hakanson *et al.*, 2009).

Equine-assisted psychotherapy is a means of equine-assisted therapy with a focus on mental health issues. Activities with the horse are performed by the participants to trigger a behavioral change and help solve emotional problems (Roberts *et al.*, 2004; Rothe *et al.*, 2005; Schultz *et al.*, 2007; Klontz *et al.*, 2007; Lentini and Knox, 2008; Selby, 2009; Keino *et al.*, 2009). Several studies show that the horse has a positive impact on children with learning disabilities and emotional problems (Kaiser *et al.*, 2006; Braat, 2008; Trotter *et al.*, 2008; Burgon, 2011). Burgon (2011) studied seven at-risk young people learning to handle the horse from the ground through different tasks during an intervention. The adolescents appeared to grow in self-confidence (measured through qualitative interviews) when able to communicate with the horse to solve the exercises.

Equine-assisted activities (EAA), like other animal-assisted activities (AAA), provide opportunities for motivational, educational, recreational, and/or therapeutic benefits to enhance quality of life, improvement in school subjects, or social behavior (Kaiser *et al.*, 2006; Ewing *et al.*, 2007; Trotter *et al.*, 2008). The horse in activities related to leisure may also have an impact on self-efficacy, although this is not the main goal of the activity (Traeen and Wang, 2006; Forsberg, 2007; Forsberg and Tebelius, 2011). Forsberg (2007) showed that 14 to 16 year old girls developed their identity through an experience of mastering activities with the horse. Forsberg used semi-structured interviews to examine girls` identity development through riding as a leisure activity, and followed the girls over time as they took part in stable work, horse-related tasks, and social interactions at the riding school. The girls in this study pointed out the importance of the horse in a task-related activity and how mastering tasks in relation to the horse influenced their beliefs in their own competence.

The farm may be used as a venue for health-promoting activities like what is termed green care, farming for health, social farming, or care farming (Sempik *et al.*, 2010). The farmer and the farm provide an environment with nature and animals with farm-related activities for a diversity of people to promote health and well-being. The focus of this study is on the experience of activities with horses in a social environment on a farm, providing a natural setting with physical activities that holds elements of task achievement, all potentially important for adolescent development (Moote and Wodarski, 1997; Barber *et al.*, 2001; Dodge and Lambert, 2009).

Self-Efficacy

Self-efficacy is described as a person`s belief that one can successfully produce the desired outcome by one's own actions (Bandura, 1997b). It is the belief that one can utilize one's abilities and skills in a certain situation to reach a specific goal (Maddux, 1995). Efficacy beliefs operate through impact on cognitive, motivational, and decisional processes (Bandura, 1997a). They determine whether a person thinks in a positive or negative way about accomplishing a task and how well they motivate themselves (Baumeister *et al.*, 2007). The most important factor in the development of self-efficacy is personal experience with trying to master a task. Mastery requires skills or knowledge that helps one master the subject, for example putting a halter on a horse. Successful performance attempts will strengthen self-efficacy related to that specific behavior, while previous failures will weaken self-efficacy beliefs about being able to perform the behavior (Maddux and Volkmann, 2010). In a social setting with people similar to oneself, seeing others succeed may also enhance one's own belief in success (Bandura, 1982).

Self-efficacy beliefs are important because they influence the goals set, the choice of goal-directed activities, and the persistence when failing at a task. High self-efficacy leads to persistence when faced with difficulties, such as mastering a difficult task, and creates a willingness to try again when faced with challenges because of the belief of success. Low self-efficacy on the other hand may lead to a less active behavior and belief that a task is harder than it actually is (Baumeister, 2004; Baumeister *et al.*, 2007). In this study the focus will be persistence on performing a task. Task-specific mastery is related to the task at hand, but experiences of mastery during a specific task may strengthen self-efficacy beliefs in general, which again might help in daring to try new tasks in other social arenas (Bandura, 2006).

Leisure activities and the relation to self-efficacy

Leisure activities in general may promote self-efficacy in adolescents (Barber *et al.*, 2001; Dodge and Lambert, 2009). Dodge and Lambert (2009) showed a positive association between participating in sports during adolescence and positive self-belief 1 year later, a concept similar to self-efficacy. Adventure-based activities may also have a positive impact on self-efficacy through task-specific activities that include cooperation, trust, communication, physical activity, and problem-solving (Moote and Wodarski, 1997). These are many of the same aspects found in activities with horses. In both qualitative and observational studies of equine-assisted activities and leisure activities with horses the participants have reported an increased belief of success and feeling of mastery through learning to master tasks with the horse. This has been interpreted as an indication that mastering tasks with the horse had an impact on self-efficacy beliefs (Forsberg, 2007; Forsberg and Tebelius, 2011; Burgon, 2011). Although studies that measured general self-efficacy directly did not find any change in the standardized measurement, the findings showed a decrease in problematic behaviors after participating in an intervention with horses and mastering tasks with the horse (Kaiser *et al.*, 2006; Ewing *et al.*, 2007). Activities with horses may therefore have an influence on specific self-efficacy beliefs through experience of mastery during potentially difficult tasks related to the horse.

Riding and interaction with horses are different from other leisure activities because of the interaction with a horse, giving a new type of interpersonal element and greater complexity to the activity. The prerequisites for mastery of tasks in relation to the horse are feeling safe in the situation and being able to communicate with the horse.

Interaction with the horse

In equine-assisted activities the qualities of the horse are used to achieve the goals of the intervention. Horses are prey animals and therefore very sensitive to their environment, responding readily to the behavior of those around them (Lentini and Knox, 2008; Smith-Osborne and Selby, 2010). Confidence in interacting with a horse is important with regards to success in getting it to cooperate (Burgon, 2011), and communication skills such as being clear and unambiguous are required (Traeen and Wang, 2006). To ensure a positive experience in relation to the horse, it is important to understand how the horse communicates (Hausberger and Muller, 2002; Birke *et al.*, 2011). The non-verbal communication often experienced with animals is especially prominent in an intervention with horses; the communication mainly happening through understanding of body language (Lentini and Knox, 2008; Krueger *et al.*, 2011). In equine-assisted psychotherapy, the behavior response from the horse is often used to improve communication and work on behavioral problems (Trotter *et al.*, 2008; Lentini and Knox, 2008; Keino *et al.*, 2009).

Physical contact, as non-verbal communication, with animals may have a calming effect on humans (Friedmann *et al.*, 1983). Stroking a dog or having a dog present is shown to give children comfort in conducting difficult tasks or in stressful situations (Gee *et al.*, 2007; Gee *et al.*, 2010; Beetz *et al.*, 2011). Studies with animals other than horses have shown that stroking an animal may have a positive physiological effect on the human (Friedmann *et al.*, 1983). Beetz *et al.* (2011) found that physical contact with a dog gave reduced cortisol level in a socially stressful situation. Stroking a dog is also shown to increase the oxytocin level through positive interaction, in both the human and the dog (Odendaal, 2000; Odendaal and Meintjes, 2003). Learning to communicate with the horse through non-verbal communication and having physical contact with the horse may be important to increase the feeling of security and calmness during the intervention.

Analysis of behaviour

Animal behavioral methodology drawn from ethology gives the opportunity to objectively study the communication between adolescents and horses and the conduct of tasks. In animal behavior studies video recording is a common method. Using animal behavior methodology, the response of the animal to the behavior of the person can be analyzed (Martin and Bateson, 2007).

Ethological methods were used previously to investigate mastery in relation to tasks with animals (Berget *et al.*, 2007; Pedersen *et al.*, 2011). These studies investigated people with psychiatric problems working with farm animals. Both studies indicated that tasks conducted with animals may be related to increased self-efficacy in some cases. Video recordings and behavior analysis have also been used to examine other activities. A study of soccer players and their relation to their coach used an ethogram with gazing and talking behaviors in a similar manner as done in the current study (Toering *et al.*, 2011). As far as we know there is no previous study using video analysis in the same context as used in the present study. Gee *et al.* (2007) conducted a study using a scale measurement on video observations to investigate the level of motor skills among children when in presence of a dog. Beetz *et al.* (2011) used video analysis of behavior in relation to cortisol measurement to investigate if stress levels were reduced during physical contact with a dog. In horse-assisted activities, a custom-designed psychosocial session form (PSF) has been used to analyze behavior of adolescents in relation to horses (Braat, 2008; Trotter *et al.*, 2008). The PSF records behaviors like looking at the horse, making contact, participating, grade of leadership, fear, and positive interactions. Measuring the interaction between horses and adolescents and investigating mastery of potentially difficult tasks with horses by use of video analysis brings a new aspect to the field, which may have implications for studies on equine-assisted activities.

Aim of the study

In this study we wanted to investigate the recorded activity patterns of adolescents during a four-month intervention with horses on a farm. The first aim was to investigate persistence during challenging tasks, by focusing on behaviors chosen after unsuccessful completion of a task after the first attempt. The second aim was to examine the communication pattern between the adolescent and the horse; that is, the interaction initiated by the adolescents towards the horse and the response of the horse to this interaction.

Methods

Participants and procedures

The participants were Norwegian adolescents between 12 and 15 years recruited through secondary schools. Ten schools were selected to be part of the project due to proximity to the selected farms where the interventions took place. The principals at the schools were contacted and informed about the project, and they approved the school's participation. The first author presented the project for each class in the first and second year at the selected schools to a total of approximately 150 adolescents at each school. The adolescents were given a brochure and information letter with detailed information about the project. A signed informed consent form was required in accordance with national ethical guidelines. Consent forms were signed by both the adolescent and their parent/guardian. Eighty students gave consent to participate in the project. All participants were offered an intervention with horses either in the spring or fall during 2009, 2010, or 2011. Five of the adolescents dropped out before the intervention started. None of the adolescents dropped out during the intervention, however. The 75 participants were randomly divided into intervention (group 1) and control (group 2) groups. Group two functioned as a waiting-list control group, beginning their intervention four months later. Four months after the end of their intervention Group one served as a follow-up group. All participants were given questionnaires before and after their intervention and either four months before or four months after their intervention. Half of the adolescents were filmed early and/or late in their intervention. The questionnaire data are not reported in this paper. In this study only those who were video-recorded both early and late in their intervention are used (N=29). The participants (10 boys and 65 girls) averaged 13.5 years old. They all attended their first (N=51) or second year (N=24) at their Junior High School (aged 13–15 years). Forty-five of the adolescents had little or no experience with handling horses while 30 had been riding before, well distributed between the two groups. All the adolescents had seen a horse before, and were motivated to learn more about handling horses. The adolescents all had access to a support system at their school, well informed about the project. The support system consisted of the principal, the contact teacher and a health professional at the school, all being aware of the project and ready to assist the participants if needed. To our knowledge, the support system was never used.

To avoid the performance pressure often found at riding schools, small farms were used as the venue for the interventions. The farms were selected by inclusion criteria related to the farm size, the country surroundings, proximity to a school, the stabling conditions for the horses, and the competence of the farmer. The criteria for the farms were: a total size of fewer than 25 acres, no more than 10 horses, situated in the country side less than 5 km from a Junior High School, and with natural surroundings. All the farms allowed their horses plenty of free movement and social contact with other horses, and most of the stables had an outdoor-stable system. The farms had cats, dogs, laying hens, goats, or sheep in addition to the horses. These were not part of the intervention, though some occasional contact occurred. None of the farms had other large animals like cows. Ten farms in Eastern Norway from the counties of Hedmark, Oppland, Buskerud, and Vestfold were selected to be part of the project.

The farmers functioned as instructors. When farms were selected the inclusion criteria for the farmers were: (a) having worked with adolescents and horses during the past year, (b) having such experience for at least a year, or (c) being a riding instructor and having some kind of education in health or pedagogy. The riding-instruction competence ranged from self-taught experience to education at a national level. The education in social work or pedagogy, ranged from fully educated teacher or health practitioner to minor courses in psychology or pedagogy. There were nine women and one man.

The horses used for the project were of Norwegian breeds or Icelandic horses, all cold-blooded horses which are known for their calm temperament and steady gaits. These were chosen due to their good temperament and steadiness often found in these breeds. All horses were kept in a herd, allowed social interaction with other horses, and were able to move freely over a large area. They were given free access to straw feed and most of them were kept bare footed. All these factors have been found to be important for the welfare and behavior of the horse and to ensure a calm and safe experience for the participants (Søndergaard and Ladewig,

2004; Lesimple *et al.*, 2011; Hartmann *et al.*, 2012).

The project was approved by The Regional Committee for Medical Research Ethics (REK) and the Norwegian Data Protection Official for Research.

Study sample

In order to be able to observe changes in task performance and communication patterns, some of the participants were filmed both early and late in their intervention. The selection of adolescents to be filmed was made using practical considerations. As many as possible were filmed, but due to simultaneous sessions at different farms and long travel distances between farms, some had to be excluded. The adolescents filmed early in their intervention were the same as those filmed late in their intervention, aiming at filming as many as possible at both time periods. The adolescents and parents were also informed about the filming when they gave their consent. They were informed that the videotapes would be used only for the purpose of analyzing behavior through a computer program, and that the video material would be deleted when the project was finished. Filming was voluntary, and the participants were free to drop out of the project at any time without giving any reasons or risking any consequences.

Of the 75 participants completing the intervention 46 participants were filmed either early or late in their intervention period. Fifteen participants were filmed early in their intervention but not late, and two of the participants were filmed late but not early. A total of 29 participants was filmed both early and late in their intervention and constituted the sample used in this study. There were four boys and 25 girls all between 13 and 15 years old, with an average of 13.5 years. A questionnaire on their prior experience with animals (N=25) showed that all but one had experience with pets (dogs, cats, or rabbits) before they started the project. Nineteen had experience with horses. Twelve of them had been riding before, while seven had sat on a horse a few times or visited friends at a stable. This sample did not differ substantially in gender, age, or experience with animals and horses from the 46 participants filmed either early or late, nor from the total group of 75 participants.

The intervention

The adolescents had a 2 hour session once a week for approximately 16 weeks. The farmers functioning as instructors were responsible for conducting the intervention guided by the researcher through an instruction letter, visits, and several conversations to make sure they understood how to conduct the intervention. The instruction letter gave specific details on how to implement the intervention, with examples of tasks in each of four categories of activities the adolescents would do in each session. The instructors decided which tasks from those categories the adolescents should carry out and when. The different activity categories were: (a) horse-related tasks, (b) riding or handling the horse from the ground, (c) grooming, and (d) stable work. Horse-related tasks ranged in difficulty from easy to hard and included putting on a halter, leading the horse, getting the horse from the field, picking up the horse's feet and cleaning the hooves, and saddling the horse. Grooming the horse consisted of saying hello to the horse, brushing the horse, and sometimes plaiting the mane and tail. Riding or handling the horse from the ground consisted of being able to steer the horse through different obstacles in the arena or going for a walk in the forest. The stable work category consisted of cleaning out the stables, giving the horse food and water, and making it tidy in the stable. Due to natural variations on a farm and varying weather conditions, the time spent on each category and activity differed, but the adolescents did some work in each category every time they came to the farm. The adolescents were assigned their own individual horses, and they kept the same horses as much as possible throughout the intervention. The intervention was all about the experience with the horse and not about becoming skilled at the activities.

Within this framework the instructors were encouraged to make an individual plan for progression during the intervention, and to provide different experiences for each adolescent for mastering tasks with the horse. This was dependent on weather conditions and the experience level of each adolescent. The progression during the intervention was accomplished by giving the adolescents more responsibility and challenges in tasks, particular during riding. The horse-related tasks varied in level of difficulty, the adolescents were therefore given the easiest tasks at the start and then progressed to tasks of increasing difficulty throughout the intervention. The level of difficulty during riding could vary from leading the horse through obstacles to riding by themselves and learning to steer the horse alone. The adolescents who had been riding prior to the study often had little experience with handling the horse in other settings, like saddling the horse, therefore all adolescents met new challenges during the intervention. When the adolescents were unable to solve the task or exercise, they were given time to figure it out by themselves. If they did not try again by themselves they either gave up or asked for help to complete the task. The instructor was with them the whole time they were at the farm, so help was available when needed.

The instructors signed an agreement with the researcher stating their responsibility for the intervention. The researcher informed the adolescents about what the intervention would be like before they consented to participate in the project. In addition, the instructors informed the adolescents about the plan for the intervention during their first visit to the farm.

Analytic approach and data collection

Video-recordings

The adolescents were video recorded for the full 2 hours when on the farm by continuous sampling. This was carried out twice during the four-month intervention; at the second or third session in the beginning of the intervention, and at the third or second to last session at the end of their intervention. This was chosen to allow the adolescents time to get to know the farmer before filming was conducted and to ensure filming at the end of the intervention. The camera used was a high definition (HD) Sony Handycam with good resolution. Extra lighting was available if the filming was to take place during darkness. The two adolescents in the same session were filmed by one observer at the same time. At times where this was not possible the observer shifted between the two adolescents by an interval of 2 min. Four different observers instructed in the same way were used in the project. The camera was hand-held to be able to follow sudden changes of movements by the participants. The distance to the participants was approximately 2–3 meters to allow both participants in the picture as much as possible when they conducted horse-related tasks, stable work or grooming, and five to fifty meters when riding.

Quantitative video-analysis

The videos were analyzed using The Observer® software (Noldus Information Technology, Wageningen, The Netherlands, www.noldus.com) which is mainly used for analyzing animal behavior. The Observer is an analyzing tool which makes analyses of behavior precise. This software utilizes an ethogram, a list of behavioral categories, made by the researcher, and adapted to the study of interest (Martin and Bateson, 2007). Each behavior has its own objective definition of what it includes and must be thoroughly evaluated before the analyzing process starts. Using The Observer behaviors may be measured as duration of time or in terms of frequency, i.e. number of occurrences per time unit.

The videos were analyzed separately for each adolescent. The activities conducted on the farm were registered as four different broad behavior categories: horse-related tasks, riding or handling the horse from the ground, grooming the horse, and stable work, as noted earlier (Table 8.1). Horse-related tasks consisted of all tasks connected to the horse. Grooming the horse was seen as an interaction and not a difficult task itself and was therefore kept as a separate category.

During horse-related tasks and riding, we scored specific secondary behaviors. We registered each occurrence (frequency of behavior) when the adolescent was unable to solve the task or exercise given to them by the instructor. The criteria for noting 'not able to

Table 8.1. Ethogram of Activities, Behaviors, and Responses as a Basis for Video Analyses.

Activity	Description
Horse-related tasks	Putting on a halter, cleaning hooves, getting the horse from the field, leading the horse, saddling the horse
Riding	Riding/leading the horse in the forest or an arena
Grooming	Grooming the horse
Stablework	Mucking out, feeding horses
Contact with other animals	Mucking out, feeding other animals
Other horse-related	Getting tack from the tackroom, cleaning up, walking to get to the field
Other not horserelated	Talking to each other or playing while waiting
No activity	Inactive while waiting
Physical contact (petting)	The adolescent initiating petting the horse
Verbal contact	The adolescent talking to the head of the horse
Eye contact	The adolescent looking at the head of the horse
Positive response from the horse	The horse turning its head toward the adolescent and/or tilting its ears forward
No response from the horse	The horse standing still and having its ears relaxed to the side
Negative response from the horse	The horse turning its head toward the adolescent and/or tilting its ears backwards
Not able to solve the task	The adolescent having problems with the task presented due to the horse or the adolescent not understanding what to do
Trying again	The adolescent trying again when not being able to solve the task
Not trying again	The adolescent not trying again when not being able to solve the task

solve the task' were if the adolescent paused in the task or exercise due to the horse or the adolescent not understanding what to do. When 'not able to solve the task' was registered we wanted to investigate the action taken next by the adolescent. One of three options was noted: (a) trying again to complete the task, (b) giving up on the task, or (c) asking for help to solve the task. The two last categories were combined into not trying again in the further analysis. The goal of the analysis was to measure persistence with the task and not success on the task, although trying again often lead to successful completion of the task. The behaviors used in the analysis are described in Fig. 8.1.

During grooming and horse-related tasks the contact initiated by the adolescents towards the horse was recorded. This was noted as physical, verbal, or eye contact. If the adolescent petted the horse physical contact was registered. If the adolescents clearly talked to the horse verbal contact was registered. If the adolescent clearly looked towards the head of the horse eye contact was registered. More than one type of these contact behaviors could be recorded at the same time. If two occurrences of the same type of contact behavior initiated by the adolescent were close in time, a minimum interval of 5 s was used before the next occurrence could be recorded. When the adolescents chose to pet, talk, or look at the horse, the response of the horse was registered using the position of ears and head. This occurred during grooming when the horse was tethered therefore other body

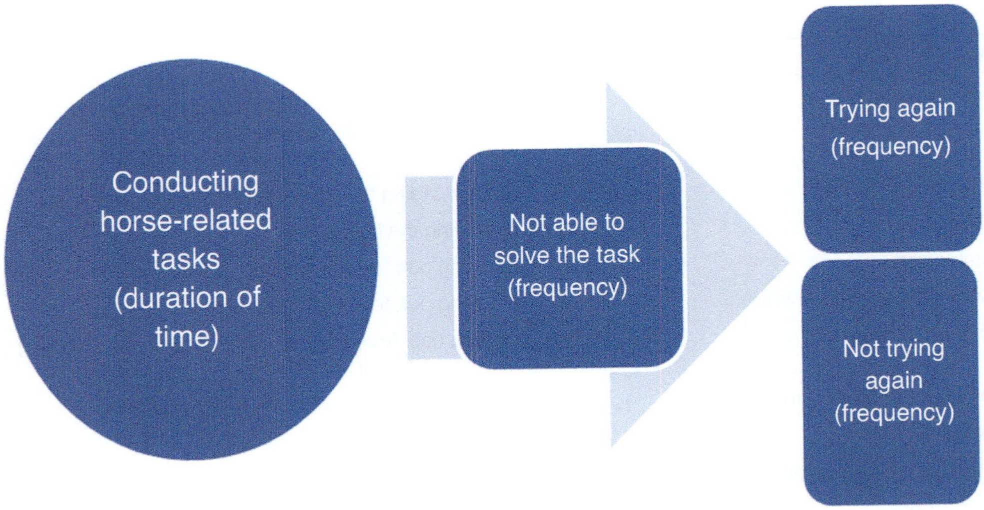

Fig. 8.1. Explanation of the different behavioral categories used in the analysis. Conducting of horse-related tasks analyzed as duration of time, not able to solve the horse-related task, and the following selection of strategy in frequency per hour.

signals were not registered. The categories for the response of the horse were: (a) ears pricked forward and/or turning the head in the direction of the adolescent, (b) ears relaxed to the side and head lowered, and (c) ears pinned backwards and/or head turned towards the adolescent raising the lips and showing teeth. The first category was used as a positive response, the second as neutral, and the third as a negative response.

Before video analyses, selected videos were viewed by two observers, the first and the third author, to discuss how 'unable to solve the task', 'trying again' and 'not trying again' should be defined. A description for each behavior was made. Videos were then analyzed separately by the observers, except in instances where there was doubt, and they were then rated by the observers together. Most of the videos were viewed in part by each of the observers, to ensure that both observers took part in the analysis of each participant.

Results

Activities on the farm

The duration of time spent in each activity category (horse-related tasks, riding, grooming, and stable work) was divided by total time recorded for each adolescent in that session to obtain the fraction of time spent in each activity for each adolescent in the video recorded session. Means and standard deviations of the percentage distribution of activities (time budget) performed during the early and the late sessions are presented in Fig. 8.2. The adolescents spent most of their time with the horse conducting horse-related tasks, through riding and grooming. Less time was spent on stable work. There was a pronounced individual variation in the distribution of work tasks during both time periods, as illustrated by the standard deviations. Some of this may be attributed to the weather conditions, as during extreme weather conditions the adolescents had to do more grooming and stable work that day. No significant differences were found between early and late parts of the intervention (See Fig. 8.2).

Contact with the horse and the response from the horse

The numbers of times the adolescent petted, looked at, or talked to the horse were divided by total time in hours spent on horse-related

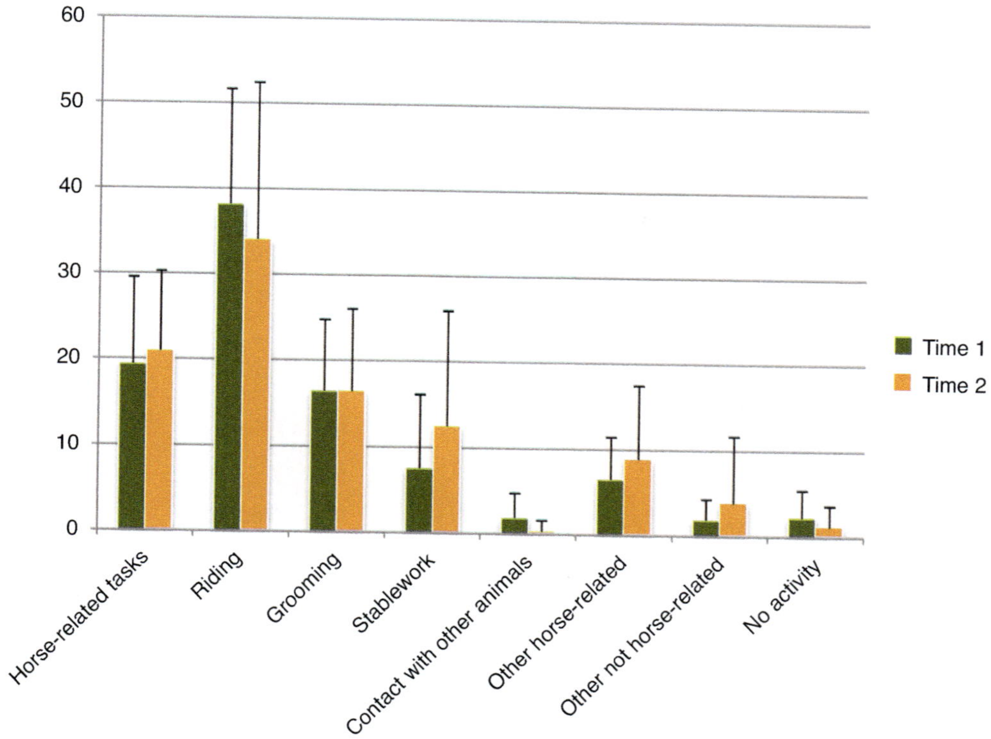

Fig. 8.2. Time budget of activities on the farm early and late during the intervention, percentages of time recorded (*Mean* + SD).

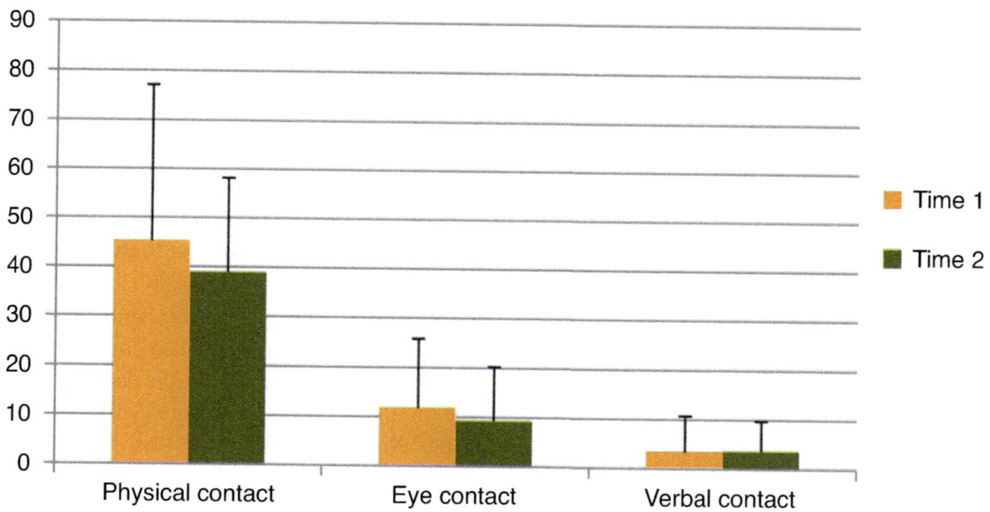

Fig. 8.3. Frequency per hour of different types of contacts the adolescents initiated with the horse, based on time used on grooming and tasks with the horse (*Mean* + SD).

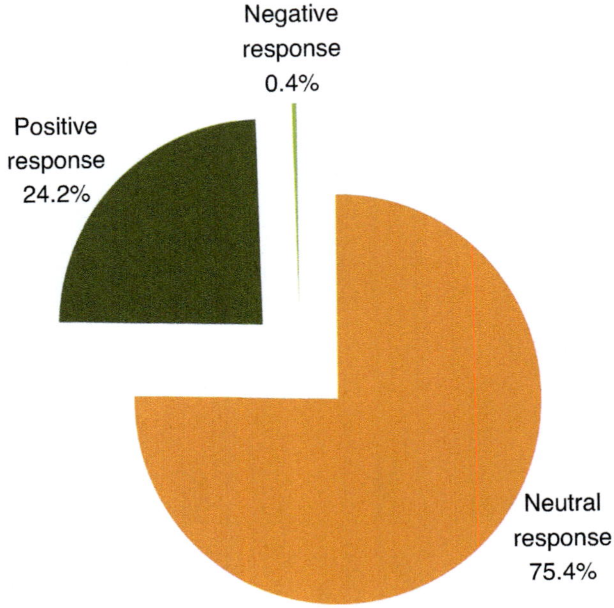

Fig. 8.4. Response from the horse to the contact initiated by the adolescents, presented as percentage distribution of frequencies per hour for early and late recordings combined.

tasks and grooming to calculate the frequency per hour of each contact variable. The different types of contact the adolescents initiated with the horse while doing horse-related activities and grooming is presented in Fig. 8.3. Petting the horse was the most frequent type of contact initiated by the adolescents, constituting 75% of all contact behavior. A paired sample t-test was conducted to compare petting with eye and verbal contact combined. Petting was performed significantly more frequent than eye or verbal contact combined during both early (T1: Petting: $M = 45.2$ $SD = 31.9$, Eye/Verbal: $M = 15.1$ $SD = 19.8$; $t(28) = 8.68$, $p<0.001$) and late recordings (T2: Petting:$M = 38.9$ $SD = 19.2$ Eye/Verbal: $M = 12.8$ $SD = 13.7$; $t(28) = 10.26$, $p<0.001$). The amount of time spent on each type of contact did not vary between early and late recordings (See Fig. 8.3).

The response from the horse was calculated for each contact variable and combined into total response from the horse for all contact variables. The distribution of responses from the horse was equal during early and late recordings. The responses from the horse in early and late sessions were therefore presented combined. The number of occurrences of each response type was divided by total time spent on horse-related tasks and grooming to obtain frequencies. The percentage distribution of these frequencies is presented in Fig. 8.4. The horse showed no response in 75.4% and a positive response in 24.2% of the cases. Only 0.4% negative responses from the horses were detected (See Fig. 8.4).

Trying again or not trying again when unable to solve the task

To calculate the frequency of being 'unable to solve the task' for each adolescent, the total number of occurrences was divided by total time (in hours) spent on horse-related tasks and riding. 'Not able to solve the task' was noted more or less equal amounts of times in early ($M = 9.9$, $SD = 7.7$) and late ($M = 8.7$, $SD = 9.7$) recordings. When 'not able to solve the task' was noted the observer waited for the action of the adolescent before recording trying again or not trying again.

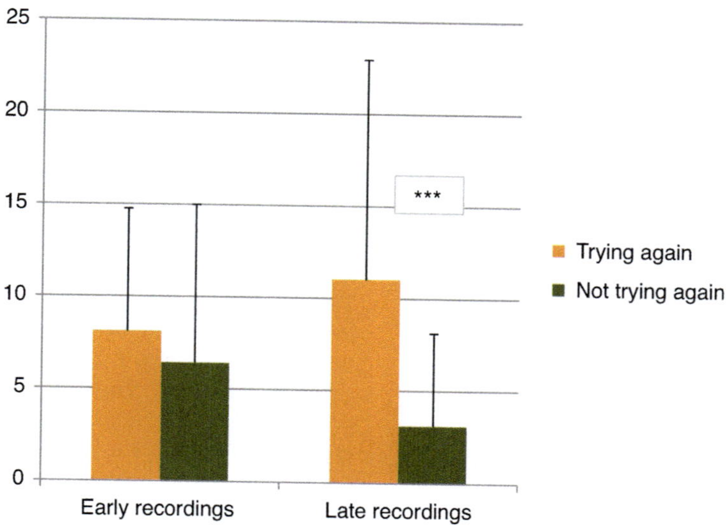

Fig. 8.5. Choice of strategy when not being able to solve the task, presented as frequencies per hour of trying again and not trying again during early and late recordings (*Mean* + SD).

'Not able to solve the task' and then 'trying again' was regarded as one strategy, while 'not able to solve the task' and then 'not trying again' was seen as another choice of strategy. The numbers of occurrences (frequency) of each of the two strategies were divided by the total duration (in hours) of horse-related tasks and riding combined for each adolescent to obtain the frequencies per hour of each strategy. The frequency per hour of trying again and not trying again in early recordings and in late recordings is presented in Fig. 8.5. Paired-sample t-tests were conducted to compare the variables 'not able to solve the task and then trying again' and 'not able to solve the task and then not trying again' in early recordings and in late recordings. In early recordings there was no difference in the frequencies of trying again or not trying again when not being able to solve the task (T1: Not trying again: $M = 6.4$, $SD = 8.6$, Trying again: $M = 8.1$, $SD = 6.7$; $t(28)= 1.15$, $p=0.259$). In late recordings, however, a significant difference between strategies was found, trying again being chosen significantly more often than not trying again when not being able to solve the task (T2: Not trying again: $M = 3.1$, $SD = 5.0$, Trying again: $M = 11.0$, $SD = 11.9$; $t(28) = 3.83$, $p<0.001$) (See Fig. 8.5).

To study potential relationships between contact with the horse and persistence during horse-related tasks, Pearson correlations were used to measure coherence between the frequencies of petting the horse and the frequencies with which the adolescents tried again. No relations were found between the variables ($r(29) = 0.21$, $p=0.28$).

Discussion

Activities and contact with the horse

The adolescents spent most of their time on horse-related tasks, riding, and grooming. This shows that that the intervention focused on activities that included contact with the horse, as the instructors were told to. The overall time used on the different activities was the same during early and late recordings, though there was variation among the individual adolescents.

Petting the horse was the main type of contact the adolescents had with the horse, both early and late in the intervention. Studies with other animals have shown that stroking an animal has positive physiological effects on e.g., the HPA axis, the major stress axis, and oxytocin Beetz, Uvnäs-Moberg, Julius, and Kotrschal, 2012; (Friedmann *et al.*, 1983; Odendaal, 2000; Odendaal and Meintjes, 2003). Gee *et al.* (2010) showed that petting a dog gave security and calmness to children in a test situation enabling them to do a test better with fewer mistakes. Non-verbal communication like petting the horse may therefore have had a calming effect on the adolescent.

Emotional bond is created through closeness, petting, and grooming the animal (Crawford *et al.*, 2006; Bachi *et al.*, 2012). The adolescents in this study showed a large interest in having contact with the horse and had the same horse each time. A feeling of connectedness with an animal might happen without owning the animal (Endenburg and Lith, 2011). The adolescents having the same horse during their intervention might have been of importance. The level of attachment was not measured, but one might consider that a bond with the horse could have developed.

Positive contact with the horse due to the response from the horse being positive or indifferent may also have been important for the feeling of security. Although no correlation was found between the frequencies of physical contact with the horse and trying again when not being able to solve a task with the horse, the positive responses received from the horse may have been important for daring to try again.

With regard to the response of the horse to the contact behavior of the adolescents, it is important that the horse's response was calm. If the horses had been aggressive, the type of and amount of contact might have been different. A review of the relationship between horses and humans points out the importance of animal husbandry for the behavior of the horse (Hausberger *et al.*, 2008). The horses we used were kept in herds, they had a good welfare, and they were mainly of cold-blooded Norwegian breeds. The positive reaction pattern of the horses may be related to these factors and indicate that choosing the right horses for the activity is important for the outcome.

The response from the horse was measured solely by ear and head positions. These are clear indicators of emotional state shown by the horse (McGreevy, 2006). In this study the horses were calm during the interaction with the adolescents, not showing much variation in behavior. This is probably due to the horse behavior being noted during grooming and horse-related tasks when the horses were held or tied-up.

Choice of strategy during tasks

Trying again when not being able to solve a task was interpreted as an act of persistence, while not trying again was a sign of less persistence of the task. The adolescents chose to try again more frequently during late recordings than they did during early recordings. As a result of this, mastery of activities with the horse might have developed during the intervention. This may have an impact on self-efficacy, as self-efficacy beliefs are related to the persistence when not being able to solve a task (Baumeister, 2004). Choosing the strategy trying again over not trying again significantly more often in the end of the intervention may therefore indicate a change in self-efficacy in relation to activities with the horse. This is in accordance with previous research that has reported that experiences of mastering tasks in relation to activities with horses may have an impact on self-efficacy (Forsberg, 2007; Forsberg and Tebelius, 2011; Burgon, 2011). This study only investigated the experiences the adolescents had during the intervention regarding contact with the horse and conducting horse-related tasks. Persistence at a task in contrast to giving up or asking others for help is an action closely related to the self-efficacy literature and might therefore enhance the self-efficacy beliefs in relation to the specific task. This might again have an influence on the experience of mastery in other arenas. This study only investigated the experiences the adolescents had during the horse-related tasks in this intervention. Though an experience of mastery at a certain task is a positive experience for adolescents, this may not be transferable to other arenas directly.

Our study did not measure if the adolescents conducted the tasks faster or more correctly late in the intervention, but the level of persistence increased while the tasks were gradually made more difficult. The level of challenges during tasks with the horse was therefore the same early and late in the intervention for all the adolescents, and therefore, possible to compare.

The importance of the setting for development of mastery

Being on a farm with a horse, the aim was not to become skilled at the activity which may be the case at a riding school (Forsberg, 2007). The instructors on the farms were told to give the adolescents time to try for themselves when attempting horse-related tasks. This setting may provide a different experience of mastery than if the focus was on conducting the task quickly and correctly, which may need to involve more active correction by the instructor.

An environment perceived to be safe and relaxed is important for receiving a positive experience of an activity (Gano-Overway *et al.*, 2009). Seven of our participants were interviewed about the impact of the environment on the experience, in a master thesis (Flatekval and Berge, 2010). This thesis concluded that the natural environment, the surroundings, and the social setting were important for the adolescents in addition to the interaction with the horse.

An experience of a social support network may have an influence on competence beliefs and the level of performing at a given activity (Ahmed *et al.*, 2010). The support from the farmer may influence the experience of an intervention on a farm (Pedersen *et al.*, 2012). The instructors in our study acted as a supportive leader, allowing the adolescents to try by themselves. The farmer might thus have played an important role for the outcome of the intervention.

The adolescents had to make an active effort in conducting horse-related tasks. Two participants were at the stable together and attempted the same type of tasks at the same time. A personal active effort and seeing other people similar to oneself succeed are important for development of self-efficacy (Bandura, 1997a). This might have been important for an increased persistence during horse-related tasks.

Methodological considerations

Using ethological methods with video analyses to investigate the relation between horses and adolescents gives a broader perspective to behavior observations, combining ethological and psychological aspects of behavior. Video recordings as a method to capture human-animal interactions have several advantages. Video analyses based on a descriptive and objective ethogram reduces subjective interpretations of the prevalence of various behaviors, and provides the possibility to analyze the behaviors in frequency per hour and average duration of each behavioral state. The possibility to control the analysis done by others, by analyzing the same situation twice by different scientists, is a clear advantage, increasing the overall reliability of the study.

The difference between analyzing animal behavior and human-animal interactions is the ability to control the recording situation. In animal behavior studies the animals are often kept in smaller compartments which enable the researcher to record the behavior at a certain angle and range. This was not possible in our study. The reliability of the analysis however depends on the behaviors chosen for the ethogram. The behaviors observed in this study were possible to see at a distance and from different positions; this was therefore not a problem for the analysis.

Other studies using The Observer® software to analyze tasks with animals in relation to mastery have mainly used duration of each behavior or task (Berget *et al.*, 2007; Pedersen *et al.*, 2011). This software also gives the opportunity to analyze the number of times a behavior occurs and reactions to this. This gives an advantage when investigating the behavior of the adolescent in relation to the task conducted, and the reaction of the horse to the behavior of the adolescent. Studies investigating non-verbal communication with horses often use these types of standardized forms to analyze the behaviors of the person (Braat, 2008; Trotter *et al.*, 2008; Keino *et al.*, 2009). Using a standardized form for analyzing behavior has its strength when comparing results between studies. Using ethological methods through video analysis gives a broader perspective of behaviors and a possibility to analyze several behaviors in relation to each other.

The aim of our study was to analyze behaviors that were easy to detect and give an objective measurement. Videos might also be used to analyze behavior on a scale measure (Gee *et al.*, 2007). The strength of using video and measuring frequency of behavior is the minimizing of a subjective interpretation. On the other hand, a more qualitative approach gives a broader perspective of the experience. In psychological studies video is often used to study emotional reactions through facial expressions (Hamm *et al.*, 2011; Kaulard *et al.*, 2012). This gives a more detailed analysis of emotions and reactions, but on the other hand does not give clear behavior categories which were important for the present study.

Further research is needed to explore this method. Studies on how mastery of tasks measured through behavior methodology may be related to other measurements of self-efficacy are relevant.

Limitations

The main aim of the intervention was to give the adolescents an experience of being on a farm with horses as close to a natural setting as possible. The farms and farmers were carefully selected to create a common experience for all participants. Despite the venue for the intervention being as similar as possible and the intervention criteria explained in detail to the instructors, being in a natural setting makes it difficult to standardize the intervention completely. A detailed manual for the intervention was not made, but detailed instructions were given to the farmers conducting the intervention. Because this intervention was about the total experience and not focused on a specific goal or learning outcome the intervention was supposed to vary in activities.

The adolescents were filmed at a relatively short distance, but the camera was a small hand held camera with a monitor, allowing the recorder to keep the camera at the chest and have contact with the adolescents to minimize intimidation made by filming. The adolescents did seem to get used to the camera and carried on as usual according to the instructor.

All participants had some kind of experience with or knowledge of horses, ranging from riding to petting a horse, before they started in the project. Experience of horses was limited to riding. Therefore many of the horse-related tasks, like putting on a halter and saddling the horse, were new to them. The tasks were also adjusted to the adolescent's level of experience. The adolescents participated in the project voluntarily, ensuring that the motivation for being around horses was present. Involvement and motivation have an influence on the outcome of interventions with animals (Melson, 2003). The results in this study are only valid for adolescents motivated to participate in such activities. Motivation is important for involvement in an activity. This might therefore have had an impact on the adolescents' persistence during horse-related tasks.

Different degrees of experience with horses did not appear to have an impact on the type of contact the adolescents had with the horse or the choice of strategy when not succeeding with the task. This indicates that the intervention itself held elements of experience of mastery for all adolescents regardless of experience level. The adolescents were filmed during the second or third time of the intervention. The first potentially insecure encounter with the horse was therefore not recorded.

Other animals on the farm could have been a distraction, but the adolescents did not have much contact with them.

The instructors had various backgrounds, but all had experience working with adolescents and horses and education both in horse-related subjects and in health or pedagogy. All the farmers acted as supportive leaders and allowed the adolescents time to try for themselves during horse-related tasks. The numbers of instructors and animals used in our study were high in relation to the number of participants data was collected for. Using different animals and different instructors conducting the intervention made it easier to distinguish the effects of an intervention from effects of the conditions (Kazdin, 2011).

The noting of 'not able to solve the task' and then trying again or not trying again during horse-related tasks and riding was thoroughly discussed by the researchers before conducting the analysis. The criteria for noting the behaviors were easy to detect, the behaviors either occurring or not occurring. Despite objective categories described in an ethogram, and discussions among the two raters on border cases, a certain bias in the results during coding of the video recordings cannot be completely ruled out. Calculation of inter-rater reliability and the use of completely independent raters were not done. This was not feasible in this study and may be seen as limitations to the conclusions. The methodology being quantitative in the observation of the behaviors is a strength, but the operationalizing of the outcome as an experience of mastery is a judgment made by the researchers. Trying to relate observed behavior through ethograms with psychological concepts brings a new aspect to the field in regards to how to investigate experiences of mastery during equine-assisted activities. This leads to opportunities for new research, but also challenges in the methodology development.

Conclusions

Adolescents that were given the opportunity to work with horses and ride once a week for four months showed the same active attitude and the same time budget of behavior both early and late in the intervention period. The adolescents showed a large interest in petting the horse. The horses provided a safe setting for the adolescents, by showing only positive or no responses to their initiation of contact. Later in the period the adolescents more persistently than earlier kept on trying to master difficult tasks with the horse rather than backing out from the challenge. This indicates that the intervention might serve to enhance task-specific self-efficacy among the adolescents during activities with horses. Even though it was not measured in this study, experiences of mastery in one arena might lead to a positive belief in oneself in other arenas, positive for adolescents' development.

Acknowledgements

We appreciate the farmers' commitment and effort in carrying out the intervention, and the master students' assistance with video-recordings.

Declaration of Interest

This research was funded by the Norwegian University of Life Sciences and the Norwegian Horse Centre. The authors report no conflict of interest. The authors alone are responsible for the content and the writing of the paper.

References

Ahmed, W., Minnaert, A., Van der Werf, G. and Kuyper, H. (2010) Perceived social support and early adolescents' achievement: The mediational roles of motivational beliefs and emotions. *Journal of Youth and Adolescence* 39(1), 36–46. DOI: 10.1007/s10964-008-9367-7.

Bachi, K., Terkel, J. and Teichman, M. (2012) Equine-facilitated psychotherapy for at-risk adolescents: The influence on self-image, self-control and trust. *Clinical Child Psychology and Psychiatry* 17(2), 298–312. DOI: 10.1177/1359104511404177.

Bandura, A. (1982) Self-efficacy mechanism in human agency. *American Psychologist* 37(2), 122–147. DOI: 10.1037/0003-066x.37.2.122.

Bandura, A. (1997a) The anatomy of stages of change. *American Journal of Health Promotion* 12(1), 8–10.

Bandura, A. (1997b) *Self-efficacy. The exercise of control New*. Freeman and Company, York.

Bandura, A. (2006) Toward a psychology of human agency. *Perspectives on Psychological Science* 1(2), 164–180. DOI: 10.1111/j.1745-6916.2006.00011.x.

Barber, B.L., Eccles, J.S. and Stone, M.R. (2001) Whatever happened to the jock, the brain, and the princess? Young adult pathways linked to adolescent activity involvement and social identity. *Journal of Adolescent Research* 16(5), 429–455. DOI: 10.1177/0743558401165002.

Baumeister, R.F. (2004) *Handbook of Self-Regulation: Research, Theory, and Applications*, 1st ed. The Guilford Press, New York.

Baumeister, R.F., Vohs, K.D. and Tice, D.M. (2007) The strength model of self-control. *Current Directions in Psychological Science* 16(6), 351–355. DOI: 10.1111/j.1467-8721.2007.00534.x.

Beetz, A., Kotrschal, K., Turner, D.C., Hediger, K., Uvnäs-Moberg, K. *et al.* (2011) The effect of a real dog, toy dog and friendly person on insecurely attached children during a stressful task: An exploratory study. *Anthrozoös* 24(4), 349–368. DOI: 10.2752/175303711x13159027359746.

Beetz, A., Uvnäs-Moberg, K., Julius, H. and Kotrschal, K. (2012) Psychosocial and psychophysiological effects of human-animal interactions: The possible role of oxytocin. *Frontiers in Psychology* 3, Article 234. DOI: 10.3389.fpsyg.2012.00234.

Berget, B., Skarsaune, I., Ekeberg, Ø. and Braastad, B.O. (2007) Humans with mental disorders working with farm animals. *Occupational Therapy in Mental Health* 23(2), 101–117. DOI: 10.1300/J004v23n02_05.

Bertoti, D.B. (1988) Effect of therapeutic horseback riding on posture in children with cerebral palsy. *Physical Therapy* 68(10), 1505–1512. Available at: http://ptjournal.apta.org/content/68/10/1505

Birke, L., Hockenhull, J., Creighton, E., Pinno, L., Mee, J. *et al.* (2011) Horses' responses to variation in human approach. *Applied Animal Behaviour Science* 134(1–2), 56–63. DOI: 10.1016/j.applanim.2011.06.002.

Braat, O.J.M. (2008) The effectiveness of equine assisted activities on positive and negative behaviours of children with learning disabilities. In: *ISES Dublin 2008, International Society for Equitation Science 4th International conference*, Royal Dublin Society Ballsbridge Dublin 4.

Burgon, H.L. (2011) Queen of the world': Experiences of "at-risk" young people participating in equine-assisted learning/therapy. *Journal of Social Work Practice* 25(2), 165–183. DOI: 10.1080/02650533.2011.561304.

Crawford, E.K., Worsham, N.L. and Swinehart, E.R. (2006) Benefits derived from companion animals, and the use of the term "attachment." *Anthrozoös* 19(2), 98–112. DOI: 10.2752/089279306785593757.

Davis, E., Davies, B., Wolfe, R., Raadsveld, R., Heine, B. *et al.* (2009) A randomized controlled trial of the impact of therapeutic horse riding on the quality of life, health, and function of children with cerebral palsy. *Developmental Medicine and Child Neurology* 51(2), 111–119. DOI: 10.1111/j.1469-8749.2008.03245.x.

Dodge, T. and Lambert, S.F. (2009) Positive self-beliefs as a mediator of the relationship between adolescents' sports participation and health in young adulthood. *Journal of Youth and Adolescence* 38(6), 813–825. DOI: 10.1007/s10964-008-9371-y.

Endenburg, N. and Lith, H.A. (2011) The influence of animals on the development of children. *The Veterinary Journal* 190(2), 208–214. DOI: 10.1016/j.tvjl.2010.11.020.

Ewing, C.A., MacDonald, P.M., Taylor, M. and Bowers, M.J. (2007) Equine-facilitated learning for youths with severe emotional disorders: A quantitative and qualitative study. *Child Youth Care Forum* 36, 59–72. DOI: 10.1007/s10566-006-9031-x.

Fine, A.H. (2010) *Handbook on Animal-Assisted Therapy: Theoretical Foundations and Guidelines for Practice*, Vol. 3. San Diego, Academic Press.

Flatekval, S. and Berge, I. (2010) *Environmental impact on youth experiences with horses in a farm environment a qualitative study of young people interacting with horses*. Master thesis, Norwegian University of Life Sciences, Ås.

Forsberg, L. (2007) *Att utveckla handlingskraft. om flickors identitetsskapande processer i stallet*. (Licentiat thesis), Luleå tekniska universitet, Luleå. Available at: http://www.ltu.se/staff/l/lenaf-1.12247?l=en

Forsberg, L. and Tebelius, U. (2011) The riding school as a site for gender identity construction among Swedish teenage girls. *World Leisure Journal* 53(1), 42–56. DOI: 10.1080/04419057.2011.552218.

Friedmann, E., Katcher, A.H., Thomas, S.A., Lynch, J.J. and Messent, P.R. (1983) Social interaction and blood-pressure - influence of animal companions. *Journal of Nervous and Mental Disease* 171(8), 461–465.

Gano-Overway, L.A., Newton, M., Magyar, T.M., Fry, M.D., Kim, M.S. *et al.* (2009) Influence of caring youth sport contexts on efficacy-related beliefs and social behaviors. *Developmental Psychology* 45(2), 329–340. DOI: 10.1037/a0014067.

Gee, N.R., Harris, S.L. and Johnson, K.L. (2007) The role of therapy dogs in speed and accuracy to complete motor skills tasks for preschool children. *Anthrozoös* 20(4), 375–386. DOI: 10.2752/089279307x245509.

Gee, N.R., Church, M.T. and Altobelli, C.L. (2010) Preschoolers make fewer errors on an object categorization task in the presence of a dog. *Anthrozoös* 23(3), 223–230. DOI: 10.2752/175303710x1275 0451258896.

Hakanson, M., Moller, M., Lindstrom, I. and Mattsson, B.. (2009) The horse as the healer – a study of riding in patients with back pain. *Journal of Bodywork and Movement Therapies* 13(1), 43–52. DOI: 10.1016/j.jbmt.2007.06.002.

Hamm, J., Kohler, C.G., Gur, R.C. and Verma, R. (2011) Automated facial action coding system for dynamic analysis of facial expressions in neuropsychiatric disorders. *Journal of Neuroscience Methods* 200(2), 237–256. DOI: 10.1016/j.jneumeth.2011.06.023.

Hartmann, E., Søndergaard, E. and Keeling, L.J. (2012) Keeping horses in groups: A review. *Applied Animal Behaviour Science* 136(2–4), 77–87. DOI: 10.1016/j.applanim.2011.10.004.

Hausberger, M. and Muller, C. (2002) A brief note on some possible factors involved in the reactions of horses to humans. *Applied Animal Behaviour Science* 76(4), 339–344. DOI: 10.1016/S0168-1591(02)00016-3.

Hausberger, M., Roche, H., Henry, S. and Visser, E. (2008) A review of the human–horse relationship. *Applied Animal Behaviour Science* 109(1), 1–24. DOI: 10.1016/j.applanim.2007.04.015.

Herrero, P., Asensio, A., Garcia, E., Marco, A., Olivan, B. *et al.* (2010) Study of the therapeutic effects of an advanced hippotherapy simulator in children with cerebral palsy: A randomised controlled trial. *BMC Musculoskeletal Disorders* 11, 71. DOI: 10.1186/1471-2474-11-71.

Kaiser, L., Smith, K.A., Heleski, C.R. and Spence, L.J. (2006) Effects of a therapeutic riding program on at-risk and special education children. *Journal of the American Veterinary Medical Association* 228(1), 46–52.

Kaulard, K., Cunningham, D.W., Bulthoff, H.H. and Wallraven, C. (2012) The MPI facial expression database - a validated database of emotional and conversational facial expressions. *PloS ONE* 7(3), e32321. DOI: 10.1371/journal.pone.0032321.

Kazdin, A.E. (2011) Establishing the effectiveness of animal-assisted therapies: Methodological standards, issues, and strategies. In: McCardle, P., McCune, S., Griffin, J.A. and Maholmes, V. (eds) *How Animals Affect Us Examining the Influence of Human-Animal Interaction on Child Development and Human Health*. Washington, American Psychological Association, pp. 35–51.

Keino, H., Funahashi, A., Keino, H., Miwa, C., Hosokawa, M. *et al.* (2009) Psycho-educational horseback riding to facilitate communication ability of children with pervasive developmental disorders. *Journal of Equine Science* 20(4), 79–88. DOI: 10.1294/jes.20.79.

Klontz, B.T., Bivens, A., Leinart, D. and Klontz, T. (2007) The effectiveness of equine-assisted experiential therapy: Results of an open clinical trial. *Society and Animals* 15(3), 257–267. DOI: 10.1163/156853007x217195.

Krueger, K., Flauger, B., Farmer, K. and Maros, K. (2011) Horses (*Equus caballus*) use human local enhancement cues and adjust to human attention. *Animal Cognition* 14(2), 187–201. DOI: 10.1007/s10071-010-0352-7.

Lentini, J.A. and Knox, M.S. (2008) A qualitative and quantitative review of equine facilitated psychotherapy (EFP) with children and adolescents. *The International Journal of Psychosocial Rehabilitation* 13(1), 17–30. DOI: 10.2174/1876391X00901010051.

Lesimple, C., Fureix, C., LeScolan, N., Richard-Yris, M.A. and Hausberger, M. (2011) Housing conditions and breed are associated with emotionality and cognitive abilities in riding school horses. *Applied Animal Behaviour Science* 129(2–4), 92–99. DOI: 10.1016/j.applanim.2010.11.005.

Maddux, J.E. (1995) *Self-efficacy, adaptation, and adjustment: Theory, research, and application*, 1st ed. Plenum Press, New York.

Maddux, J.E. and Volkmann, J. (2010) *Self-Efficacy Handbook of Personality and Self-Regulation*. Malden MA: Wiley-Blackwell, pp. 315–331.

Martin, P. and Bateson, P. (2007) *Measuring behaviour. An introductory guide*, 3rd ed. Cambridge University Press, Cambridge.

McGreevy, P. (2006) *Equine Behavior. A Guide for Veterinarians and Equine Scientists*, 3rd ed. Saunders, Philadelphia PA.

Melson, G.F. (2003) Child development and the human-companion animal bond. *American Behavioral Scientist* 47(31), 31–39.

Moote, G.T. and Wodarski, J.S. (1997) The acquisition of life skills through adventure-based activities and programs: A review of the literature. *Adolescence* 32(125), 143–167.

Odendaal, J.S.J. (2000) Animal-assisted therapy - magic or medicine? *Journal of Psychosomatic Research* 49(4), 275–280.

Odendaal, J.S. and Meintjes, R.A. (2003) Neurophysiological correlates of affiliative behaviour between humans and dogs. *Veterinary Journal* 165(3), 296–301. 2003/04/04. eng.

Pedersen, I., Nordaunet, T., Martinsen, E.W., Berget, B. and Braastad, B.O. (2011) Farm animal-assisted intervention: Relationship between work and contact with farm animals and change in depression, anxiety, and self-efficacy among persons with clinical depression. *Issues in Mental Health Nursing* 32(8), 493–500. DOI: 10.3109/01612840.2011.566982.

Pedersen, I., Ihlebaek, C. and Kirkevold, M. (2012) Important elements in farm animal-assisted interventions for persons with clinical depression: A qualitative interview study. *Disability and Rehabilitation* 34(18), 1526–1534. DOI: 10.3109/09638288.2011.650309.

Roberts, F., Bradberry, J. and Williams, C. (2004) Equine-facilitated psychotherapy benefits students and children. *Holistic Nursing Practice* 32–35.

Rothe, E.Q., Vega, B.J., Torres, M.R., Soler, S.M.C. and Pazos, R.M.M. (2005) From kids and horses: Equine facilitated psychotherapy for children. *International Journal of Clinical and Health Psychology* 5(2), 373–383.

Schultz, P.N., Remick-Barlow, G.A. and Robbins, L. (2007) Equine-assisted psychotherapy: A mental health promotion/intervention modality for children who have experienced intra-family violence. *Health & Social Care in the Community* 15(3), 265–271. DOI: 10.1111/j.1365-2524.2006.00684.x.

Selby, A. (2009) *A systematic review of the effects of psychotherapy involving equines*. The University of Texas at Arlington, Texas.

Sempik, J., Hine, R. and Wilcox, D. (2010) *Green Care: A Conceptual Framework. A Report of the Working Group on the Health Benefits of Green Care. COST 866, Green Care in Agriculture*. Loughborough University, Loughborough.

Smith-Osborne, A. and Selby, A. (2010) Implications of the literature on equine-assisted activities for use as a complementary intervention in social work practice with children and adolescents. *Child and Adolescent Social Work Journal* 27(4), 291–307.

Søndergaard, E. and Ladewig, J. (2004) Group housing exerts a positive effect on the behaviour of young horses during training. *Applied Animal Behaviour Science* 87(1–2), 105–118. DOI: 10.1016/j.applanim.2003.12.010.

Toering, T., Elferink-Gemser, M., Jordet, G., Jorna, C., Pepping, G.J. *et al.* (2011) Self-regulation of practice behavior among elite youth soccer players: An exploratory observation study. *Journal of Applied Sport Psychology* 23(1), 110–128. DOI: 10.1080/10413200.2010.534544.

Traeen, B. and Wang, C.E. (2006) Perceived gender attribution, self-esteem, and general self-efficacy in female horseback riders. *Journal of Equine Veterinary Science* 25(10), 439–444. DOI: 10.1016/j.jevs.2006.08.007.

Trotter, K.S., Chandler, C.K., Goodwin-Bond, D. and Casey, J. (2008) A comparative study of the efficacy of group equine assisted counseling with at-risk children and adolescents. *Journal of Creativity in Mental Health* 3(3), 254–284. DOI: 10.1080/15401380802356880.

9 Equine Assisted Social Work as a Means for Authentic Relations Between Clients and Staff

Catharina Carlsson[1]*, Daniel Nilsson Ranta[2] and Bente Traeen[3]

[1]Department of Social Work, Linnaeus University, Kalmar, Sweden; [2]Department of Social Work, Mid Sweden University, Östersund, Sweden; [3]Department of Psychology, University of Oslo, Oslo, Norway

Abstract

The purpose of this study was to explore, by qualitative methods, the meaning of equine-assisted social work (EASW) both to young women with self-harm problems and their staff in residential treatment. Data were collected by in-depth interviews with eight staff members and nine clients. Human-horse interaction was observed subsequently in three of the staff and four of the clients. Based on the narratives of the staff members and the clients, who agreed, the horse's ability to read human emotions made the staff and clients aware of their own emotions. In order not to alienate the horse or make the situation dangerous, which intensified their sense of being in the present, both the clients and staff had to regulate their emotions. As a result the resistance for change decreased, made the participants more likely to regulate their emotions and change behavior rather than avoid the situation or the information. The horse seemed to set the framework for the interaction between the staff and young women. The fact that even the staff could fail in the interaction with the horse, affected the interaction between staff and clients making it more informal and dynamic. A relationship based on empathy, trust, respect and negotiation, where clients shared private matters, resulted in the perception of a more authentic relationship. Further research is needed to see which conditions are favourable for achieving an authentic relationship and which conditions possibly fail.

Keywords: equine-assisted social work, mirror of emotions, be in the presence, emotional awareness and regulation, authentic relationship

Introduction

Social work practice is interactive and process-oriented. Social workers treat problems by building therapeutic relations and identifying needs, goals, and resources that unfold over time, representing unique combinations for every client (Adams *et al.*, 2009). Even so, clients and social workers may disagree on the cause of a problem as well as on the choice of treatment (Nevonen and Broberg, 2000), adding to the complexity of the work. Introducing a horse as part of treatment may seem to make social work even more complex. Previous research (Bower and MacDonald, 2001; Bizub *et al.*, 2003; Macauley and Gutierrez, 2004; Rothe *et al.*,

*Corresponding author: catharina.carlsson@lnu.se

2005; Smith-Osborne and Selby, 2010; Holmes *et al.*, 2011; Burgon, 2012) defines this type of work with horses as different although the way the staff work and what clients they emphasize seem to overlap. This provokes the question: are there differences in the essence of what the staff, clients and the horses do? To find out, this paper therefore asks what the horse's role might be in a treatment context?

In the first place there is evidence that Equine-Assisted Psychotherapy seems to facilitate the restoration of trust and building relationships between staff members and their clients (Frewin and Gardiner, 2005; Foley, 2008). In this context, it becomes pertinent to highlight that professional skills also affect the possibilities of establishing a relationship with the client (Hasenfeld, 2010). The discretion of the staff also varies, for example in a bureaucratic organisation, the work may contain elements of incitement, intimidation and persuasion. The reverse could be defined as emotional labour were the work is based on establishing trust and thereby motivate the client (Guy *et al.*, 2010).

If we return to research about horses in care, generally, there are indications that this kind of therapy is associated with restoring the client's self-esteem and self-image. It also seems to develop communication skills, emotional awareness, and regulation, as well as reducing anxiety (Beck and Katcher, 2003; Gress, 2003; Bizub *et al.*, 2003; Macauley and Gutierrez, 2004; Kaiser *et al.*, 2004; Karol, 2007; Dell *et al.*, 2008; Holmes *et al.*, 2011; Dell *et al.*, 2011; Rothe, Vidrin *et al.* (2002); William (2005); Rothe *et al.* (2005); Yorke *et al.*, 2008; Walsh (2009)). These outcomes indicate that Equine Assisted Social Work (EASW) may be fruitful for the particular client groups in the present study, namely, young women with eating disorders, self-injury problems and tendencies to suicidal behavior.

Moving back to the field of social work, there is lack of research on treatment methods and the effects of treatment. Much of the work carried out in institutions does not build on existing theories or models (Sallnäs, 2000; Andreassen, 2003). One variation of treatment that has emerged in recent years, includes horses (Statens offentliga utredningar Statens offentliga utredningar SOU, 2000:109:109). In Sweden today, nearly twenty organisation units are using horses in treatment and care, and most of these are under private management (Håkansson *et al.*, 2008). The research about the horse's role for staff and clients is very limited, and more research is needed to establish whether EASW can be defined as a separate method or a tool.

Considering aspects of definition, we know that the horse serves different purposes for people, and is given different meanings in different contexts. Some claim that the horse is a partner or a friend, whereas others state that the horse is a tool, a master, a motivational factor, or a mirror (Ewing *et al.*, 2007; Klontz *et al.*, 2007; Vidrine *et al.*, 2002). Norwegian therapists involved in horse-assisted therapeutic interventions for clients with prolonged, severe eating disorders conclude that the horse provides the clients with an opportunity to experience intimacy in the relationship, and to train their relational skills in a clear, effective and less threatening way (Traeen *et al.*, 2012). The staff believe that dealing with the horse increases the clients' awareness of their own and others' emotional reactions, body language, and intentions. However, that particular study did not interview the clients themselves. The present study takes the experience of the Norwegian therapists one step further and explores the role of the horse in a social-work-treatment setting, as experienced by both clients with eating disorders, other self-harm problems and staff members on the premises.

Equine assisted social work and therapy

To begin with, horses are flight animals. They relate to the environment they live in as well as to the emotional status of the humans they interact with (Ekesbo, 2011). In the wild, a horse's survival depends on its ability to sense fear and to interpret other animals' intentions as expressed in body language and pheromones. Accordingly, the horse's survival lies in the knowledge of recognizing other horses' emotions as well as those of other species, including humans (Ekesbo, 2011). Studies show that when a person has negative feelings towards a horse, the horse's heart rate increases. A person with neutral or positive feelings toward the horse does not affect the horse's heart rate. An individual's attitude toward the horse affects the behavior of the horse (Hama *et al.*, 1996; Chamove *et al.*, 2002; Henry *et al.*, 2005). The horse's sensitivity to emotions could make it suitable for use in working with specific groups of clients.

EASW represents an alternative or a supplement to conventional treatment and social work and involves a triangular relationship between staff, client, and horse. EASW is sometimes inaccurately called 'riding therapy.' The primary emphasis in EASW is not riding as in therapeutic riding for the disabled. The activities in EASW can involve care of the horse, training with the horse, riding, carriage driving, and vaulting (gymnastics on horseback), similar to other equine assisted interventions. A possible difference could be that in EASW there is no manual or program to follow. Instead, the work will always depend on the staff's skills and educational backgrounds, the horse's abilities, and the client's skills and wishes.

Only recently researchers have begun to focus on the mechanisms involved in the horse-human interaction where the horse assists therapy and social work. Several methodological difficulties arise (Pauw, 2000; Smith-Osborne and Selby, 2010), and qualitative and quantitative studies reach different conclusions about the effects (Pauw, 2000). The research about therapy and social work with horses is often hypothesis-generating (Wilson and Barker, 2003; Kaiser *et al.*, 2004; Johansen, 2008). Based on observations, client reports, and the reports from staff who practice equine assisted work, there seem to be potential benefits for clients with mental illnesses, including eating disorders (Cumella and Marx, 2003). In a program for adolescents with emotional, behavioral, or learning problems where horses assist, Holmes *et al.* (2011) found a significant reduction in self-reported anxiety. The stable can be regarded as a social arena where young girls/women in particular are given the opportunity to increase their perception of self-efficacy (Koren and Træen, 2003). In a sample of Norwegian adolescents, Hauge *et al.* (2013), no effects were found regarding self-efficacy, but a relationship between low self-efficacy and increased persistence during tasks with the horses could be observed. Other studies among risk populations of adolescents and young adults reported improved self-confidence, self-esteem, self-efficacy, sense of mastery, empathy, and the opening of positive opportunities, thanks to the interaction with the horses (Bower and MacDonald, 2001; Burgon, 2012).

Clients under treatment

The clients in this study were in residential treatment for eating disorders, self-injury behavior as well as suicidal tendencies. These conditions often develop during adolescence, and known factors involve a combination of the individual's perfectionism, high standards, and low self-esteem (Lunner *et al.*, 2000; Kåver and Nilsone, 2003; Lundh *et al.*, 2007; Holmqvist *et al.*, 2007; Abrahamsson *et al.*, 2007; Lundh and Bjärhed, 2008; Jablonska *et al.*, 2009; Hagqvist, 2010). Another aspect relevant to the present study is emotional regulation, which may play a significant role in the maintenance of eating problems (Gianini *et al.*, 2013). Emotional regulation could be defined as an activity to regulate either the magnitude or the duration of an emotional response (Gross, 2013). Emotional regulation skills in a social context are shown to be of importance for well-being in adolescence (Silvers *et al.*, 2012). Those who have not learned emotional self-regulation are more likely to have difficulties in school and with friendships. Furthermore, attachment orientations, cognitive-emotional complexity and mindfulness seem to play a role in emotional regulation (Coats and Blanchard-Fields, 2008; Hill and Updegraff, 2012; Naim *et al.*, 2013). The potentially damaging impact of negative emotions on the processing of emotional information can be prevented by high emotional awareness or with the implementation of reappraisal as an emotional regulation strategy (Szcygiel *et al.*, 2012).

Residential treatment is a last resort when other methods of treatment have failed and the choice of therapy seems random (Frensch and Cameron, 2002; Andreassen, 2003; Sunseri, 2004). The client's own experiences should be regarded as important in the process of investigating residential treatment (Rutter, 2000; Nevonen and Broberg, 2000; Holleran Steiker, 2008). Clients, particularly young clients, may perceive institutional environments as stressful

(Bickman *et al*., 2004; Zegers *et al*., 2006). Faced with this stress, they tend to turn to the staff to manage the institutional environment (Bickman *et al*., 2004; Zegers *et al*., 2006). Young clients who have a positive relationship with staff have fewer problems in general, better school performance, and more friendships (Walter and Petr, 2008). The staff's mutual respect, genuine listening, and openness toward clients are required to achieve a trusting relationship (Richmond and Padgett, 2002; Manso *et al*., 2008; Hasenfeld, 2010), which is associated with empathy, honesty, respect, sensitivity, kindness, helpfulness, creativity, commitment, responsibility, reliability, and patience. What is perceived positively and negatively may differ between the individual client and members of staff, however (Richmond and Padgett, 2002; Manso *et al*., 2008).

Framing the study

An ideal of our time is for an individual to act in a way that increases the likelihood of others' perception of one as a 'true' or 'real' person, as opposed to a 'false' or 'untrue' person. We emphasise the value of acting in accordance with our inner qualities. These descriptions indicate that our self-presentation can be analysed in terms of Giddens (1991) concepts 'nature,' 'culture,' and 'authenticity'. Neither 'nature' nor 'culture' are given qualities. The concepts are given a meaning and content. A researcher may call something 'natural' or 'cultural' after having defined the concepts with the intent to use them for a particular purpose. However, these definitions may not overlap or correspond to those of the informant. If an informant should use the concept, the researcher must work to separate his or her analytical use from the common language of the informant. In this study, we will understand the concept of 'nature' as something 'raw,' not socialized or cultivated. 'Nature' will primarily be used as the opposite of the concept 'culture.' There is a long tradition of analyzing human responses within these categories. 'Culture' is the human, the formed, the socially accepted, and the non-random. When we talk about 'culture' we talk about it as opposed to 'nature,' as categories we assign to something informants have expressed using other words. It is the same with 'authenticity'. According to Giddens (1991) understanding, the 'authentic' is both 'cultural' and 'natural.' Authenticity is not necessarily opposed to culture, but it stands in contrast to the imagined, the fake, and the insincere. To be 'authentic' is to be true to yourself and show true thoughts and feelings. Non-authentic, on the other hand, refers to hiding oneself from others and from one's own self. An 'authentic' individual is the one who lives out her/his inner, true self in a world where many do not reach this inner state. The most authentic individual is she/he who expresses the natural in a cultural manner. In addition, according to Jordan (2010), people are born social and have an urge to have social relationships. People strive for social acceptance and recognition for who they are in their relationships. People are, to varying degrees, insecure about themselves, however, and fear rejection. Those who fear this more than average, or what may be considered 'normal' may tend to devalue themselves and hide the qualities of which they are ashamed. In turn, they may perceive themselves as less authentic in their social relationships, and, accordingly, the relationships they enter into may not fully fulfil their desires and needs (Jordan, 2010). The relationships are not authentic. When a person has been hurt severely and rejected in a previous relationship, he or she may have a greater need to build new relations rather than to avoid further social contact. Simultaneously, he or she may perceive that authentic relationships are not safe (Jordan, 2010). This is called the *relational paradox*.

Therefore staff cannot take a position of power, distance, and objectivity to reach an 'authentic relationship' in a therapeutic setting (Jordan, 2010). Staff need to be present in the moment, take in the differences that arise, and be open to change themselves (Jordan, 2010). Present in the moment could be explained as not focusing on doing but focusing on being or a state of active, open, intentional attention on the present. To become an observer from moment to moment without judging - a form of non-judgemental awareness. To handle the behaviors and the feelings of clients, staff members need to (1) know the client's feeling by intuition and communication and at the same time adapt their feelings to the situation;

(2) elicit the desired emotional response, and (3) understand that the client may not want to show the emotions necessary to create interpersonal relationships according to Guy *et al.* (2010). This form of emotional labour requires the staff to manage their own emotions as well as the client's emotions, a task that active listening, responsiveness, empathy and the ability to negotiate can achieve (Guy *et al.*, 2010). Although it is often characterised as unprofessional for staff to show their emotions to the clients, it may be of value in some interventions (Guy *et al.*, 2010).

The client's thoughts and emotions and their ability to deal with them are also of importance for the treatment outcome. To clarify, according to the theory of *cognitive dissonance* the participants would try to reduce the dissonance and achieve consonance, or avoid situations and information, which could increase dissonance (Festinger, 1957). However, the way the dissonance is treated is affected by whether the message is perceived as moderately uncommon. If not moderately uncommon defense mechanisms could appear and make it more likely that avoidance is chosen according to Bateson (1972).

Method

The purpose of this study was to explore by means of qualitative methods the meaning of EASW for young women with self-harm problems in residential treatment. Seen from the eyes of the clients and the staff members, what role and meaning are assigned to the horse as used in EASW? How does the interaction with the horse contribute to the client's and staff's perceptions of authenticity and the possibilities for being in an authentic relation?

Participants

There were no exclusion criteria except for those under 15 years that could not attend because of difficulties in getting consent from their parents. Those who attended the first interview, nine out of eleven female clients aged 15–21 years, were those who wanted to participate in the study. Four of these were observed and interviewed since they had experience of EASW for more than one semester. Two other clients with the same experience were not observed because they dropped out of the treatment. Those were the girls with experience with horses prior to their inclusion in the EASW. The other girls had met horses in the company of friends a couple of times except for one that had no prior experience of horses or of other animals kept as pets. EASW is an individual treatment in this setting and the clients were therefore observed individually. The clients came from different regions in Sweden, and all but one had Swedish ethnicity.

Furthermore, we interviewed eight staff members who worked with horses as a complementary treatment or activity. Only those who worked in the complementary treatment EASW were also observed and interviewed in depth, others such as the staff that only worked with horses as an activity, were excluded. All staff had previous experience with horses and all had owned one or more horses. The staff had different educational backgrounds; one was a social pedagogue, one was a social worker, two were psychotherapeutically trained, two were nurses, one was a riding instructor, and one was a treatment assistant. Even if the staff in the study came from different educational backgrounds they all have specific education and experience in Cognitive Behavioral Therapy (CBT) and Dialectic Behavioral Therapy (DBT), which is the theoretical base in the work with the clients. In addition all three staff, who were observed and interviewed in depth, had undergone a 3 year riding-therapy educational program.

Moreover, the horses observed represented different races and age, gender, experience, temperament and size varied. Everything from Shetland ponies, Icelandic horses, Lusitanos, Dutch warm-bloods and Norwegian fjord horses were represented in the study.

Setting

To reduce the risk of stigma, the clients' treatment was based on a combination of cognitive, behavioral, and personality approaches (Lunner *et al.*, 2000; Nevonen and Broberg, 2000; Norring *et al.*, 2002). EASW was offered

1 hour weekly and used as a complement to the ordinary program. Client participation was voluntary. Because EASW is flexible, an individual design was made for each client based on her treatment goals and desires. EASW was carried out individually with each client based on knowledge of this client group's needs for perfection and previous experiences that their clients were not comfortable working in a group with horses. The EASW was designed to increase the client's self-esteem, modify her behavior, reduce her fears and phobias, and give her the possibility to relax and have fun.

In addition the staff set the boundaries for what could be done in each session based on the safety of the client and the horse. The horse's temperament, behavior, and willingness to interact were assessed on a daily basis. The client's assessment included a mental status examination. If the client exhibited a negative emotional status, signs of aggression, and could not contain their anger for example, the session with the horse was cancelled for safety reasons. Other caveats were if there were indications that the clients were acting as if the horse was a potential sex-object. It was unusual that the staff needed to cancel the session. Usually the clients' empathy for the horse made themselves say that they needed to wait until they were okay to be with the horse. There was no requirement for previous experience with animals or horses prior to EASW participation. The clients continuously learned about the horses so that they would be able to interact safely with the horses. Several of the clients were initially frightened, but this often changed to an appreciation of the horses.

Procedure/Design

The regional Ethical Review Board in Linköping, Sweden, approved the study for both humans and animals, and it was conducted in accordance with ethical standards both considering humans and horses (Andersson *et al.*, 1999; Saslow , 2002; McGreevy, 2004; Minero and Canali, 2008). First in-depth interviews that lasted between 40 and 60 min based on six to seven themes were conducted separately with the clients and the staff members at the treatment setting, except for one staff member who was interviewed at a hotel. The goal of the first interview was a freedom for the informants to decide for themselves what they wanted to say as the study is exploratory. Conducting interviews in the same way with both staff and clients was based on the wish to obtain the perspectives of both groups and make comparisons. Next observations were video taped and conducted three times with each pair of staff and client who worked together in EASW. Conducting video taped observations was based on the hypothesis that there could be aspects in the interaction with the horse that clients and staff were not aware of. In addition the study is exploratory, not all questions were ready in advance. From an ethical standpoint the participants were spared from numerous observations if additional issues would be raised along the way. The videotape also provided opportunities for other scientists to see the observations and review analysis and participants had the opportunity to comment on their experience.

After the observations, the clients and the staff members were interviewed separately, with open questions based on an initial analysis from the first interviews and observations. These in-depth interviews were conducted in direct conjunction with one of the observations made with clients and staff members, thus giving the informants a greater ability to provide rich descriptions of their experiences. All the interviews were recorded digitally and transcribed word for word (Kvale, 1997; Hallberg, 2002).

Analysis

The material from the interviews was grouped by clusters of meaning with the data program MACQDA. The data was analyzed through an empirical phenomenological analysis, which focused on the clients' and staff members' experiences of their personal and social worlds, and in accordance with Smith *et al.* (2012), moving from the particular to the shared and from the descriptive to the interpretative. The first in-depth interviews and observations were analyzed before the second in-depth interviews were carried out. This analysis resulted in new questions, almost the same for the staff and the clients, which were raised in the second in-depth interview. The videos were studied in the light of the answers from the interviews, theoretical framework, and earlier research. The empirical data was compared between clients and between staff and finally even between client and staff to look for patterns across cases (Smith *et al.*, 2012). This resulted in the re-labelling of themes and when combining interviews with observations the figure developed which illustrates the relationships between themes (Smith *et al.*, 2012). Field notes written by the researcher accompanied with constant dialogue with reference persons; students, practitioners and researchers in the field of human–horses in interaction for health both national and internationally were made to consider validity and reliability in the study.

Results

Based on a method of constant comparison between clients and clients, staff and staff as well as clients and staff, three clusters emerged from the empirical data analysis. The clusters which emerged from this analysis were: (1) the horse mirroring emotions - an opportunity for emotional communication, (2) emotional awareness and regulation through presence in the moment, and (3) authentic relationships despite relational paradox.

The horse mirroring emotions - an opportunity for emotional communication

The horse reflected the emotions, intentions, attitudes, and mental status of both the client and staff. According to both staff members and clients you could lie to each other about how you felt but it was impossible to lie to the horse.

> *We have had girls who did not want to open up and said that this is easy. I have no anxiety. I have no problem. I have nothing I want to talk about and then they may not even come close to the horse. I use the horse as the mirror and it gets pretty un-dramatic. You cannot lie to the horse and the horse gives me the answer. You can lie to me but you cannot lie to the horse. (Staff L)*

The clients and the staff members agreed that the horse gave an immediate response to emotions and intentions and did not care how they appeared otherwise. The clients perceived the horse as mirroring their inner state.

> *If I'm worried and have anxiety in my body the horse gets worried and does not know where to go. The horse mirrors my emotions. (Client G)*

In addition several clients stated that the relationship with the horse was the most important. Some clients considered quitting the program if they could not continue working with a particular horse. They all agreed that horses communicated with greater clarity and were more honest than people were. An important aspect in this context, which contributed to their perception that the relationship with the horse was authentic, was the horse's ability to sense the client's feelings.

> *It feels like the animals can feel how you are and that is something humans do not always do. So if you feel bad and then the human may not perceive it, then you can feel even worse when you know that the person does not even notice it. The animals do not lie, deceive, or betray—not as long as you do not betray them, or actually not even then. (Client I)*

Consequently the staff stated that it was easier to talk about the clients' feelings as the behavior of the horse reflected them. The clients were calmer and did not find it so threatening to talk about their feelings.

> *It is true that you cannot disguise yourself, a horse knows all about you, if you are afraid, angry, stressed, or so on. We can read the horse and give that information back to the client. (Staff L)*

According to both clients and staff members, the treatment with the horses was about the client getting to know herself and about improving her ability to empathize, improving her self-esteem and self-confidence, and overcoming her fears and phobias.

> *It's not just about a horse. You learn to know yourself. The course is not all about the horses. It's about working with yourself. (Client G)*

Furthermore the clients looked upon the horse as a close and trustworthy friend. This perception of the horse's trustworthiness derived from clients' feelings that the horse read their emotions and intentions and from their opportunity to cuddle them.

Emotional awareness and regulation through presence in the moment

Because the interaction with the horse could be dangerous, the humans had to develop awareness. Both the clients and the staff were aware that a horse could be dangerous if the human was not calm and did not behave properly. For this reason, a person must always be present at the moment when working with horses.

> *There must be a calm environment. You cannot run around in the barn or in the pasture because the horse gets stressed. You must feel calm when you go to the horse, otherwise it will not end up well. (Client G)*

Therefore it was important to avoid frightening the horse. A scared horse will show avoidance behavior, will not cooperate, or may even show aggression. The horse could turn its head high toward the client and tilt its ears backwards showing that it did not appreciate what was happening. To prevent this, the staff members as well as clients had to monitor their feelings and intentions at all times. The horse responded to their emotions, intentions, and energy. A positive response could be turning its head toward the client, tilting its ears forward. Neutral response could be ears relaxed to the side, head lowered and the eyes soft and relaxed. Both staff and clients had to reflect upon themselves and, as seen in the observations, be mindful about what they brought into the situation with the horse. The clients took more account of the horses than they did of the staff. The clients sometimes acted aggressively inside the treatment centre but never when they were among the horses. Nothing was thrown or broken in the stable. They seemed to become more self-regulating in terms of feeling less aggression, fear, or stress.

> *You learn to regulate your emotions. It was great that I could stay focused and calm with a horse like that, because I have difficulties with being calm. You have to be calmer when you are having riding therapy than you have to be when you are in the house. (Client D)*

In addition the staff members talked about the possibility to create unpredictable situations that could occur when the clients worked with the horses. These situations could not be re-created in a normal situation in the treatment facilities. The horse gave the clients opportunities to develop their abilities for empathy, stress management, mental training, and relaxation; they found it was easier to get close to a horse than to a person. Being present in the moment made the clients more open to learn new experiences. Because the horse did not accuse them of anything, wrongdoing could be forgiven, which lowered the clients' urge to take a defensive stand. It seemed from the observations the clients did not perceive a message from the horse—in contrast to a message from the staff—as threatening. The work with the horses seemed to provide a calm bodily sensation in the clients and an opportunity to learn and practice—for example, putting up limits and saying no, which they otherwise found difficult to do. To be present in the moment also made the clients more relaxed. Feeling relaxed was partly associated with the physical contact with the horse. The clients also expressed that their anxiety levels decreased in the company of horses. They thought this made it possible to act more independently outside the stables as well, in school, for example, where they dared to say no to peers.

> *I've learned to say no with the help of the horse. I have learned to say no to other people. (Client I)*

As a result the opportunity for self-development was a common motive underlying work with the horses. The horse gave the clients opportunities to develop their abilities for empathy, stress management, mental training, and relaxation; they found it easier to get close to a horse than to a person.

Moreover all staff members emphasized the importance of knowing the horse you work with in therapy. It was important to rely on the horse. At the same time, staff believed that there might be problems working with the staff's own horse because this tended to be more personal, and their own horses did not always offer the calmness and size that the clients needed (the clients preferred small and quiet horses). All staff members also agreed that their view of the horses had changed over time. The horse had gone from being a competitive sports tool or a hobby to being a friend and a partner. They learned that the horse gave them an opportunity to develop such personal qualities as approach, reaction patterns, attitudes, values and ways of thinking and acting. The staff members more often reflected upon how they themselves affected the horse depending on their emotional status, such as being stressed or in a bad mood. The staff also explained that the situation with the horse and the client was for real; neither the client nor the staff could put on an act. Staff shared their own experiences with the clients, and staff and clients talked more about what happens here and now than about possible future events and happenings.

Authentic relation despite the relational paradox

The informants were asked about their views on what a good relationship should be like, about their own inter-personal relationships, and about their relationship with the horse. Both the staff and the clients said that you could not lie to the horse. This meant that the relationship with the horse was built on trust and respect. As a consequence, they tended to perceive the relationship with the horse as more authentic than relationships with humans.

> *When you work with the girls out here, we meet on a more equal level and you can give each other more respect in some ways. Many have real difficulties in trusting others. When we talk about trust, they say, Yes, you may be able to trust me, but I will never trust anyone. I never trust anyone, I rely only on myself. But when we are doing this kind of job together, the relations are different. (Staff M)*

Furthermore all staff members and most of the clients agreed that the relationship between the staff and the clients in the presence of the horse was less strict and based more on mutual respect. The observations supported this notion and showed a shift in power balance between the staff and the clients, which made them act on more equal terms. One factor that contributed to this was that the staff members did not always have the correct answer. This provided the clients with a possibility to show their competence and to succeed where the staff had shortcomings.

> *We might succeed and we may not succeed. Then we can have a discussion about it; was it me or the patient who had the stress? Just a thing that it is not even certain that I succeed. Sometimes it is me, sometimes it is the opposite. (Staff B)*

The clients often mentioned that they had bad previous experiences from close relationships. They wished for relationships where the other would stay close to them no matter what. Not all of the clients had a close relationship with the staff. One client, who initially was very interested in horses, stopped the treatment with horses because she claimed she had bad chemistry with the staff and that the staff expected too much of her. Another client said that she did not like the staff person prior to meeting her with the horse, but this changed after seeing how the staff member interacted with the horses. After this observation, the client perceived the staff member to be a more multidimensional individual.

> *The staff told me about her emotions and then she became a person for me not only the boss. (Client A)*

Similarly the staff believed that working with the horses gave the staff and the clients a more authentic picture of one another. They had observed that it was consistently easier to communicate with the clients in the presence of the horses.

> *They see another side of us who engage in equine-assisted social work. To raise difficult topics, which are unpleasant to talk about. It does not hurt so much when you have mutual experiences with the clients from being with the horse. I feel it might also be me that feels it is safer to talk about difficult things or conflicts, because I have experienced things with the client and the horse, and then you establish a relationship faster. (Staff H)*

Furthermore, the staff and the clients were not directly facing each other, and this seemed to make the conversation more casual. They did not have the direct eye contact of regular verbal therapy, where staff member and client sit opposite each other, which may prevent the client from talking.

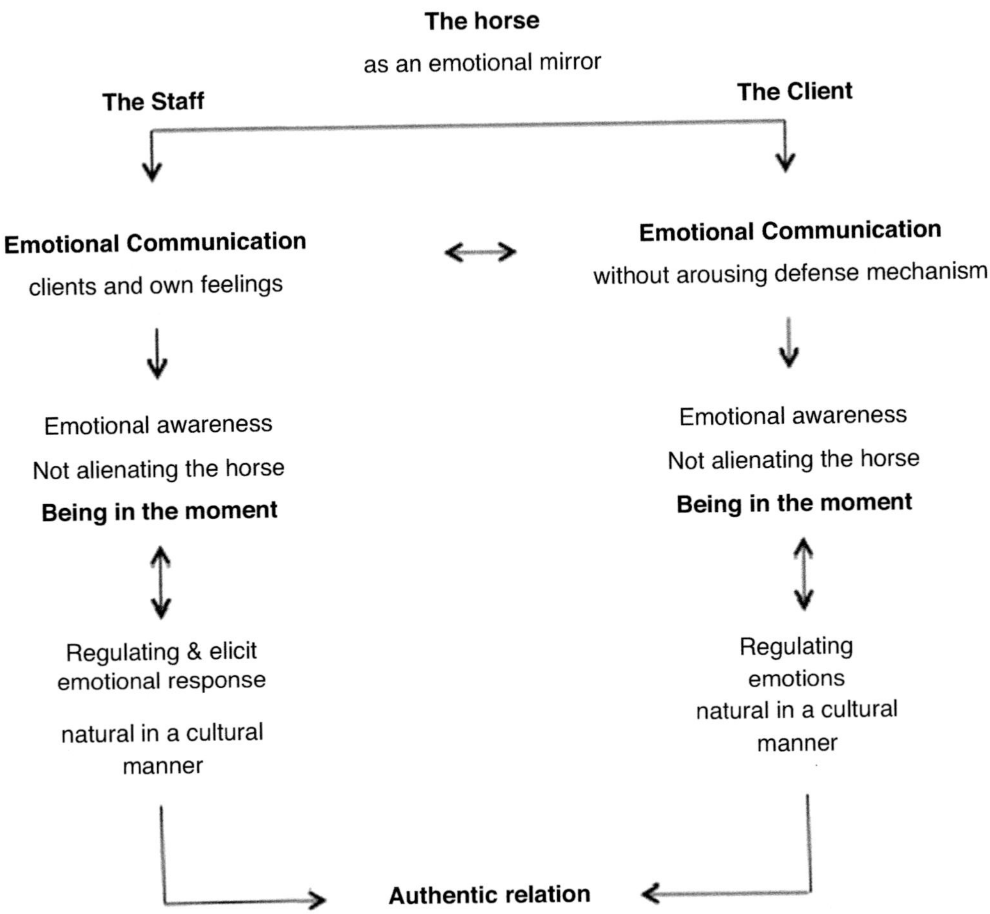

Fig. 9.1. The horse as an emotional mirror in Equine-Assisted Social Work (EASW) providing opportunities for an authentic relationship between staff and clients.

Discussion

We have seen that the horse's ability to read human emotions made the staff and clients aware of their own emotions. The horse has the capacity to be an 'emotional mirror' for the participants according to the concept of 'causal capacity' (Brante, 2001) (Fig. 9.1). In order not to alienate the horse or make the situation dangerous, the clients and staff said that they had to detect and recognize their own emotions and subsequently regulate them, which intensified their sense of being in the present. It has been claimed that the horse's distinctive combination of fearfulness and power, allied with its natural ability to read and mirror human emotions, intentions, non-verbal communication and behavior, seems to be important in explaining why EASW can be effective (Wipper, 2000; Vidrine *et al.*, 2002; Frewin and Gardiner, 2005; Rothe *et al.*, 2005; Cody *et al.*, 2011). Although the effect of EASW was not the topic of this study, our findings lead in the same direction. Because the horse can be dangerous, our participants said they needed a high degree of awareness of their own emotions, intentions, and attitudes in the present moment when interacting with the horse. Klontz *et al.* (2007) report similar findings. Consequently, as observed in the video and confirmed by the staff and the clients themselves, the clients felt calmer, had less anxiety, and were less unfriendly or aggressive toward others when they were dealing with the horse on the stable premises. They thus experienced a higher degree of emotional self-regulation, which had been found in other studies (Vidrine *et al.*, 2002; Kaiser *et al.*, 2004; Klontz *et al.*, 2007; Holmes *et al.*, 2011). The clients were not the only ones influenced by the horse's presence. Similarly the staff members had to focus on the self-regulation of their own emotions, however, in addition to considering the emotional needs of the client, who may not want to share thoughts and feelings with them.

In light of the clients' experiences of being hurt and rejected in earlier relationships, they highlighted the horse as more honest and trustworthy than humans are. The horse's role in reflecting emotions seemed to open up topics the clients were reluctant to talk about because they feared rejection. The clients did not always want to show their emotions to the staff. However, the interaction with the horse did not seem to raise defence mechanisms or resistance for change in the clients because the horse was seen as non-judgmental. Rather the opposite, the interaction with the horse produced feelings of trust and respect, which in turn made it possible for the client to show her 'true emotions' and 'authentic self.' She could escape her 'false self' and 'take off the mask,' which otherwise protected her from her environment. This scenario has also been described in earlier research (Christian, 2005; Buswell and Leriou, 2007; Cody *et al.*, 2011; Traeen *et al.*, 2012). Hence, the relationship with the horse may be seen as an 'authentic relationship.'

In addition the staff did not always have the right solution, it could be the client instead, and therefore the client could move from a self-perception of 'being nothing' to one of 'being someone', which made them feel like a real, true, or 'authentic' person. Also the staff felt more authentic in the stable as they stepped out of traditional frameworks for how to be a professional. In a western society the professional social worker is supposed to have a certain distance to their client according to Brydon (2012). In contrast to this culture the staff members observed that they were more direct in their feedback to the clients in the company of horses. Not only did the usual verbal communication change; the horse seemed to allow staff and clients to make physical contact, both with the horse and themselves, for example in the form of adjustments to the client's body movements. For staff members to use physical contact, show emotions, give direct feedback, and share private matters could be seen as unprofessional in a clinical context. In the stable arena, however, other rules seemed to apply. Emotional expressions by both the staff and clients are likely to produce a sense of being real and true, and the interaction is likely to be perceived accordingly as a more 'authentic relationship.' Because of the negotiation necessary to handle the horse and make the situation predictable, the staff-client relationship seemed to be informal, dynamic, equal and power balanced in the stable arena. The risk of alienating the horse and the possibility of getting close to the horse seemed to set the framework for the staff and the clients meeting in the stable arena. Stable-based relationships experienced as based on trust, respect, and negotiation can be interpreted accordingly as representing more authentic relationships compared to relationships outside of the stable.

As a result, the horse altered the perception of what is considered socially acceptable and appropriate behavior by the staff. Furthermore the staff explained that the interaction with the horse was perceived as 'real' because neither the client nor the staff could interact according to the prescribed roles in a traditional treatment setting. This facilitated the possibility of achieving the authentic relationship.

Implications for Clinical Work and Research

According to the informants, empathy, honesty, respect, sensitivity, responsibility, patience, active listening, ability to negotiate, and responsiveness are the main ingredients in the work with the horse. These aspects are connected with a good therapeutic alliance and may contribute to a successful treatment outcome (Martin *et al.*, 2000; Wampold, 2001). The human-horse interaction seems to create feelings of trust. The clients find it easier to bond with the horse because it has no expectations of the clients, unlike the staff who are perceived to have expectations. The clients perceive the horse as non-judgmental, honest, reliable, and without prejudices; it offers physical contact and gives direct feedback. In relation to the relational paradox it seems as if the horse has something to teach the staff in terms of attitude to clients. Consequently the staff should consider working with non-judgemental awareness, active listening and constant negotiation.

Nevertheless the horse provides an opportunity for both clients and staff to mirror their emotions. The interaction with the horse gives options other than in traditional social work, where it is easier for both staff and clients to hide their emotions (Karol, 2007). The emotional awareness in the moment changes the framework for the meeting between staff members and clients, making it more natural in a cultural manner and therefore more authentic according to Giddens (1991). Including emotions in social work research is more than adding emotions to the inventory of social research topics according to Gausel (2011), Wetherell (2012) and Morrison (2007). There is a need for a change in ontological and epistemological views, a paradigm change. Social constructionism has treated feelings as rules and discourses but perhaps that is over-simplifying in a complex world? Affective practice is about the emotional as it appears in social life and to follow what participants do. There are no simple lines or causations or emotion categories according to Wetherell (2012). The stable provides opportunities for interaction other than the therapy room because other rules apply and staff and clients becomes more equal and perceive their relations as more 'authentic'. Furthermore altered frames of the conversation, the absence of direct eye contact, might look similar in horticultural therapy and green care which is why comparisons with research in those fields should be done.

In addition the horse opened the door for clients to get rid of anxiety and aggression, which resulted in calm and relaxation. This opened up communication between staff and client about emotions, which in turn could increase insights. Not only client's resistance but also professionals' resistance is highlighted and focus should be directed on relationships that produce mutually beneficial identities according to Juhila and Abrams (2011). Other implications regarding emotional awareness and emotional regulation are that the work together with the horse, containing the opportunity to be in the present, could be a social context required for providing learning and well being for this particular client group (Lundh and Bjärhed, 2008; Hill and Updegraff, 2012; Szcygiel *et al.*, 2012; Silvers *et al.*, 2012; Naim *et al.*, 2013; Gianini *et al.*, 2013). Considering the definition of EASW as a method or a tool, the staff and the clients agree to define the horse as a friend or a partner. The observations indicating a strong link with CBT and DBT combined with defining the horse as a friend or a partner leads to the conclusion that EASW becomes more of a supplement to existing practices and methods.

Finally considerations are needed about the small sample, but the results correspond with previous research, reference persons and staffmembers and clients agree with each other indicating transferability. The staff seems to influence whether the mirroring of emotions, emotional awareness and regulation reproduced here go ahead or not, which in turn affects the event of an authentic relationship. This could for example explain the inconsistency about increasing perception or no effects regarding self-efficacy (Bower and MacDonald, 2001; Koren and Træen, 2003; Burgon, 2012; Hauge *et al.*, 2013). Therefore further research is needed to see which conditions are favourable for achieving an authentic relationship and which conditions possibly fail.

Funding

This research was supported in part by grants from the Children´s Welfare Foundation Sweden

References

Abrahamsson, E., Torbiornsson, A. and Hägglöf, B. (2007) Causes of adolescent onset anorexia nervosa: Patient perspectives. *Journal of Eating Disorders* 15(2), 125–133.

Adams, K., LeCroy, C. and Matto, H.C. (2009) Limitations of evidence-based practice for social work education: Unpacking the complexity. *Journal of Social Work Education* 45(2), 165–186.

Andersson, M.K., Friend, H.T., Evans, W.-J. and Bushong, D.M. (1999) Behavioral assessment of horses in therapeutic riding programs. *Applied Animal Behavior Science* 63, 11–24.

Andreassen, T. (2003) *Institutionsbehandling av ungdomar vad säger forskningen?* Stockholm, Centrum för utvärdering av socialt arbete, Statens institutionsstyrelse och Förlagshuset Gothia AB.

Bateson, G. (1972) *Steps to an Ecology of Mind: Collected Essays in Anthropology, Psychiatry, Evolution, and Epistemology*. University of Chicago Press, Chicago, Illinois.

Beck, A.M. and Katcher, A.H. (2003) Future directions in human-animal bond research. *American Behavioral* 79–93.

Bickman, L., Vides de Andrade, A.R., Lambert, E.W., Doucette, A., Sapyta, J. *et al.* (2004) Youth therapeutic alliance in intensive treatment settings. *Journal of Behavioral Health Service and Research* 31(29), 124–148.

Bizub, A.L., Joy, A. and Davidson, L. (2003) It's like being in another world": Demonstrating the benefits of therapeutic horseback riding for individuals with psychiatric disability. *Psychiatric Rehabilitation Journal* 26, 377–384.

Bower, M. and MacDonald, P. (2001) The effectiveness of equine-facilitated psychotherapy with at-risk adolescents. *Journal of Psychology and Behavioral Sciences* 15, 62–76.

Brante, T. (2001) Consequences of realism for sociological theory-building. *Journal for the Theory of Social Behavior* 31(2), 167–195.

Brydon, K. (2012) Promoting diversity or confirming hegemony? In search of new insights for social work. *International Social* 55(2), 155–167.

Burgon, H.L. (2012) Queen of the world": Experiences of "at-risk" young people participating in equine-assisted learning/therapy. *Journal of Social Work Practice: Psychotherapeutic Approaches in Health, Welfare and the Community* 25(2), 165–183.

Buswell, D. and Leriou, F. (2007) Perceived benefits of students service-learning experiences with hippo therapy 23(1), 20–26.

Chamove, A.S., Crawley-Hatrick, O.J.E. and Stafford, K.J. (2002) Horse reactions to human attitudes and behavior. *Anthrozoös* 15, 323–331.

Christian, J.E. (2005) All creatures great and small: Utilizing equine-assisted therapy to treat eating disorders. *Journal of Psychology and Christianity* 24, 65–67.

Coats, A.H. and Blanchard-Fields, F. (2008) Emotional regulation in interpersonal problems; The role of cognitive-emotional complexity, emotion regulation goals and expressivity. *Psychology And* 23(1), 39–51.

Cody, P., Holleran Steiker, L. and Szymandera, M.L. (2011) Equine therapy: Substance abusers healing through horses. *Journal of Social Work Practice in the Addictions* 11(2), 198–204.

Cumella, E.J. and Marx, R.D. (2003) Questions & answers. *Eating Disorders: The Journal of Treatment* 11(2), 143–147.

Dell, C.A., Chalmers, D., Dell, D., Sauve, E. and MacKinnon, T. (2008) Horse as healer: An examination of equine assisted learning in the healing of first nations youth from solvent abuse. *A Journal of Aboriginal and Indigenous Community Health* 6(1), 81–106.

Dell, C.A., Chalmers, D., Bresette, N., Swain, S., Rankin, D. *et al.* (2011) A healing space: The experiences of First Nations and Inuit youth with equine-assisted learning (EAL). *Child Youth Care Forum* 40, 319–336.

Ekesbo, I. (2011) *Farm Animal Behavior Characteristics for Assessment of Health and Welfare*. Cambridge University Press, UK.

Ewing, C.A., MacDonald, P.M., Taylor, M. and Bowers, M.J. (2007) Equine-facilitated learning for youths with severe emotional disorders: A quantitative and qualitative study. *Child Youth Care Forum* 36, 59–72.

Festinger, L. (1957) *A Theory of Cognitive Dissonance*. Stanford University Press, Stanford, CA.

Foley, A.J. (2008) Conflict and connection: A theoretical and evaluative study of an equine-assisted psychotherapy program for at risk and delinquent girls. Thesis, Oklahoma State University, Whitehurst USA.

Frensch, K.M. and Cameron, G. (2002) Treatment of choice or a last resort? A review of residential mental health placements for children and youth. *Child & Youth Care Forum* 31(5), 307–339.

Frewin, K. and Gardiner, B. (2005) New age or old sage? A review of equine assisted psychotherapy. *Journal of Counselling* 6, 13–17.

Gausel, N. (2011) The importance of emotions in social work: Thematic directions from a national conference. *European Journal of Social* 14(2), 287–291.

Gianini, L.M., White, M.A. and Masheb, R.M. (2013) Eating pathology, emotion regulation and emotional overeating in obese adults with binge eating disorders. *Eating Behaviors* 14, 309–313.

Giddens, A. (1991) *Modernity and self-identity: Self and society in the late modern age*. Polity Press, Cambridge.

Gress, K. (2003) Animals helping people. People helping animals. Interview by Shirley A Smoyak. *Journal Psychosocial Nursing Mental Health Services* 41(8), 18–25.

Gross, J.J. (2013) Emotional regulation: Taking stock and moving forward. *Emotion* 13(3), 359–365.

Guy, M., Newman, M.A., Mastracci, S.H. and Maynard-Moody, S. Hasenfeld, Y. (ed.). (2010) *Emotional Labor in the Human Service Organization*. Sage, London.

Hagqvist, C. (2010) Discrepant trends in mental health complaints among younger and older adolescents in Sweden: An analysis of WHO data 1985-2005. *Journal of Adolescent* 46, 258–264.

Håkansson, M., Karlsson Palmgren, C. and Sandgren, V. (2008) *Kartläggning av verksamheter som använder hästar i vård och behandling i Sverige 2008*. Stiftelsen Svensk Hästforskning.

Hallberg, L.R.-M. (ed.) (2002) *Qualitative Methods in Public Health Research - Theoretical Foundations and Practical Examples*. Studentlitteratur. Lund.

Hama, H., Yogo, M. and Matsuyama, Y. (1996) Effects of stroking horses on both humans and horses heart rate responses. *Japanese Psychological. Research* 38(2), 66–73.

Hasenfeld, Y. (2010). Worker-Client Relations: Social Policy in Practice. In Y. Hasenfeld *Human Services as Complex Organizations* (405-427) London: Sage.

Hauge, H., Kvalem, I.L., Pedersen, I. and Braastad, B.O. (2013) Equine-assisted activities for adolescents: ethogram-based behavioral analysis of persistence during horse-related tasks and communication patterns with the horse. *Human-Animal Interaction Buletin* 1(2), 57–81.

Henry, S., Hemery, D., Richard. M.A., Hausberger, M. (2005). Human-mare relationships and behavior of foals toward humans. Applied Animal Behavior Science, 50, 341-362.

Hill, C.L.M. and Updegraff, J.A. (2012) Mindfulness and its relationship to emotional regulation. *Emotion* 12(1), 81–90.

Holleran Steiker, L.K. (2008). Cultural Considerations for Residential Treatment of Children and/or Adolescents. *Residential Treatment for Children & Youth*, 23(1), 61-74.

Holmes, C.M.P., Goodwin, D., Redhead, E.S. and Goymour, K.L. (2011) The benefits of equine assisted activities: An exploratory study. *Child Adolescent Social Work Journal* 29, 111–122.

Holmqvist, R., Carlberg, M. and Hellgren, L. (2007) Deliberate self-harm behavior in Swedish adolescent girls reports from public assessment and treatment agencies. *Child Youth Care Forum* 37, 1–13.

Jablonska, B., Lindberg, L., Lindblad, F. and Hjern, A. (2009) Ethnicity, socio-economic status and self-harm in Swedish youth: A national cohort study. *Psychological Medicine* 38, 87–94.

Johansen, S.G. (2008) Equine facilitated psychotherapy in mood disorders. *Journal of Affective Disorders* 107, 53–12.

Jordan, J.V. (2010) *Relational-Cultural Therapy*. United Book Press, Inc, Baltimore, MD.

Juhila, K. and Abrams, L.S. (2011) Special issue editorial: Constructing identities in social work settings. *Qualitative Social Work* 10(3), 277–292.

Kaiser, L., Spence, L.J., Lavergne, A.G. and Vanden Bosch, K.L. (2004) Can a week of therapeutic riding make a difference? – a pilot study. *Anthrozoös* 17(1), 63–72.

Karol, J. (2007) Applying a traditional individual psychotherapy model to equine-facilitated psychotherapy (EFP): Theory and method. *Clinical Child Psychology and Psychiatry* 12(1), 77–90.

Kåver, A. and Nilsone, Å. (2003) *Dialektisk beteendeterapi vid emotionellt instabil personlighetsstörning-teori, strategi och teknik*. stockholm natur och kultur.

Klontz, B.T., Bivens, A., Leinart, D. and Klontz, T. (2007) The effectiveness of equine-assisted experiential therapy: Results of an open clinical trial. *Society And* 15, 257–267.

Koren, E. and Træen, B. (2003) Jenter og hest. stallen som arena for sosialisering og mestring. *Tidsskrift for Ungdomsforskning* 3(2).

Kvale, S. (1997) *Interviews: An Introduction to Qualitative Research Interviewing*. Sage Publications, Thousand Oaks, CA.

Lundh, L.-G. and Bjärhed, J. (2008) Deliberate self-harm in 14-year-old adolescent: how frequent is it, and how is it associated with psychopathology, relationship variables, and styles of emotional regulation? *Cognitive Behavior Therapy* 37(1), 26–37.

Lundh, L.-G., Karim, J. and Quilisch, E. (2007) Deliberate self-harm in 15-year-old adolescent: A pilot study with a modified version of the deliberate self-harm inventory Scandinavian. *Journal Of Psychology* 48, 33–41.

Lunner, K., Werthem, E.H., Thompson, J.K., Paxton, S.J., McDonald, F. *et al.* (2000) A cross-cultural examination of weight-related teasing, body image, and eating disturbance in Swedish and Australian samples. *The International Journal of Eating Disorders* 28, 430–435.

Macauley, B.L. and Gutierrez, K.M. (2004) The effectiveness of hippo therapy for children with language – learning disabilities. *Communication Disorders Quarterly* 25(4), 205–217.

Manso, A., Rauktis, M.E. and Boyd, A.S. (2008) Youth expectations about therapeutic alliance in residential setting. *Residential Treatment for Children & Youth* 25(1), 55–72.

Martin, D.J., Garske, J.P. and Davis, M.K. (2000) Relation of the therapeutic alliance with outcome and other variables: A meta-analytic review. *Journal of Consulting and Clinical Psychology* 68(3), 438–450.

McGreevy, P. (2004) *Equine Behavior: A Guide for Veterinarians and Equine Scientists*. UK Elsevier Limited.

Minero, M. and Canali, E. (2008) Welfare issues of horses an overview and practical recommendations. *Italian Journal Animal Science* 8, 219–230.

Morrison, T. (2007) Emotional intelligence, emotion and social work: Context, characteristics, complications and contribution. *British Journal of Social* 37, 245–263.

Naim, B., Hirschberger, S., Ein-Dor, G. and Mikulincer, M. (2013) An experimental study of emotion regulation during relationship conflict interactions: the moderating role of attachment orientations. *Emotion* 13(3), 506–519.

Nevonen, L. and Broberg, A.G. (2000) The emergence of eating disorders: An exploratory study. *European Eating Disorders Review* 8, 279–292.

Norring, C., Engström, I. and Enzell, K. (2002) Eating disorders across Europe Swedish eating disorders services-a brief modern history. *European Eating Disorders Review* 10, 161–167.

Pauw, J. (2000) Therapeutic horseback riding studies: Problems experienced by researches. *Physiotherapy* 86(10), 523–527.

Richmond, P.A. and Padgett, D.L. (2002) Finding the line: Boundary decisions in residential treatment. *Residential Treatment for Children* 20(1), 53–66.

Rothe, E.Q., Vega, B.J., Torres, R.M., Campos Soler, S.M. and Molina Pazos, R.M. (2005) From kids and horses: Equine facilitated psychotherapy for children. *International Journal of Clinical Health Psychology* 5(2), 373–383.

Rutter, M. (2000) Children in substitute care: Some conceptual considerations and research implications. *Children and Youth Services Review* 22, 685–703.

Sallnäs, M. (2000) *Barnavårdens institutioner – framväxt, ideologi och struktur*. Stockholms Universitet, Rapport i Socialt Arbete nr 96.

Saslow, C.A. (2002) Understanding the perceptual world of horses. *Applied Animal Behavior* 78, 209–224.

Silvers, J.A., Mc Rae, K., Gabrieli, J.D.E., Gross, J.J., Remy, K.A. *et al.* (2012) Age-related differences in emotional reactivity, regulation and rejection sensitivity in adolescence. *Emotion* 12(6), 1235–1247.

Smith, J.A., Flowers, P. and Larkin, M. (2012) *Interpretative Phenomenological Analysis Theory, Method and Research*. SAGE Publications Ltd, London.

Smith-Osborne, A. and Selby, A. (2010) Implications of the literature in equine-assisted activities for use as a complementary intervention in social work practice with children and adolescents. *Child Adolescent Social Work Journal* 27, 291–307.

Statens offentliga utredningar SOU (2000:109) *En svensk hästpolitik Betänkande från Hästpolitiska utredningen*. stockholm, Nordstedts tryckeri.

Sunseri, P.A. (2004) Family functioning and residential treatment outcomes. *Residential Treatment for Children* 22(1), 33–53.

Szcygiel, D., Buczny, J. and Bazinska, R. (2012) Emotional regulation and emotional information processing: The moderating effect of emotional awareness. *Personality and Individual Differences* 52(3), 433–437.

Traeen, B., Moan, K.-A. and Rosenvinge, J.H. (2012) Terapeuters opplevelser av hestebasert behandling for pasienter med spiseforstyrrelser. *Tidskrift for Norsk Psykologforening* 49, 348–355.

Vidrine, M., Owen-Smith, P. and Faulkner, P. (2002) Equine facilitated group psychotherapy applications for therapeutic vaulting. *Issues in Mental Health Nursing* 23, 587–603.

Walsh, F. (2009) Human-animal bonds: The relational significance of companion animals. *Family Process* 48(4), 462–480.

Walter, U.M. and Petr, C.G. (2008) Family-centered residential treatment: knowledge, research, and values converge. *Residential Treatment for Children & Youth* 25(1), 1–16.

Wampold, B.E. (2001) *The great psychotherapy debate: Model, methods, and findings*. Lawrence Erlbaum Associates, Mahwah, NJ.

Wetherell, M. (2012) *Affect and Emotion – A new social science understanding*. SAGE, London.

William, B. (2005) Horsemastership part 1: Therapeutic components and link to occupational therapy. *International Journal of Therapy & Rehabilitation* 12(2), 83–83.

Wilson, C.C. and Barker, S.B. (2003) Challenges in designing human-animal interaction research. *American Behavioral Scientist* 47(16), 16–28.

Wipper, A. (2000). The Partnership: The Horse-Rider Relationship in Eventing. *Symbolic Interaction* 23(1), 47-70.

Yorke, J., Adams, C., and Coady, N., (2008). Therapeutic Value of Equine-Human Bonding in Recovering from Trauma. *Anthrozoös* 21(1), 17-30.

Zegers, M., Schuengel, C., Van Ijzendoom, M. and Janssens, J. (2006) Attachment representations of institutionalized adolescents and their professional caregivers: Predicting the development of therapeutic relationships. *American Journal of Orthopsychiatry* 76(3), 325–333.

Appendix

The questions in the first interview focused more specifically on themes and were the same for clients and staff (except for one question to the staff about theoretical foundation) and were as follows:

1. Earlier experience with animals/horses, attitude to animals/horses today
2. Reasons for participating in Equine Assisted Social Work (EASW)
3. Description of experiences EASW similarities differences other experiences of treatment
4. Requirements for participation
5. Experienced relationship client staff/role models relationship/similarities differences relationship staff, horses
6. Pictures about future/goals

Questions for the in-depth interview with clients after and in direct conjunction with the observations.

a. Can you describe EASW for me in your own words?
b. How did you experience the horse in EASW? How would you define the horse?
c. Does EASW influence your view of the horse/horses?
d. Could you focus on the present or were your thoughts elsewhere?
e. How did you experience the staff in EASW?
f. How would you describe the relationship with the horse? Similarities, differences other relationships?
g. How would you describe the relationship with the staff? Similarities, differences other relationships?
h. Do you think/feel that you have learned anything from today's practice?
i. Were there any thoughts or feelings awakened during EASW?
j. What are your thoughts on performance in relation to EASW and the staff you work with?
k. Do you feel more or less motivated to participate in EASW after today's practice? Motivate?
l. Are there any differences or similarities between EASW and traditional treatment work?

Questions for the in-depth interview with staff members after and in direct conjunction with the observations.

a. Can you describe EASW for me in your own words?
b. How did you experience the horse in EASW? How would you define the horse?
c. Does EASW influence your view of the horse/horses?
d. How would you describe the goal of EASW?
e. How did you experience the client in EASW? How do you think she experienced the EASW?
f. How would you describe the relationship with the horse? Similarities, differences other relationships?
g. How would you describe the relationship with the client? Similarities differences other relationships?
h. Do you think that the client has learned something today from EASW?
i. Were there any thoughts or feelings awakened during EASW?
j. What are your thoughts on performance in relation to EASW and the clients you work with?
k. Do you feel that EASW affects you as a professional in the meetings with the client?
l. Are there any differences or similarities between EASW and traditional treatment work?

10 The Helping Horse: How Equine Assisted Learning Contributes to the Wellbeing of First Nations Youth in Treatment for Volatile Substance Misuse

Cindy Adams[1], Cheryl Arratoon[2], Janice Boucher[3], Gail Cartier[3], Darlene Chalmers[4], Colleen Anne Dell[5]*, Debra Dell[6], Dominique Dryka[3], Randy Duncan[7], Kathryn Dunn[8], Carol Hopkins[9], Loni Longclaws[10], Tamara MacKinnon[3], Ernie Sauve[10], Serene Spence[7] and Mallory Wuttunee[7]

[1]University of Calgary, Faculty of Veterinary Medicine; [2]Canadian Centre on Substance Abuse; [3]Cartier Equine Learning Centre; [4]University of Regina, Faculty of Social Work; [5]Department of Sociology & School of Public Health, University of Saskatchewan; [6]Youth Solvent Addiction Committee; [7]Department of Sociology, University of Saskatchewan; [8]University of Saskatchewan, School of Public Health; [9]National Native Addictions Partnership Foundation; [10]White Buffalo Youth Inhalant Treatment Centre

Abstract

There has been recent interest in Canada exploring the benefits of equine assisted interventions in the treatment of First Nations youth who misuse volatile substances. Using the richness of an exploratory case study involving the White Buffalo Youth Inhalant Treatment Centre and the Cartier Equine Learning Center, our community-based study examined the question of how an Equine Assisted Learning (EAL) program contributes to the wellbeing of First Nations female youth who misuse volatile substances. Both programs are grounded in a holistic bio-psycho-social-spiritual framework of healing. Our study shares how the EAL horses, facilitators and program content contributed to youths' wellbeing in each area of the healing framework (bio-psycho-social-spiritual), with emphasis on the cultural significance of the horse and its helping role. The horse is a helper in the girls' journeys toward improved wellbeing—the horse helps through its very nature as a highly instinctive animal, it helps the facilitators do their jobs, and it also helps put the treatment program activities into practice. In addition, the role of First Nations culture in the girls' lives was enhanced through their encounters with the horses. The findings support the limited literature on equine assisted interventions and add important insights to the youth addictions treatment literature. Key implications to consider for EAL and volatile substance misuse policy, practice and research are identified.

*Corresponding author: colleen.dell@usask.ca

A dearth of scholarship exists that examines the relationship between humans and horses, and how these bonds may contribute to human wellbeing (Chalmers, 2014). Horses have also received relatively limited scholarly attention in the literature, despite increased interest in incorporating them into human health and welfare services (Bachi, 2013a). Animal assisted intervention studies have mainly focused on understanding small companion animals within the context of therapy and therapeutic relationships between animals and humans (Nimer and Lundahl, 2007). The horse has unique characteristics that are related to its cultural role in some Indigenous populations (Lawrence, 1983; Lawrence, 1998; Dell *et al.*, 2008) and also its status as a flight animal, which enables it to respond instinctively to human behaviour and intent through its heightened sensitivity to its surrounding environment (Goodwin, 2002; McDonnell, 2003; Feh, 2005). As such, human interactions with horses can be a distinct avenue for exploring social relationships and human wellbeing (Gonzalez and Sutton, 2011). This paper shares the key findings of our study, which asked how an Equine Assisted Learning (EAL) program contributes to the wellbeing of First Nations youth who misuse volatile substances. We found that the horse is a helper in the youths' journeys toward improved wellbeing—the horse helps through its very nature as a highly instinctive animal, it helps the facilitators do their jobs, and it also helps put the treatment program activities into practice. In addition, the role of First Nations culture in the girls' lives was enhanced through their encounters with the horses.

Equine Assisted Learning

In the past decade, equines have been included in a range of therapeutic human service contexts, including counselling and learning programs within corrections (Deaton, 2005; Bachi, 2013a), mental health facilities (Bizub *et al.*, 2003), social services (Burgon, 2003), women's and youth addiction treatment (Pollack, 2009) and with high-risk youth and veterans who have experienced emotional and behavioural trauma (Basile, 1997; Burgon, 2011; Schultz, 2005; Foley, 2007; Jarrell, 2009; Duncan *et al.*, 2014; Carlsson *et al.*, 2014). Much of the research focuses on therapeutic riding programs with hybrids that include both riding and non-riding activities. These approaches are mainly informed by outcomes and theories from the broader human-animal bond knowledge base (Kruger and Serpell, 2006) and descriptive literature (Frewin and Gardiner, 2005; Brooks, 2006; Kakacek and Ottens, 2008; Chardonnens, 2009).

Equine Assisted Learning (EAL), a non-riding program, is a relatively new field within the area of equine assisted interventions and draws primarily on the tenets of experiential or 'hands-on' learning. While resonating with some of the core values (MacKinnon, 2007) found within other equine guided interventions (e.g. equine assisted psychotherapy, therapeutic riding), in general EAL is an educational program that is facilitated within a group format and focuses on ground activities. In EAL participants engage in structured, facilitator-led sessions with constant feedback related to the participants' experiences. The horse is generally considered the teacher and a human facilitator is there to help guide the participants on a journey of learning and understanding. The group sessions provide opportunities for participants to become engaged in situations that require interaction with the horse and group members, and to reflect on these experiences. The overall intent is to create opportunities for participant self-development and to internalize this awareness within the sessions and generalize it to other life situations (Equine Assisted Growth and Learning Association, 2010).

Volatile Substance Misuse

Research indicates that volatile substances are often the first mood-altering substance used by children and youth because they are readily available, inexpensive and easily concealed (Dell *et al.*, 2011). Volatile substances are a large and diverse group of chemical compounds located in hundreds of households and industrial products, including paint thinner, glue, gasoline and correctional fluid (Dell and Beauchamp, 2006). Volatile substance misuse (VSM) is the deliberate inhalation of fumes or vapours given off from a substance for its intoxicating and mind altering effect (National drug abuse information centre, 1998). The health effects of inhaling can be acute, and include frostbite and burns (Albright *et al.*, 1999), brain and nerve cell damage (Basu *et al.*, 2004) and sudden heart failure (Ballard, 1998). The social effects are equally destructive and include poor academic performance (Basu *et al.*, 2004), spiritual harm (Honouring Our Strengths: Honouring our strengths: Indigenous culture as intervention in addictions treatment, 2014), and problem behaviour, such as delinquency (Best *et al.*, 2004).

Epidemiological and other international research indicates that there are higher rates of VSM among youth experiencing disenfranchised life conditions (Dell *et al.*, 2011). This has been documented among street youth, inner city youth, and some First Nations and Inuit youth living in select rural and remote areas of Canada (Bone *et al.*, 2011). VSM has been linked to high rates of poverty, boredom, loss of self-respect, unemployment, family breakdown and poor social and economic structures. In Canada, these conditions are connected to the historic impact of the colonization of First Nations people, systemic racism and discrimination, residential schooling and multi-generational losses of land, language and culture. Canada's First Nations people comprise approximately 2.6 percent of the country's population (Métis 1.4 percent & Inuit 0.2 percent) (Statistics Canada, 2011).

Since a widely-played media clip in 1993 of Innu youth in Davis Inlet, Labrador getting high by sniffing gasoline, there has been ongoing interest in effective ways to treat this health issue. In 1996, a major response to VSM was undertaken on the part of First Nations people and Health Canada's First Nations and Inuit Health Branch; eight residential youth solvent treatment centres were established. Canada is now considered an international leader in providing residential treatment to First Nations youth who misuse volatile substances.

The Canadian Experience

The White Buffalo Youth Inhalant Treatment Centre (White Buffalo) is one of the established VSM centres. It is a six-month residential program for female, First Nations adolescents who misuse volatile substances. It is located on the Sturgeon Lake First Nation, near Price Albert, Saskatchewan. The program is based on the concept of living therapy, which integrates four corner-stones of treatment that parallel teachings of the First Nations' medicine wheel—spiritual, emotional, mental, and physical. The majority of youth who enter the program have extensive histories of multiple forms of abuse (Youth Solvent Addiction Committee (YSAC), 2008). A foundational concept of the White Buffalo program is nurturing the inner spirit, which is practiced through traditional First Nations teachings and holistic healing (e.g. fasting, prayer). Alongside this, through structured programming, White Buffalo attempts to realign the youths' association with and reliance on their greater community. With belief that EAL compliments its approach, White Buffalo expanded its treatment to include the Cartier Equine Learning Center EAL program in the past decade.

Located north of Prince Albert, Saskatchewan, the Cartier Equine Learning Center has been a leader in establishing industry standards in the area of EAL certification and program development in Canada (Saskatchewan Horse Federation). This program offers its participants a structured curriculum that incorporates a variety of 'building block' lessons and learning opportunities. The program is a learner-centred educational experience with horses that focuses on the animal's non-verbal communication as a teaching modality for cognitive and behavioural change. A fundamental guiding philosophy of the program is understanding that '[b]y their intuitive nature and innate sensitivity, horses can provide facilitators with a window into the participant's personality creating opportunities for immediate outcomes and feedback' (MacKinnon, 2007, p.4). The Center has developed an 'EAL Formula' that is considerably different from other equine guided programs, and is designed to produce consistent, desired and predictable outcomes when all six components of its formula are present. The components are: relationship, curriculum, formula (experiential learning focus), horse, facilitation and partnerships. Each exercise is custom designed to maximize progressive learning potential and focuses on developing individual skills as participants work through interactive challenges.[1]

Our current study is informed by three peer-reviewed articles in which various members of our team examined EAL in the treatment of VSM among First Nations youth for the first time in Canada. The first, a 2008 article titled 'Horse as Healer: An Examination of Equine Assisted Learning in the Healing of First Nations Youth from Solvent Abuse' assessed the theoretical intersections between the White Buffalo and Cartier programs (Dell *et al.*, 2008). This work illustrated that the culture-based model of resiliency underlying White Buffalo's program, which accounts for an individual's inner spirit and relations with their collective community, is complementary to Cartier's EAL program. Both support a holistic bio-psycho-social-spiritual framework of healing. This understanding contributed to a 2011 study that examined the benefit of EAL to First Nations youths' healing from VSM at the Nimkee NupiGawagan Healing Centre in Muncey, Ontario (Dell *et al.*, 2011). Similar to White Buffalo, this VSM program adheres to a culture-based model of resiliency. The resulting publication, titled 'Creating Healing Spaces: The Experiences of First Nations and Inuit Youth with Equine-Assisted Learning' found that the youths' healing was aided through the availability of a culturally-relevant space, comprised of spiritual exchange (i.e. between the youth, horses and sometimes EAL facilitators), complementary communication (i.e. relaying upon intuition with the horses and sometimes transferring this to humans) and authentic occurrence (i.e. overcoming barriers to understanding imposed by Western ways of knowing).

Extending this work, the third publication was released in 2011 and is a scoping review of 13 EAL studies (Chalmers and Dell, 2011). It concluded that Western epistemologies or ways of knowing guided the limited existing research through post-positivism (which suggests that a shifting truth can be located outwardly) and constructivism (which claims that knowledge is co-created by those experiencing it) and to the exclusion of Aboriginal epistemology (which asserts that knowledge is ever-present and internal to the self and external in relation to others). Knowledge in this sense is understood to be a way of life and is grounded in a history of cultural practices and traditional understanding. Recognizing this, the Western-derived participatory paradigm was identified as a potential bridge to guide future EAL research, and specifically with First Nations. As a method for constructing knowledge, the participatory paradigm is emancipatory in its focus on bringing about change through action, initiated by community identification, direction and owning of the research process (Davis and Reid, 1999).

Taking this cumulative understanding into account, the overarching question originally set out for the current study was if and how the Cartier Equine Learning Center EAL program contributed to the wellbeing of First Nations youth who misuse solvents. With congruence already established between the bio-psycho-social-spiritual framework of healing applied by the White Buffalo Youth Inhalant Treatment Centre and Cartier's program, it was determined that in theory the EAL program does contribute to youths' wellbeing. The validity of the cultural foundation of the White Buffalo program was not questioned; to do so would be a colonizing act. Specifically examined in this paper then is the remaining question of – *how do youth experience wellbeing based on their participation in the EAL program?*

The Present Study

A community-based research approach was adopted in this study for several reasons. Foremost, it emphasizes relationship building throughout the research process with all stakeholders to ensure the study is relevant to each (Scott, 1992; Smith, 1999; Martial *et al.*, 2003). This focus supports the prioritization of First Nations knowledge in the research (Smith, 1999) and self-determination by ensuring jurisdiction over the data at all times and directing the research to serve First Nations interests (Sakai *et al.*, 2004). This is imperative given the history of research with Indigenous communities is fraught with disrespect and the appropriation of power, which is reflective of the history of colonization of First Nations people in Canada (Smith, 1999). As a method for constructing knowledge, this approach also parallels principles found within the participatory paradigm.

The make-up of our community-based team facilitated our approach to the research such that it was 'by, for and in balance with' the research population and not 'on' it (Kirby *et al.*, 2006, p. 32). Our study is a collaborative effort of front line workers, community leaders, Elders, researchers, decision makers, the White Buffalo Youth Inhalant Treatment Centre, the Cartier Equine Learning Center, the Youth Solvent Addiction Committee (YSAC), the National Native Addictions Partnership Foundation (NNAPF), the Canadian Centre on Substance Abuse (CCSA), and researchers at the Universities of Calgary, Saskatchewan and Regina. YSAC provides theoretical direction to Canada's solvent abuse treatment centres based upon a culture-based model of resiliency, NNAPF guides the First Nations and Inuit Addictions System in Canada towards holistic, culturally-relevant and seamless community-driven addiction services, and CCSA is Canada's non-governmental addictions agency. Our team's research practices are in adherence with the Tri-Council Policy Statement on research involving the First Nations, Inuit and Métis people of Canada (Canadian Institutes of Health Research, Natural Sciences and Engineering Research Council of Canada, and Canadian Institutes of Health Research, 2010). Consent was received from all participants, including parents/guardians for the youth, as part of the information package when they entered the White Buffalo treatment facility; everyone was offered the opportunity to decline participation with no repercussions. Ethics approval for the study was granted by the Universities of Saskatchewan, Calgary and Regina Behavioural and Animal Research Ethics Boards.

With this study situated in a holistic bio-psycho-social-spiritual framework of healing, as applied by White Buffalo and compatible with Cartier's EAL program, the beginning point for understanding wellbeing was this framework alongside the peer-reviewed literature (including our team's studies referenced above and where possible the equine-specific literature), and the experiential and cultural expertise of our diverse team members. Taking this all into account, it was established that *biological* healing refers generally to physical health and the physiological requirements to have a healthy body. This can include, for example, food, water, fresh air, exercise and connection to and respect for the land. In this study we focused on three measures of biological healing, defined as physical wellbeing and identified through (a) physical status, (b) importance of physical touch with the horse, and (c) acknowledging the role of the land in healthy being. *Psychological* healing refers generally to mental status. This can include mood/feelings, cognition, perception, thoughts, self, intellect, emotion, judgment and identity. In this study we applied the most common understanding of psychological healing located in the equine literature, characterized it more inclusively as mental wellbeing and identified it through perceptions of self. *Social* healing generally refers to an individual's relationships or associations with friends, family, community and the universe. We focused on social healing for First Nations in this study, defining it as community involvement and relations with others, and it was identified through quantity and quality of interactions with fellow participants and staff in the residential treatment program and family members. And last, *spiritual* healing refers to an individual's inner system of beliefs (e.g. purpose, meaning, value, higher power) and their essence, being or inner spirit. We focused on two measures of spiritual healing, defined as development of an equine-human relationship (trust) and spiritual bond, and it was identified through (a) development of respect and trust with horses, (b) spiritual bond to horses, and (c) increased spiritual functioning (see Fig. 10.1).

Key to this understanding of wellbeing is that the health of an individual is comprised of their subjective and relational interactions within four inter-related quadrants: biological, psychological, social and spiritual. This coincides with First Nations knowledge and understandings of health that focuses on a balance of mental, physical, social and spiritual wellbeing (Youth Solvent Addiction Committee (YSAC), 2008). To illustrate, the work of Laframboise

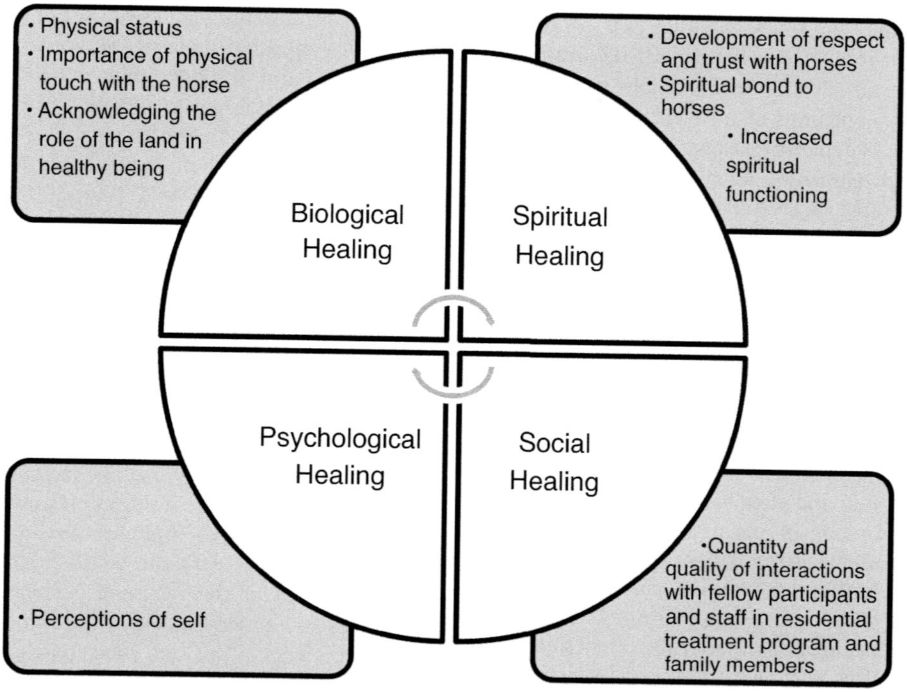

Fig. 10.1. Holistic bio-psycho-social-spiritual framework of healing.

and Sherbina (2008) supports that '[w]ithin First Nations culture the Medicine Wheel reflects these same principles of holistic healing. These teachings are among the oldest teachings of First Nations people. The teachings found on the Medicine Wheel create a bio-psychosocial and spiritual foundation for human behaviour and interaction' (2008). Applying a holistic framework of wellness to this exploratory study is necessary because the field 'currently lacks a unified, widely accepted, or empirically supported theoretical framework for explaining how and why relationships between humans and animals are potentially therapeutic' (Kruger and Serpell, 2006, pp. 25–26). This understanding frames how we collected and analyzed our data in the study.

Case study

Considering the exploratory nature of this study, our team chose a case study design. The case study in this project is not an individual but rather the Cartier EAL program. This expanded approach to conventional case study research is used to explore topics in their context using detailed and in-depth data gathered through multiple sources (Baxter and Jack, 2008). This design is particularly useful when questions pertaining to 'how' and 'why' are explored; when a detailed account of the context is needed to fully comprehend the topic; when the stories of the research participants are crucial to attaining understanding; and when a collaborative approach to the research process and knowledge building is essential (Creswell, 2013). As shared by Hodkinson and Hodkinson (2001), key strengths of a case study design are that they are useful for understanding complex inter-relationships (e.g. between White Buffalo and Cartier), are grounded in lived reality (e.g. youths' experiences with EAL), facilitate exploration of the unusual and unexpected (e.g. undertaking an exploratory study with a limited empirical base), use multiple cases that facilitates focus on the ideosyncratic (that is, accounting for different experiences as well as similar), and can facilitate rich conceptual and theoretical development (Hodkinson and Hodkinson, 2001). And possibly most important, the openness of a case study allowed our team to adapt the design to account for our research approach – applying community-based research principles that facilitate the prioritizing of First Nations knowledge. Holistic sharing of dynamic and complex knowledge in this study reflects a First Nations culture-based storytelling approach to knowing.

Our case study is designed to be both in-trinsic and instrumental. In general, an instrumental case study is undertaken when an understanding of a particular issue is of interest and an intrinsic case study explores cases where description and detail are required for a full understanding of the topic area and how the case is unique (Stake, 1995). Our study is instrumental in that it focuses on understanding how participation in an EAL program contributes to the wellbeing of First Nations youth in treatment for misusing volatile substances. It is also intrinsic because gaining such insight requires an in-depth understanding of the EAL program itself; that is, it accounts for the context within which the phenomenon of youth wellbeing occurs.

A single case, the Cartier EAL program, was identified for this study based on several factors. First, the Cartier program, or case, has an established curriculum and is nationally recognized for offering EAL training certification and is distinct in this way from other EAL programs. Second, the program has provided EAL to youth in treatment for VSM over the past 6 years, and as such has an established relationship with White Buffalo. As a result, purposeful sampling was used to access a group of 'information rich' participants based on the commonality of their experiences with EAL and VSM. Third, with the instrumental focus of our case study, we looked exclusively at Cartier's EAL program within a specified time period (each cohort of youth attended the EAL program for 20 weeks) and within a specified population (First Nations, female youth, 12–18 years of age). Last, the EAL and VSM programs are in close geographic proximity to each other and the majority of the researchers' institutions. It was important for our team to be clear from the study outset about the case sample and design. Establishing boundaries around the individual unit of study was necessary as this delineated the case (Cartier EAL program) and the case context (the horse, the facilitators, the program activities and First Nations' youth in treatment for volatile solvent misuse), even though 'phenomenon and context are not always distinguishable' (Yin, 2003, pp. 13). In keeping with the case study design, this step provided clarity and direction to our data collection strategies and analysis.

Data collection

The contextual units of analysis within the case in the study included the horse, the facilitators, the program content and wellbeing. Data were gathered from multiple and diverse sources, including interviews with the youth to gain a detailed account of their general perceptions, experiences and understanding of the four identified components of healing; youth participant journaling about experiences each week on the day following the EAL program and during classroom time for approximately 20 min; interviews with the Cartier EAL program facilitators to gain a detailed account of their understandings; Cartier EAL program facilitator field notes on participants in the Cartier program areas of curriculum, objectives, and intended outcomes and responses; interviews with White Buffalo staff to gain a detailed account of their insights and observations on the youths' encounters with the EAL program; and White Buffalo staff recordings in participant files to document observations of youths' attitudes and behaviors.

Data were collected from four White Buffalo six-month treatment cohorts, starting in June 2010 through to June 2012. A total of 66 girls participated in the study, with 26 completing it (39%). All youth who discontinued their participation in the study left the White Buffalo program for varying reasons, and at times new youth attended part-way through an established cohort. The youth participated in the Cartier EAL program a maximum of two times a week for 2 hours and for 20 weeks total. A number of differences were experienced within and between the youth cohorts that posed challenges for the data collection, with the most limiting being drop-out. The average age of the girls was 15. In addition to speaking English a quarter spoke their native language (Cree, followed by Dene, Saulteaux, and Innu), and the majority were from Saskatchewan, followed by the provinces of Manitoba, Labrador, Ontario and Alberta.

Collecting data over time allowed our team to start off by gaining an in-depth and isolated picture of the EAL program from the first cohort and apply what was learned to the remainder of the data collection. It allowed us to collect descriptive and insightful information on contextual factors and research process issues that were not anticipated by our team at the start of the study. For example, it was quickly learned that involving community-based researchers from both White Buffalo and Cartier in the pre-interviews with the girls was helpful to establish rapport with them. We revised and honed our data collection techniques based on such experiences. This was consistent with a case study approach, where necessary data collection processes are revised while the data is collected. It is also important to note that the multi-month span of time between the measures (e.g. pre-post interviews) helped us to avoid a testing effect.

Data analysis

In case study research 'the analysis is rich in the context of the case or setting in which the case presents itself' (Merriam, 1998, pp. 41). Through our data analysis processes we detailed aspects of the history of the case, the chronology of events, and day-by-day rendering of the activities of the case (Stake, 1995). Following this description, we focused on key issues for understanding the complexity of the case (Creswell, 2013). We used categorical aggregation as proposed by Stake (1995). We examined a collection of instances from the data that revealed issue-relevant meanings. We used a process of data reduction, that is, we identified themes and we cross-compared them between cohorts (Yin, 2003). We arrived at a meaning for the case in terms of the EAL program in its totality and its contributions to the wellbeing of First Nations youth in treatment for VSM. Practically, this resulted in the identification of key themes in answer to our research question. The participatory paradigm that framed our approach to the study and was especially paramount to the data analysis.

To arrive at the results, our team followed an integrated data analysis process. We developed our initial coding tree based on the work of others and our own understandings, most prominently informed by the holistic bio-psycho-social-spiritual framework of healing applied by White Buffalo. We began with our Indigenous understanding of wellbeing (deductive) and then added detail to it based on our data analysis (inductive) of the first cohort, both verifying and generating themes and explanations for each of the four components of wellbeing. Interview transcripts, observer recordings, White Buffalo staff reflections, and Cartier EAL session debriefing notes were analyzed and searched for patterns and possible new codes. As Stake (1995) articulates, analysis tends to translate to 'a search for patterns, for consistency, for consistency with certain conditions, which we call correspondence' (Stake, 1995, pp. 78). Our community-based researchers from White Buffalo and the Cartier EAL program were the leads on this work. A refined coding tree

Table 10.1. Initial coding tree of youth wellbeing themes derived from the literature, team expertise & analysis of cohort one.

Biological/ physical	Social	Psychological mental/emotional	Spiritual
• Physical health • Physical touch • Nature/natural setting	• Get along • Develop relationships • Importance of community • Importance of family • Importance of friends	• Identity • Self-esteem	• Just being • Bond • Spiritual functioning

emerged from our team's analysis of the first cohort's data (see Table 10.1).

We thereafter applied a simultaneous deductive and inductive approach to theme identification. It was an iterative process. The White Buffalo and Cartier researchers and the larger research group continually revisited the coding tree. As such, all perspectives and insights were shared, discussed and agreements arrived at. This process helped to not only trim and refine our themes but also served three key functions. First, by returning to the team members vis-à-vis research participants member-checking occurred. Second, the process ensured attention to interrater reliability as several of the team members were involved in both the data collection and analyses. And finally, these steps contributed to the overall validity of our findings. Although all themes were identified to a notable extent in the multiple data sources collected, and are therefore considered findings, our team also undertook a process in which each theme was identified for saturation, and demarcated as low, medium or high. The two community-based researchers undertook this categorization based on their indepth experiences of coding the data sources and as a validation check an independent researcher on our team did the same with a careful review of the coded data. Whenever there was disagreement, a discussion was held between the three individuals and a decision arrived at. This was necessary for three themes – touch (decided upon high), nurturing (decided upon low) and spiritual functioning (decided upon low).

Drawing on all of our data sources, a comprehensive file was created for each youth and a narrative written. Although the findings in this paper only focus on the identified themes and the illustrative words of the youth, the original analysis is situated within the youths' complete narratives. The detailed and time intensive process of creating files for the youth resulted in 21 individual files; a file was not developed if only limited data was available for a youth (e.g. youth left the program early).

Results

The EAL program's contribution to youths' wellbeing

As shared, this study is framed within the holistic bio-psycho-social-spiritual framework of healing applied by the White Buffalo Youth Inhalant Treatment Centre. The combined experiences, observations, reflections and analyses in the study led to the finding that the Cartier EAL program contributes to the wellbeing of the First Nations youth in the White Buffalo Youth Inhalant Treatment Program in several ways. The research question asked was: how does an Equine Assisted Learning (EAL) program contribute to the wellbeing of First Nations female youth who misuse volatile substances? The

Table 10.2. Final coding tree (findings) of youth wellbeing themes.

Culture	Biological/ physical	Social	Psychological mental/emotional	Spiritual
• Knowledge • Moontime (menstruation)	• Physical touch • Physical health • Anatomy • Nature/Natural setting	• Develop relationships • Communication • Get along • Experience • Behavior change • Importance of friends • Importance of family • Importance of community • Team player • Leader	• Identity • Self-esteem • Feelings/Mood • Problem solving • Attitude • Participation • Nurturing	• Just being • Bond • Cultural activities • Spiritual functioning

horse is a helper in the girls' journeys toward improved wellbeing—the horse helps through its very nature as a highly instinctive animal, it helps the facilitators do their jobs, and it also helps put the treatment program activities into practice. In addition, the role of First Nations culture in the girls' lives was enhanced through their encounters with the horses.

Table 10.2 outlines 28 themes in the four areas of wellbeing, plus the addition of the category termed culture as an area of wellbeing. Thirteen of the themes originated from pre-existing work (refer back to Table 10.1) and were all supported in this study and the remaining 15 were newly identified in this study. Findings within each of the five areas of wellbeing are discussed in order of greatest significance or most saturation. The addition of the culture category was in response to the youth speaking specifically to the relationship between culture, the horse and their wellbeing. This should not be a surprise given that the horse is a prominent animal in the culture of many First Nations. It was concluded that the horse has a helping role in facilitating youth wellbeing most prominently through the sharing of cultural knowledge. One participant commented that 'Knowing that those horses are part of my culture made me realize that I can believe in a Creator' (Cohort 2, Youth 3). Another shared: 'it [the horse program] taught me to stay with my family and learning about culture that I will take back home' (Cohort 2, Youth 10). The girls also spoke about becoming familiar with the power of the female spirit of First Nations through the horses. For example, they learned about the cultural teaching that they should not be near the horses when menstruating. One girl shared '[T]hey say you aren't supposed to go on your time, so respect yourself' (Cohort 2, Youth 7).

The most significant finding regarding biological or physical wellbeing is experiencing healthy or safe touch (i.e. brushing the body, braiding the mane, petting the horse, hugging the horse, standing close to allow for body contact). For example, one youth commented 'whenever he would do something that is right he would put his head on my chest and wait for me to hug him' (Cohort 2, Youth 3) and another shared 'the horses know they feel appreciated when you do that [brushing] and tell them they did a good job and petting them they will know' (Cohort 3, Youth 4). Another youth commented 'certain horse I liked brushing [because] I felt calm' (Cohort 3, Youth 11). This finding was followed by improved physical health, that is, taking part in other physical activities because of EAL participation. The youth offered: 'it made me feel more active' (Cohort 1, Youth 13) and 'I can actually run around now' (Cohort 2, Youth 10). Third, the youth shared that they learned about anatomy/how the body works for horses and by extension humans and their own bodies. The youth shared: 'I learned that a horse's heart beats way faster than a human and they can like run really fast and that they are really strong and they can carry babies and all that' (Cohort 3, Youth 4) and 'yeah, when a horse is mad their ears pin back, their eyes go big, their muscles tighten up and their tails start swishing. It's pretty interesting. I never knew that' (Cohort 3, Youth 2). And last, and as the least saturated category, the youth shared that they connected with the EAL setting by being at the farm and in nature. One youth shared: 'when you go in the door and you are in there [the arena] it smells like a million bucks' (Cohort 1, Youth 5). And another said 'I just like it over there because there are horses over there' (Cohort 1, Youth 11) and another simply commented 'peaceful' (Cohort 3, Youth 11).

There were five prominent findings attributed to psychological (mental/emotional) wellbeing. One of the most significant findings was that the horse helped the youth to learn about their identity. For instance, one youth mentioned 'I can actually show my real me when I am at the horse program. It [horse program] made me think smartly, helped me with courage and stuff like that' (Cohort 2, Youth 10). Another meaningful finding illustrated how the horse increased the youths' sense of self-esteem and selfworth. One youth stated 'yeah, the teacher once said the horses know how you feel. This one girl they had a few intakes ago was feeling angry and not good, and she was at Cartier and the horse wouldn't listen to her because he knew how she felt, and she started crying and that horse hugged her. So when I go to Cartier's I feel good about myself so the horses can feel good' (Cohort 1, Youth 3). Another shared: 'I found out that I am not, I don't know how to put it, mean sometimes, I am actually really nice' (Cohort 2, Youth 3). The youth also reported to develop a greater sense of self-awareness about their feelings and moods. For example, one youth stated 'at the horse program I learned that I was actually happy inside and I didn't know that before' (Cohort 2, Youth 10). Moreover, the youth increased their ability to problem solve: "I found that I didn't really have good boundaries with some people. I will use this at home with my friends and my family' (Cohort 2, Youth 3). Another youth reflected: 'if I see someone doing a bad thing I know they will get mad at me for doing the right thing but that is the way I was taught, to do the right thing and tell the truth, not living in lies' (Cohort 3, Youth 2). The youth also established a more positive attitude. For instance, one youth said 'the horse program helped me feel good cause it made me, when I am depressed it makes me happy' (Cohort 2, Youth 7).

In addition, though to a less extent, the youth experienced psychological healing through increased participation and effort on their part. One youth expressed 'I didn't really respect anyone, then I got with the horses and I needed to respect them or they would kick me, and then I came back here [treatment centre] and started to respect the others' (Cohort 1, Youth 12). Another youth shared: 'well, we were practicing walking and he [the horse] would walk in front of me and I would say whoa and I would have to back him up and put boundaries in place for him. It was a long process' (Cohort 3, Youth 2). The least saturated finding was experiencing or witnessing nurturing. One youth described the experience of observing a pregnant mare and her foal as 'exciting, we saw a pregnant horse and then later on saw the baby horse' (Cohort 1, Youth 3).

There were five key findings specific to social wellbeing. First, and most prominent, relationships were developed specifically with the horse facilitators. Some youth expressed their relationships with the facilitators as: 'I like them talking to me' (Cohort 2, Youth 6), 'I feel awesome around them' (Cohort 2, Youth 7) and 'they are like my parents' (Cohort 2, Youth 9). Another prominent finding was a greater sense of communication, for instance understanding of space, body language and dialogue. One youth described her ability to communicate with horses 'yeah, they are like humans (the horse), they know when you are feeling scared and they know when you are feeling open with them' (Cohort 3, Youth 10). Not only did the youth get along with the facilitators, they also reported developing communication skills with each other. For example, two different youth shared the sentiment that 'when I first got here I didn't know those girls and then I got paired up with other girls and got to know them and the horses' (Cohort 1, Youth 12) and 'when I first met [name of youth], we were partnered up and I taught her some things and we were relationship building and from there we started talking' (Cohort 3, Youth 2). Another social wellbeing aspect identified in the study was the importance of a new experience. The youth had a chance to try new roles and participate in a new activity. For instance, one youth said 'something [work with the horses] I wouldn't do at home [it] was something new' (Cohort 1, Youth 3). Another youth shared: 'I felt, I really didn't want to go to it at first. I just felt, I don't know. Now there is this horse Maverick, and when you take the halter off him and the lead rope he just follows me everywhere' (Cohort 3, Youth 11). Finally, there was a positive change in behavior, for instance, an increased awareness of self, others and one's own learning. One youth was able to describe her experience with boundaries as 'like a horse has boundaries, where they want a person to be. The distance and their reaction when you cross those boundaries, so I look for those things in people too. Where their boundaries are' (Cohort 1, Youth 3).

Although to a less extent, the youth were also able to improve their social wellbeing with increased acknowledgement of the importance of friends and family in their lives. For example, the youth developed an increased awareness of the importance of what it means to have a friend. One individual spoke about friendship as 'someone who is just there for you' (Cohort 1, Youth 3). Expanding this understanding to family, another youth commented 'I will respect my mom more after participating in this treatment' (Cohort 1, Youth 12) and another shared 'I need to show them [family] more respect' (Cohort 2, Youth 3). In addition, youth participants began to think differently about what their home community means to them. In other words, they acknowledged the importance of community. One youth said 'hmm, it got me to think about like all the, when I took a look at the horses and my friends here, everybody was like stressed out when they came and then now they are just bonding and just all stressed out again and I am scared to go back too [to home community]' (Cohort 1, Youth 5). Youth also worked on being a better team player. One youth shared, 'when I first got here [another youth] started talking to me, then she stopped talking to me and I stopped talking to her, and then she was my teammate at the Cartier Farm, and then we had to talk' (Cohort 1, Youth 11). Another youth shared: 'yeah, when you support each other, for example, say if one of us got on each other's bad side then we had to be partnered up, we had to put our differences aside and work together' (Cohort 3, Youth 2). Finally, the youth stated that they started to become better leaders by being a positive influence, as one example. One youth mentioned that 'I learned a lot about the horse, and you know how you say to be a role model here? And then we did an exercise how to be a leader, well I gained confidence there' (Cohort 3, Youth 2).

Focussing on spiritual wellbeing, the youth identified the importance of just being and spending time with the horse. An illustration of this is in one youth's statement that 'the best part of the program was getting to know the horses better and just spending time with them' (Cohort 1, Youth 3). The advancement of spiritual wellbeing was also illustrated through the youths' development of a relationship or special bond with the horse. One youth stated 'he listens well and pays attention and he turns my day around and when I am having a sad day he puts his nose on my shoulder and then he smells my hand' (Cohort 2, Youth 9). To a less extent, the youth identified increased participation in cultural activities. One youth shared '[cultural activities] keep you busy to not go back to the way you were before' (Cohort 1, Youth 11) and '[cultural activities] are important because it's me, part of my identity and beliefs' (Cohort 1, Youth 10). The youth also shared about their increased spiritual functioning, that is feeling more connected to their spirit. One youth shared 'I didn't really think lots about God at the time but when I started going to the horse program, it got me thinking lots and now I just pray every night and like I just pray that everything goes alright with my family when they come here [for a visit at the Treatment Centre]' (Cohort 2, Youth 10).

It should be noted that our team also examined the challenges youth identified with the program. For the most part, the challenges dissipated through the course of the program. The key challenges were associated with the youth attending the White Buffalo Youth Inhalant Treatment program, including most prominently: coming into a new environment, sorting out dynamics with peers, meeting new individuals, residing in a new community, and meeting White Buffalo as well as Cartier program staff. There were also complaints associated with the equine facility, including it being 'cold' and 'stinky like a farm'. These too declined as the EAL and White Buffalo programs progressed.

Discussion

The most significant finding for biological/physical wellbeing was experiencing healthy or safe touch. The importance and healing quality of physical touch to human wellbeing is well documented in the literature (Yorke *et al.*, 2008). Given the high rate of physical and sexual abuse among First Nations youth in treatment for VSM, the horses can offer a 'safe' and 'non-sexual' mechanism for physical touch to occur (Held, 2006). Robin and Bensel's related work explains that pets can 'satisfy the child's need for physical contact and touch without the fear of the complications that accompany contact with human beings' (Robin and Bensel, 1985, p. 71). The literature also supports our finding that there may be physical health improvements following participation in an equine program. According to the Cartier program, this may be due in part to equine programs contributing to increased hand/eye coordination, balance and mobility. This is likewise identified in the general animal assisted intervention research (e.g. exercise and increased balance by walking a dog) (Friedmann *et al.*, 2010). Regaining motor coordination is an area of concern for youth who have misused volatile substances (Lubman *et al.*, 2008). The third key finding of this study, although not identified in the equine assisted intervention literature, was learning about anatomy and how the body works through the EAL program and being with the horses. This has been identified however in emerging scholarship specific to AAIs with youth in treatment for solvent misuse (Dell *et al.*, 2011). And last, connection with the EAL setting, that is, being at the farm and in nature, was identified in this study as well as the health literature. The work of Crofoot-Graham, for example, notes that connecting with nature is an important part of healing for Indigenous people (Crofoot-Graham *et al.*, 2001). The land is intimately related to culture, spirit and other dimensions of healthy being. The biophilia hypothesis similarly proposes that as humans we have a genetic and fundamental need to affiliate with other living organisms, including nature (Wilson, 1984; Kahn, 1997; Melson, 2000; Melson, 2008).

We applied the most common measure of psychological wellbeing located within the equine-assisted interventions literature in our study, and defined it through perceptions of self, and specifically identity and self-esteem. Our findings meaningfully supported both. For example, a study by Toukonen Cuffari (2011) explored the perceived benefits of female youth interacting with horses in recreation and therapeutic programs. Her findings suggest that perceived reciprocity surfaced between the study participants and their horses regardless of the type of program, and that this contribution to self-perception and self-esteem was a potential contributor to the youths' wellbeing. Self-identity and esteem are key areas of concern for First Nations youth in Canada given the historic impacts of colonization which attempted to 'kill the Indian in the child' (Harper, 2008).

Although not found specifically in the equine literature, related literature to this study has identified the importance of the other measures of psychological wellbeing in this study, including some specific to Indigenous youth healing from addictions and AAI research: feelings/mood (Roberts *et al.*, 2001), problem solving, attitude (Poole and Dell, 2006), participation (Hawkins *et al.*, 2004) and nurturing (Dell *et al.*, 2011). This is also supported in studies of atrisk youth and equine-assisted interventions in participant self-reporting of decreased depressive feelings and decreased feelings of loneliness (Bowers and MacDonald, 2001; Ewing *et al.*, 2007), problem solving and nurturing (Toukonen Cuffari, 2011), emotional regulation (Carlsson *et al.*, 2014), and improvements in mental health and psychosocial functioning (Hayden, 2005; Schultz *et al.*, 2007).

The EAL program was able to offer social wellbeing through interpersonal relationships, comprised foremost of developing relationships and getting along better with others, and secondary as acknowledging the importance of community, family and friends. Since the early 1990s, extensive medical literature has emerged confirming a strong, positive link between social support through involvement in community activities and improved human health and survival (Kruger and Serpell, 2006). Our own research in Ontario with First Nations youth participating in an EAL program and healing from VSM also identified the important relationship that develops between the youth and horse facilitators (Dell *et al.*, 2011). Although not EAL specific but related to equine assisted interventions, the literature also supports an increase in future community involvement with participants in therapeutic horseback riding programs (Iannone, 2003). Some equine assisted interventions have also demonstrated an increase in feelings of unconditional love and acceptance among participants (Iannone, 2003). One of the earliest animal assisted intervention studies by Cobb (as cited in Kruger and Serpell, 2006) referred to this as social support, that is, 'information leading the subject to believe he is cared for and loved, esteemed, and a member of a network of mutual obligations' (p. 27). A sense of belonging and community is key to the identity and wellbeing of First Nations youth in Canada (Honouring Our Strengths: Honouring our strengths: Indigenous culture as intervention in addictions treatment, 2014). We also know that peers are an important role in the initiation of VSM (Beauvais *et al.*, 2002).

Findings were also identified that are not specific to the EAL literature, but again, are important to youth and addictions treatment, including for youth who misuse volatile substances as well as link with the AAI literature generally. The key findings were communication (Fergus and Zimmerman, 2004; Botvin and

Griffin, 2004), impact of a new experience and behavior change (Rollnick and Miller, 2009; Bone *et al.*, 2011). Secondary were being a better team player (Roberts *et al.*, 2001) and a leader (Hawkins *et al.*, 2004). These findings likewise resonate with the equine-assisted interventions scholarship related to communication and mastery in a new experience (Toukonen Cuffari, 2011), and behavior change (Trotter *et al.*, 2008; Carlsson *et al.*, 2014).

One of the most potentially meaningful and unique findings of this research is the impact of the EAL program on spiritual wellbeing. There were three main findings that connect to the existing animal assisted intervention literature, and specifically with equines. The first is the human-equine relationship, and the merits of the youth just spending time with the horse. The horse-human relationship is based on the development of mutual respect and trust. In equine-guided activities, positive interactions with horses are most often observed when participants approach the horse, and thus the experience, from a stance of mutual respect and trust (Parelli and Parelli, 2012). The literature supports that opportunities to interact with animals provides a starting place to begin to explore and develop trust and a relationship with another living being (Latella, 2003; Wilkes, 2009). McNicholas and Collis (2006) explain, for example, that '[s]ocial signals from animals are less complex than from humans, and the reduced processing load may permit a greater degree of social understanding and social interaction than would be otherwise possible' (2006: 69). We know from trauma-related research that the '[r]estoration of the trauma victim's capacity for recovery hinges on provision of safety and development of trust...' (Yorke *et al.*, 2008, p. 22).This is of particular significance for First Nations in consideration of the traumatic effects of colonization on individuals, families and communities and resulting mistrust. We know that youth who misuse volatile substances are more likely to have histories of abuse than youth who abuse other substances (Sakai *et al.*, 2004).

Second, was the development of a relationship or special bond with the horse. According to the available literature, the bond between equine and human spirits is a 'holding environment' (Bachi, 2013b, p. 4) that allows for a bond to develop with animals and specifically with horses. Some refer to a meeting of spirits within this context. Individuals who have experienced spiritual trauma, for example, may find that being with animals provides an empathic space whereby 'having someone witness their pain [may] bring hope through responding therapeutically to their suffering [and] that clients can begin to heal' (Ashbrooke in Wilkes, 2009, p. 103). The work of Yorke *et al.* (2008) suggests that a strong intimacy/nurturing bond can form between humans and horses. They concluded that '...unique elements of equine-human relationships may foster deep, intimate connections' (pp. 18–19). Again, this is an important area of need for First Nations youth in treatment for VSM. A third finding in this study, which is identified in the addictions literature but not as a key finding in this study although present, was increased spiritual functioning (Arnold *et al.*, 2002). It is also identified as key to the healing of First Nations people (Honouring Our Strengths: Honouring our strengths: Indigenous culture as intervention in addictions treatment, 2014).

And last, to a less extent, the youth identified increased participation in cultural activities, and this is not identified in the existing literature. Within an Indigenous worldview and the treatment centre's culture-based model of resiliency, the inter-connectedness of all living things as a community of beings is viewed as essential to an individual's sense of wellbeing. Within some First Nations communities, the role of other beings, including horses, holds significance as a part of identity, culture and spirit. Since the introduction of the horse to some First Nations people and communities, it has had significant working (e.g. hunting and gathering) and ceremonial (e.g. traditional Horse Dance) roles in the culture and their lives. The horse has historically been viewed by some with a profound sacredness, just as there is sacredness believed to be in all living things (Bastien, 2004). The horse is identified as having a strong spiritual power, and is seen to be a 'teller of truth' who desires to 'do the right thing'. It is believed that a horse's spirit will lead individuals in the 'right direction' and will assist them in understanding their place in the circle of life (Dell *et al.*, 2008).

Today's global economy is placing unparalleled value on knowledge. Our current state of EAL and VSM knowledge, including AAIs and addictions generally, was supported. New understandings were also uncovered. The creation of new knowledge and its transference into practice is especially required in the substance abuse field (Simpson, 2002), and specific to VSM (Dell *et al.*, 2011). This is likewise true in the Equine Assisted Learning area (MacKinnon, 2007). Similarly, in the context of Indigenous health, unique consideration is being allotted to the development of understanding through multiple perspectives and its transference within specific social and political contexts (Estey *et al.*, 2007). By designing our study with these key points in mind, awareness of our findings is already being shared by our knowledge using and practice-oriented team members. Several of the key policy and practice implications they are focussing on include the need to:

- acknowledge the lack of research in EAL, and with youth who misuse volatile substances, and that the findings of this study show promise to study the area further;
- recognize that the horse and First Nations culture are historically linked, and that there is room for further understanding about this in the context of EAL;
- value a multi-disciplinary, community-based team approach to researching the diverse areas of EAL and VSM. A case-study design had not been conducted before and this study showed promise for its ability to offer a contextual understanding of EAL;
- practice traditional First Nations ceremony to ensure a 'good' and respectful research process. There is growing recognition of the importance of this, as was experienced in this study;
- formally acknowledge the impact of the EAL program facilitator, who up to this point in the literature, has not received due credit for their unique and essential role in assisting participants in EAL programs; and
- understand that there is significant variation in how EAL is applied across programs and therefore there is a need for further exploratory and evaluation studies designed specific to individual programs.

Limitations

There are two key limitations to the results of this study. First, the research findings are specific to White Buffalo and Cartier and it is difficult to transfer the data to additional cohorts. Further research is needed to establish the findings to be able to generalize beyond the studied cohorts. Related, a case-study design has important merits, and especially in an under-researched field, but it also has its limitations. The most significant is that it is not generalizable.

A second limitation of the present study relates to data collection. During the initial stages of the EAL program each cohort fluctuated due to drop out rates. Additionally, the youth had difficulty leaving their families and home community for treatment and their adaption to the treatment centre environment, staff, other youth, routines and their own detoxification upon entering treatment. These issues surfaced during the initial few weeks of treatment and therefore the data collected during this time period is limited. These limitations are important considerations when considering the implications of this study and for future research.

Conclusion

The linkage between what is commonly termed 'horse therapy' or equine assisted interventions and treating youth who misuse volatile substances has gained increased attention in Canada over the past decade. However, a large gap remains in the literature explaining the outcomes. Our team's study applied a multidisciplinary, community-based, exploratory case study approach to better understand how an EAL program contributes to the wellbeing of First Nations youth in treatment for volatile substance misuse. The findings suggest that the youth enhanced their bio-psycho-social-spiritual wellbeing as outlined in White Buffalo's framework for healing. The horses, facilitators and program content supported the youths' wellbeing in each area of the holistic framework, with emphasis on the cultural significance of the horse and its helping role. Understanding from the existing EAL literature was likewise supported in this study and new knowledge identified as well. The results of this study offer a meaningful understanding and contribution to the literature, as studies are rare in the EAL and VSM fields and youth addictions treatment generally. Further, our team's blending of traditional Indigenous knowledge and an academic approach has allowed us to premise our work on Indigenous ways of knowing. We believe this model can also help fill a gap in the literature and practice. And finally, as shared, the findings of this study offer important implications for both policy and practice.

Note

[1] The program is now offered in Alberta.

Acknowledgements

We would like to acknowledge the generosity of our project funder, the Alberta Centre for Child, Family & Community Research and participants at the White Buffalo Youth Inhalant Treatment Centre, Meegwich.

References

Albright, J., Lebovitz, B., Lipson, R. and Luft, J. (1999) Upper aero-digestive tract frostbite complicating volatile substance abuse. *International Journal of Pediatric Otorhinolaryngology* 49(1), 63–67.

Arnold, R., Avants, K., Margolin, A. and Marcotte, D. (2002) Patient attitudes concerning the inclusion of spirituality into addiction treatment. *Journal of Substance Abuse Treatment* 23, 319–326.

Bachi, K. (2013a) Equine-facilitated prison-based programs within the context of prison-based animal programs: State of the science review. *Journal of Offender Rehabilitation* 52(1), 46–74. DOI: 10.1080/10509674.2012.734371.

Bachi, K. (2013b) Application of attachment theory to equine-facilitated psychotherapy. *Journal of Contemporary Psychotherapy* 43(3), 187–190. DOI: 10.1007/s10879-013.

Ballard, M. (1998) Inhalant abuse: A call for attention. *Journal of Addictions and Offender Counselling* 19(1), 28–32.

Basile, R.B. (1997) The psychological effects of equine facilitated psychotherapy on behavior and self-esteem in children with attention deficit/hyperactivity disorder (ADHD). *Scientific Journal of Therapeutic Riding* 3, 10–15.

Bastien, B. (2004) *Blackfoot ways of knowing: The worldview of the Siksikaitsitapi*. University of Calgary Press, Calgary, AB.

Basu, D., Jhirwal, O., Singh, J., Kumar, S. and Mattoo, S. (2004) Inhalant abuse by adolescents: A new challenge for Indian physicians. *Indian Journal of Medical Sciences* 58(6), 245–249.

Baxter, P. and Jack, S. (2008) Qualitative case study methodology: Study design and implementation for novice researchers. *The Qualitative Report* 13(4), 544–559. Available at: http://www.nova.edu/ssss/QR/QR13-4/baxter.pdf

Beauvais, F., Wayman, J.C., Jumper-Thurman, P., Plested, B. and Helm, H. (2002) Inhalant abuse among American Indian, Mexican American, and non-Latino white adolescents. *The American Journal of Drug and Alcohol Abuse* 28(1), 171–187.

Best, D., Manning, V., Gossop, M., Witton, J., Floyd, K. *et al.* (2004) Adolescent psychological health problems and delinquency among volatile substance users in a school sample in South London. *Drugs: Education, Prevention and Policy* 11(6), 473–482.

Bizub, A., Joy, A. and Davidson, L. (2003) It's like being in another world": Demonstrating the benefits of therapeutic horseback riding for individuals with psychiatric disability. *Psychiatric Rehabilitation Journal* 26(4), 377–384.

Bone, R., Dell, C., Koskie, M., Kushniruk, M. and Shorting, C. (2011) The lived experience of volatile substance misuse: How support contributes to recovery and sustained wellbeing. *Substance Use & Misuse: An International Interdisciplinary Forum* 47(S1), 119–127.

Botvin, G. and Griffin, K. (2004) Life skills training: Empirical findings and future directions. *Journal of Primary Prevention* 25(2), 211–232.

Bowers, M. and MacDonald, P. (2001) The effectiveness of equine-facilitated psychotherapy with at-risk adolescents. *Journal of Psychology and the Behavioral Sciences* 15, 62–76.

Brooks, S.M. (2006) Animal-assisted psychotherapy and equine-facilitated psychotherapy. In: Webb, N.B. (ed.) *Helping Traumatized Youth in Child Welfare: Perspectives of Mental Health and Children's Services Practitioners*. Guildford Press, New York, NY, pp. 196–218.

Burgon, H. (2003) Case studies of adults receiving horse-riding therapy. *Anthrozoos* 16(30), 263–276.

Burgon, H. (2011) Queen of the world': Experiences of "at-risk" young people participating in equine-assisted learning/therapy. *Journal of Social Work Practice: Psychotherapeutic Approaches in Health, Welfare and the Community* 25(2), 165–183. DOI: 10.1080/02650533.2011.561304.

Canadian Institutes of Health Research (2010) Natural sciences and engineering research council of canada, and social sciences and humanities research council of Canada. In: *Tri-Council Policy Statement: Ethical Conduct for Research Involving Humans*. Canadian Institutes of Health Research, Ottawa, ON.

Carlsson, C., Ranta, D. and Traeen, B. (2014) Equine assisted social work as a mean for authentic relations between clients and staff. *Human-Animal Interaction Bulletin* 2(1), 19–38.

Chalmers, D. (2014) Fulfilling the fundamentals of life: A grounded theory of the process of human-horse relationship development, University of Calgary, Calgary, Canada.

Chalmers, D. and Dell, C. (2011) Equine-assisted therapy with first nations youth in residential treatment for volatile substance misuse: Building an empirical knowledge base. *Native Studies Review* 20(1), 59–87.

Chardonnens, E. (2009) The use of animals as co-therapists on a farm: The child-horse bond in person centered equine-assisted psychotherapy. *PCEP Journal of the World Association for Experiential and Person-Centered Experiential Psychotherapy and Counseling* 8(4), 319–332.

Creswell, J. (2013) *Qualitative inquiry and research design: Choosing among five approaches*. Sage, Thousand Oaks, CA.

Crofoot-Graham, T., Cellarius, K., Clothier, P., Moore, L. and Hawkins, J. (2001) *Transition programs in Indian Country*. Casey Family Programs, Seattle, WA.

Davis, S. and Reid, R. (1999) Practicing participatory research in American Indian communities. *The American Journal of Clinical Nutrition* 69(4), 7555–7595.

Deaton, D. (2005) Humanizing prisons with animals: A closer look at '"cell dogs"' and horse programs in correctional institutions. *Journal of Correctional Education* 56(1), 46–62.

Dell, C. and Beauchamp, T. (2006) *Youth Volatile Solvent Abuse: Frequently Asked Questions*. Canadian Centre on Substance Abuse, Ottawa.

Dell, C., Chalmers, D., Dell, D., Sauve, E. and MacKinnon, T. (2008) Horse as healer: Applying equine assisted learning to uncover and strengthen the spirit of first nations youth who abuse solvents. *Pimatisiwin: A Journal of Indigenous and Aboriginal Community Health* 6(1), 81–106.

Dell, C., Gust, S. and MacLean, S. (2011) Global issues in volatile substance misuse. *Substance Use & Misuse: An International Interdisciplinary Forum* 47(S1), 1–7.

Duncan, R., Critchley, S. and Marland, J. (2014) Can Praxis: A model of equine assisted learning (EAL) for PTSD. *Canadian Military Journal* 14(2). Available at: http://www.journal.forces.gc.ca/vol14/no2/page64-eng.asp

Equine Assisted Growth and Learning Association (2010). Available at: http://www.eagala.org

Estey, A.E., Kmetic, A.M. and Reading, J.L. (2007) Innovative approaches in public health research: Applying life course epidemiology to aboriginal health research. *Canadian Journal of Public Health* 98(6), 444.

Ewing, C., MacDonald, P., Taylor, M. and Bowers, M. (2007) Equine-facilitated learning in youths with severe emotional disorders: A quantitative and qualitative study. *Child & Youth Care Forum* 36, 59–72.

Feh, C. (2005) Relationships and communication in socially natural horse herds. In: Mills, D. and McDonnell, S. (eds) *The Domestic Horse: The Origins, Development and Management of Its Behaviour*. Cambridge University Press, Cambridge, UK, pp. 83–93.

Fergus, S. and Zimmerman, Z. (2004) Adolescent resilience: A framework for understanding healthy development in the face of risk. *Annual Review of Public Health* 26, 399–419.

Foley, A.J. (2007) A theoretical and evaluative study of an equine-assisted psychotherapy program for delinquent girls. In: *Annual Meeting of the American Society of Criminology*.

Frewin, K. and Gardiner, B. (2005) New age or old sage? A review of equine assisted psychotherapy. *The Australian Journal of Counselling Psychology* 6, 13–17.

Friedmann, E., Son, H. and Tsai, C. (2010) The human/animal bond: Health and wellness. In: Fine, A. (ed.) *Handbook on Animal-Assisted Therapy: Theoretical Foundations and Guidelines for Practice*. San Diego, CA: Elsevier, pp. 85–107.

Gonzalez, G. and Sutton, P. (2011) Horse and human personalities: A deep interaction at work. In: Leffhalm, J. (ed.) *Horses: Biology, Domestication, and Human Interactions*. Nova Science Publishers, Inc, New York, NY, pp. 121–129.

Goodwin, D. (2002) Horse behavior: Evolution, domestication and feralisation. In: Waran, N. (ed.) *The Welfare of Horses*. Kluwer Academic Publishers, Secaucus, NJ, pp. 1–18.

Harper, S. (2008) Prime minister offers full apology on behalf of Canadians for the Indian residential school system. *Ottawa*. Available at: http://www.pm.gc.ca/eng/news/2008/06/11/prime-minister-harper-offers-full-apology-behalf-canadians-indian-residential

Hawkins, E., Cummins, H. and Marlatt, G. (2004) Preventing substance abuse in American Indian and Alaska Native youth: Promising strategies for healthier communities. *Psychological Bulletin* 130(2), 304–323.

Hayden, A. (2005) An exploration of the experiences of adolescents who participated in equine facilitated psychotherapy: A resiliency perspective, Alliant International University, San Diego.

Held, C.A. (2006) Unpublished doctoral dissertation. In: *Horse Girl: An Archetypal Study of Women, Horses, and Trauma Healing*. Pacifica Graduate Institute, California.

Hodkinson, P. and Hodkinson, H. (2001) The strengths and limitations of case study research. In: *Paper presented to the "Learning and Skills Development Agency conference Making an Impact on Policy and Practice", Cambridge*, pp. 5–7.

Honouring our strengths: Indigenous culture as intervention in addictions treatment (2014) Indigenous wellness framework and outcomes. Saskatoon: University of Saskatchewan. Canadian Institutes of Health Research. Funding Reference Number AHI-120535.

Iannone, V. (2003) Evaluation of a vocational and therapeutic riding program for severely emotionally disturbed adolescents, The Catholic University of America, Washington.

Jarrell, N. (2009) A healing triangle: Clients learn much about themselves through equine-assisted therapy. *Addiction Professional* 7(1), 15–20.

Kahn, J.R.P. (1997) Developmental psychology and the biophilia hypothesis: Children's affiliation with nature. *Developmental Review* 17, 1–61.

Kakacek, S. and Ottens, A. (2008) An arena for success: Exploring equine-assisted psychotherapy. *Michigan Journal of Counseling* 35(1), 14–23.

Kerston, G. and Thomas, L. (1997) Straight from the horse's mouth: The truth about equine-assisted therapy. *The Counsellor* 25, 18–24.

Kirby, S., Greaves, L. and Reid, C. (2006) *Experience research social change: Methods beyond the mainstream*. University of Toronto Press, Ontario.

Kruger, K. and Serpell, J. (2006) Animal-assisted interventions in mental health: Definitions and theoretical foundations. In: *Handbook on Animal-Assisted Therapy: Theoretical Foundations and Guidelines for Practice*. San Diego, CA: Academic Press, pp. 33–48.

Laframboise, S. and Sherbina, K. (2008) The Medicine Wheel. Dancing to Eagle Spirit Society. Available at: http://www.dancingtoeaglespiritsociety.org/medwheel.php

Latella, D. (2003) *Animals as A Therapeutic Modality: A Curriculum Model for Occupational Therapy*. University of Bridgeport, Connecticut.

Lawrence, E.A. (1998) Human and horse medicine among some native American groups. *Agriculture and Human Values* 15, 133–138.

Lawrence, E.A. (1983) The human-horse bond in crow Indian life and culture. In: Katcher, A. and Beck, A. (eds) *New Perspectives on Our Lives with Companion Animals*. University of Pennsylvania Press, Philadelphia, PA, pp. 82–86.

Lubman, D.I., Yucel, M. and Lawrence, A.J. (2008) Inhalant abuse among adolescents: Neurobiological considerations. *British Journal of Pharmacology* 154(2), 316–326.

MacKinnon, T. (2007) At the Heart of Equine-Assisted Learning. M.E.S. Ltd. North American Riding for the Handicapped Association, Inc. Available at: http://www.narha.org

Martial, R., Macaulay, A. and Freeman, W. (2003) *A conversation: Communities and cathedrals. Pimatiswin: A Journal of Aboriginal and Indigenous Community Health*. Alberta ACADRE Network, Alberta.

McDonnell, S. (2003) *A practical guide to horse behavior: The equid ethogram*. The Blood-Horse, Inc, Lexington, KY.

McNicholas, J. and Collis, G. (2006) Animals as social supports: Insights for understanding animal-assisted therapy. In: *Handbook on Animal-Assisted Therapy: Theoretical Foundations and Guidelines for Practice*. pp. 49–71.

Melson, G. (2000) Companion animals and the development of children: Implications of the biophilia hypothesis. In: Fine, A. (ed.) *Handbook on Animal-Assisted Therapy: Theoretical Foundations and Guidelines for Practice*. Academic Press, San Diego, CA, pp. 375–383.

Melson, G. (2008) Children in the living world: Why animals matter for children's development. In: Fogel, A., King, B. and Shanker, S. (eds) *Human Development in the Twenty-First Century. Visionary Ideas from Systems Scientists*. Cambridge University Press, New York, NY, pp. 147–154.

Merriam, S. (1998) *Qualitative research and case study applications in education*. Jossey-Bass Publishers, San Francisco, CA.

National Drug Abuse Information Centre (1998) Deaths due to volatile solvent abuse. In: *Stat Update Number 8*.

Nimer, J. and Lundahl, B. (2007) Animal-assisted therapy: A meta-analysis. *Anthrozoos* 20(3), 225–238.

Parelli, P. and Parelli, L. (2012) Looking at equine assisted counseling from the horse's perspective.

Pollack, S. (2009) *Equine-assisted psychotherapy with women with addictions*. Wilfred Laurier University, Ottawa, ON.

Poole, N. and Dell, C. (2006) *Girls, women and substance use*. Canadian Centre on Substance Abuse, Ottawa.

Roberts, G., McCall, D., Stevens-Lavigne, A., Anderson, J., Paglia, A. *et al*. (2001) *Preventing substance use problems among young people: A compendium of best practices*. Health Canada, Ottawa.

Robin, M. and Bensel, R. (1985) Pets and the socialization of children. *Marriage and Family Review* 8(71), 63–78.

Rollnick, S. and Miller, W. (2009) What is motivational interviewing. *Behavioural and Cognitive Psychotherapy* 23(4), 325–334.

Sakai, J.T., Hall, S.K., Mikulish-Gilbertson, S.K. and Crowley, T.J. (2004) Inhalant use, abuse, and dependence among adolescent patients: Commonly comorbid problems. *Journal of American Academy of Child & Adolescent Psychiatry* 43(9), 1080–1088.

Schultz, B. (2005) The effects of equine-assisted psychotherapy on the psychosocial functioning of at-risk adolescents ages 12-18, Denver Seminary, Denver.

Schultz, P., Barlow, N., Remick, A. and Robbins, L. (2007) Equine-facilitated psychotherapy: A mental health promotion/intervention modality for children who have experienced intra-family violence. *Health and Social Care in the Community* 15(3), 265–271.

Scott, K. (1992) *Indigenous Canadians. A substance use background paper*. Prepared for the National Native Alcohol and Drug Abuse Program.

Simpson, D. (2002) A conceptual framework for transferring research to practice. *Journal of Substance Abuse Treatment* 22(4), 171–182.

Smith, L. (1999) *Decolonizing methodologies: Research and Indigenous peoples*. St Martin's Press, New York.

Stake, R. (1995) *The art of case study research*. Sage, Thousand Oaks.

Statistics Canada (2011) Peoples in Canada: First Nations People, Métis and Inuit. National Household Survey, 2011. Catalogue no. 99-011-X2011001. Available at: http://www12.statcan.gc.ca/nhs-enm/2011/as-sa/99-011-x/99-011-x2011001-eng.pdf

Toukonen Cuffari, M. (2011) The relationship between adolescent girls and horses: Implications for equine-assisted therapies, Kent State University, Cleveland, OH.

Trotter, K., Chandler, C., Goodwin-Bond, D. and Casey, J. (2008) A comparative study of the efficacy of group equine assisted counseling with at-risk children and adolescents. *Journal of Creativity in Mental Health* 3(3), 254–284.

Trotter, K. (ed.) (n.d.) *Harnessing the power of Equine Assisted Counseling: Adding Animal Assisted Therapy to your practice*. Routledge, New York, pp. 17–26.

Wilkes, J. (2009) *The role of companion animals in counseling and psychology: Discovering their use in therapeutic process*. Charles C Thomas Publisher, Ltd, Springfield, PA.

Wilson, E.O. (1984) *Biophilia: The human bond with other species*. Harvard University Press, Cambridge, MA.

Yin, R. (2003) *Applications of case study research*, 2nd ed. Sage, Thousand Oaks, CA.

Yorke, J., Adams, C. and Coady, N. (2008) Therapeutic value of equine-human bonding in recovery from trauma. *Anthrozoos* 21(10), 17–30.

Youth Solvent Addiction Committee (YSAC) (2008) YSAC Annual Report 2008-2009. Saskatoon, SK: YSAC.

11 Psychotherapy Incorporating Horses to Address Anxiety in Women College Students: A Pilot Study

Erika L. Berg[1*], Amber R. Bach-Gorman[2] and Carolyn J. Hammer[1]
[1]Department of Animal Sciences, North Dakota State University; [2]Counseling Center, North Dakota State University

Abstract

An annual 2019 survey of directors of college counseling centers reported that 60.7% of students seeking counseling present with the predominant concern of anxiety. Furthermore, counseling centers across the United States have seen a rise in the number of students requesting services in recent years, frequently resulting in wait lists for students to receive counseling. Group psychotherapy sessions may mitigate wait times; however due to student anxiety and negative stigma surrounding mental health treatment it can often be challenging to attract students to participate in groups compared to individual sessions. This pilot study aimed to investigate the efficacy of group psychotherapy incorporating equines for undergraduate women college students presenting with concerns of anxiety. Implementing the Equine Assisted Growth and Learning Association's model of psychotherapy, six women participants attended one 90-minute group session per week for 7 weeks on the campus of North Dakota State University. Students completed the Self Compassion Scale (SCS) and the Depression Anxiety Stress Scales (DASS) at weeks 0 and 7 as pre and post measures of the intervention. A significant increase was observed in the common humanity subscale of the SCS; however, no other differences were found in the remaining SCS subscales nor in the DASS subscales. Anecdotally, multiple students expressed that they would not have persisted in coming to group without the horses present. Furthermore, there was markedly improved attendance rates in the psychotherapy incorporating equine group compared to other group therapies at the counseling center. We recommend that future research examine whether psychotherapy integrating both horses and the tenets of self-compassion can help woman college students improve their anxiety management skills and become more resilient.

Keywords: psychotherapy, equines, anxiety, college counseling, undergraduate students

Mental Health on College Campuses

College counseling centers have been experiencing a steady increase in the utilization of counseling services for the past decade (Reetz *et al.*, 2015). Of the numerous mental health concerns of college students, those presenting with Generalized Anxiety Disorder (GAD) continue to rise to the top. GAD is characterized by excessive worry that is difficult to control and significantly impacts daily functioning (American Psychiatric Association, 2013). Nordberg *et al.* (2013) found that college students who report distress, including GAD, are more likely to seek counseling services. In its 2019 annual survey, the Association for University and College

*Corresponding author: erika.berg@ndsu.edu

Counseling Center Directors (AUCCCD) found that 10% of university students nation-wide seek counseling services and that women are twice as likely as men to seek services. Anxiety has risen steadily as the predominant concern among those students presenting to college counseling centers, increasing 23.9% from 2009 to 2019 (Association of University and College Counseling Center Directors (AUCCCD), 2010; Association of University and College Counseling Center Directors (AUCCCD), 2019; Association of University and College Counseling Center Directors (AUCCCD) (2019)). This increase is consistent with national and global data, which suggest that anxiety disorders comprise the most prevalent mental health diagnoses (Bandelow and Michaelis, 2015). Furthermore, the tremendous economic cost of anxiety disorders in the United States is estimated to be upwards of $42 billion annually (Kessler and Greenberg, 2002).

In addition to students presenting with GAD, 17.5% of all students also indicate feelings of social isolation and distress associated with loneliness. Pascarella and Terenzini (2005) discussed that students coping with internal stressors associated with mental health conditions such as anxiety, social anxiety and depression may be particularly vulnerable to adjustment problems that compromise their longevity at institutions. Women in college report a higher incidence of mental health problems than men (Eisenberg *et al.*, 2009). Social anxiety and other extenuating impacts of GAD have been shown to predict academic problems and to increase risks that students will not attach to their academic institutions, which may negatively impact their eventual retention and academic success (Strahan, 2003; Nordstrom *et al.*, 2014). Moreover, students with mental illness are less engaged in college, report poorer relationships, and have lower graduation rates than students who do not identify as having a mental illness (Salzer, 2012). Thus, there is an imperative need to find comprehensive treatments that support college students who struggle with mental illness.

Despite the increased number of students seeking counseling services, many college mental health centers lack the staff necessary to successfully manage students' needs (Kraft, 2011). College counseling centers continue to have difficulty managing the influx of clients, and 34% of directors have reported using waitlists to manage client flow (Association of University and College Counseling Center Directors (AUCCCD), 2018).

Consequently, establishing effective therapeutic interventions that provide for a more efficient provision of mental health services is desperately needed.

Mental health professionals use group therapy as an approach to treating multiple students simultaneously. Group therapy has also been found to be an effective intervention for GAD and social anxiety (Yalom and Leszcz, 2005). Bernhardsdottir and Vilhjalmsson (2013) reported that psychologically distressed women university students reported 'feeling more self-confident and in control' after taking part in cognitive behavioral group therapy (CBT). Van Der Kolk (2014) discussed the importance of varying interventions that include creative and experiential therapies for individuals who may not be able to process emotions like anxiety, negative thought patterns, and adverse experiences through simple language due to challenges such as trauma. Furthermore, there is very little research focusing on the impacts of group therapy targeting anxiety symptoms for college students, and no current literature exploring group therapy that incorporates equines with this population.

One potential insulator against debilitating anxiety for college students is self-compassion. Self-compassion, as defined by Neff (2003a), refers to having a warm and accepting stance towards aspects of oneself that are disliked or painful. Research has demonstrated a correlation between higher levels of self-compassion with reports of greater life satisfaction, social connectedness, as well as reductions in self-criticism, anxiety and depression. Werner *et al.* (2012) report findings that individuals with social anxiety disorder report lower levels of overall self-compassion, potentially contributing to reduced academic success and difficulty establishing meaningful relationships in higher education (Nordstrom *et al.*, 2014).

Psychotherapy Incorporating Equines

Evidence suggests that some types of animal-assisted interventions may benefit adults physically and psychologically, and help to reduce stress, anxiety, and depression (Fine, 2015). Psychotherapy incorporating equines is a form of experiential therapy that is appropriate for individuals or groups (Notgrass and Pettinelli, 2014). Research investigating benefits of psychotherapy incorporating equines have cited increases in self-confidence, communication, and problem solving for individuals of varying ages (Kersten and Thomas, 2004). Trotter *et al.* (2008) reported that adults and children participating in psychotherapy incorporating equines had lower anxiety scores, as well as reduced anxiety and depressive symptoms associated with trauma-related stress. Additionally, individuals have been found to focus less on fear and guilt and more on independence and self-support when participating in psychotherapy incorporating equines (Klontz *et al.*, 2007). Participation in psychotherapy incorporating equines has also been found to alleviate some post-traumatic stress disorder symptoms in maltreated youth (McCullough *et al.*, 2015) and veterans (Burton *et al.*, 2019).

The Equine Assisted Growth and Learning Association Model

The Equine Assisted Growth and Learning Association (Equine Assisted Growth and Learning Association (EAGALA), 2019) has developed a model of psychotherapy to provide a structural framework for conducting psychotherapy sessions incorporating equines that includes four key components. The first key component stipulates a collaborative, team approach. More exactly, it is expected that a licensed mental health professional and qualified equine specialist, both of whom are trained in the EAGALA model, work in partnership with each other, equines and clients. A second key component is the solution-focused, experiential nature of the therapy sessions that empowers clients to create the best solutions for themselves when given the opportunity to discover them. The third key component is that the sessions are entirely ground-based with no mounted horseback riding involved, allowing horses the freedom to interact with clients as they choose. The fourth key component is that mental health and equine professionals, who are certified in the EAGALA model, follow a strict code of ethics to ensure best practices and adherence to the highest industry standards. The EAGALA Code Of Ethics includes twelve principles designed to uphold the safety and well-being of clients and horses; maintain professional integrity; demonstrate honesty and respect among colleagues; and seek ongoing education and training of best industry practices.

The EAGALA model invites clients to engage in a process with the horses and the practitioners are there to observe (not interpret) and provide feedback to clients based on 'SPUD'S' (Equine Assisted Growth and Learning Association (EAGALA), 2019). SPUD'S—an abbreviation that stands for Shift, Pattern, Unique, Discrepancy, and Self—is the framework for the experiential learning that takes place in sessions following the EAGALA Model. Morgan (2017) reported that equine-assisted activity sessions following the SPUD'S framework of the EAGALA model were as effective as yoga and mindfulness in reducing stress in undergraduate students. Limited research exists on the implementation of psychotherapy incorporating equines – and more specifically the EAGALA model – to address mental health needs of students in higher education. Consequently, the purpose of the present study was to determine whether participants experienced a shift in self-compassion or anxiety levels after once weekly 90-min group psychotherapy incorporating equines sessions for 7 weeks.

Method

General Research Approach

This pilot study used a one group, pre-and post-test design.

Participants

Human subjects' participation was approved by North Dakota State University's Institutional Review Board (protocol #AG17234). Participants for this pilot study were recruited through the NDSU Counseling Center. To be considered for the study, individuals must have been: 1) a full-time woman undergraduate student between the ages of 18 and 24 years; 2) a current NDSU Counseling

Center client presenting with the predominant concern of anxiety; and 3) available for the scheduled time of psychotherapy that would incorporate equines. Potential participants self-selected by either responding to a general 'all-student' email informing them of the group opportunity or through a counselor at the Center who informed them of the group opportunity. They were then referred to the study investigators for additional information about the group opportunity. A total of seven individuals met the criteria and were recruited for study participation. Potential participants met with the investigating mental health professional in a private room at the counseling center where a hard copy of the informed consent document was reviewed. Potential subjects had the opportunity to ask questions about the study at that time. Additionally, the option to participate in the psychotherapy incorporating equines sessions without use of individual data for the study was given as an alternative; however, no one selected this option. All individuals signed the consent document and were sent home with a copy for their information.

Practitioners

The same mental health professional (MHP) and equine specialist (ES) facilitated all sessions with equines and clients for this study. Both are EAGALA certified professionals and have participated in a PATH Intl. Equine Specialist in Mental Health and Learning workshop (2011). In addition, the MHP has 20 years of clinical counseling experience; 15 of which she has spent in collegiate mental health. She has a PhD in Counselor Education with her dissertation focused on canine-assisted therapy in counseling for youth in a residential childcare facility. She has attended numerous AAI continuing education opportunities hosted by the University of Denver School of Social Work and the International Institute of Play Therapy. In addition, she has served as a volunteer handler and therapy team evaluator for Pet Partners and was recently invited to participate as a content expert on a panel to create AAI credentialing for Pet Partners. The equine specialist is a veterinarian and equine science program director at a land-grant institution and has over 35 years horse experience in management, horsemanship, behavior and veterinary care. She has also participated in Eponaquest trainings, teaches an equine guided leadership course based on the teachings of Linda Kohanev (2015) , and actively pursues continuing education in the field of EAS.

Equines and facilities

Daily Management

The equine partners in this study were protected under North Dakota State University's Institutional Animal Care and Use Committee (protocol #A20031). A total of five program horses met the criteria for working in the psychotherapy incorporating equines setting. These criteria included soundness; responsiveness to human movement and noises; tolerant of being handled by different individuals of varying skill levels; willing to load; traveling and unloading in a trailer; desensitized to a multitude of props used during sessions, not displaying fear or repeated distress during sessions of psychotherapy that incorporated horses; and being able to work in a herd situation.

All daily management practices for horse care fell within the recommendations published in the *Guide for the Care and Use of Agricultural Animals in Research and Teaching* (Federation of Animal Science Societies, 2010). Horses were maintained at the university equine center as a herd, observed daily for injury or illness and turned out together from 0700 and 1630 daily in a 24m by 122m dry lot enclosed with a 1.5m high pipe fence. Horses had ad libitum access to fresh water, trace mineral and salt blocks, and a round bale of mixed brome, timothy and orchardgrass hay. Because the dry lot has no shelter, horses were kept inside during inclement weather and turned out in groups of three to four for free exercise in the 30m by 36m indoor arena enclosed with a 1.7m high pipe fence. Between 1630 and 0700, horses were maintained in individual 3m by 3m steel stalls with pine shavings that were cleaned daily. They were fed a.m. and p.m. grain meals (sweet feed or senior feed) with access to fresh water and grass and/or alfalfa hay to meet their dietary requirements. Horses were groomed and exercised one to three times each week, depending on individual needs, and more frequently when programming was in session. They were vaccinated for rabies, tetanus, West Nile Virus, eastern and western encephalitis, rhinopneumonitis, and influenza; de-wormed every 12 weeks; had their teeth floated annually at a minimum and more frequently if needed; and hooves trimmed every 6 to 8 weeks and were shod if required. In addition, horses received alternative treatments (i.e. chiropractic, acupuncture or massage), as well as joint medications or injections if indicated.

Psychotherapy sessions incorporating equines

All sessions were conducted within the framework of the American Counseling Association's Animal-Assisted Therapy in Counseling Competencies of knowledge, skills and attitudes (Stewart *et al.*, 2016). Each domain includes competency subthemes which are described hereafter. 'Knowledge' specifies that a practitioner has formal training in animal-assisted therapy in counseling (AATC), in-depth animal knowledge and knowledge of existing ethical requirements. 'Skills' refers to mastery of basic counseling skills, intentionality of including animals in a treatment session, and having a specialized skill set with regard to including animals in treatment sessions. 'Attitudes' includes animal advocacy, professional development in AATC, and demonstration of professional values.

Psychotherapy sessions incorporating equines took place from 0930 to 1100 in the campus livestock pavilion, which is within walking distance to the dorms and easily accessible via public transportation. The pavilion arena is located behind doors that can be closed during sessions to help ensure client privacy. No more than three horses were transported 6.5 km (one-way) via a five-horse slant load trailer from the university equine center to a campus livestock pavilion for weekly psychotherapy sessions incorporating equines. All horses were familiar with the hauling routine and the facility. To allow horses the opportunity to re-familiarize themselves with the arena space, they were turned out for 60–90 min once a week for 2 weeks prior to sessions beginning. During psychotherapy sessions incorporating equines weeks, horses arrived at the pavilion arena approximately 45 min prior to the start of the session and were hauled back to the university equine center 30 min after the session ended. The pavilion arena is 13.7m by 27.4m with dirt footing that was watered as needed to keep dust levels down. The arena perimeter was a solid 1.5m concrete wall with an additional 0.9m of pipe fencing on top of the wall. Horses had access to clean, fresh water at all times in the arena.

Horses were observed for signs of distress before, during and after sessions. Potential distressed or negative behaviors that have been identified in our horse herd include bite, bite threat, kick, kick threat, lip lick/chew (indicative of return to a calmer parasympathetic state from a distressed sympathetic state), ears pinned, head shake, head toss, startle, and eliminative behaviors. Positive observed behaviors include mutual grooming, play behavior, investigative behavior, resting/relaxed body posture, and ingestive behaviors. Weekly horse selection was based on the planned activity for the day (e.g. some horses may be more or less willing to participate in a given activity so this was considered when deciding which horses to bring), in addition to how the horse(s) responded during the session the week prior and how they were behaving the day of sessions. For example, it became apparent over two sessions that one of the mares and one of the clients did not work well together. The mare consistently pinned her ears each time this particular client would approach or put her hands on the horse. For the safety of the client and the well-being of the mare, the decision was made to no longer include that mare in sessions as she was communicating a significant level of discomfort.

Instruments

Self-compassion scale

Self-compassion entails three basic components: self-kindness, common humanity, and mindfulness (Neff, 2003a). The SCS is a 26-item self-report questionnaire that aims to measure the construct of self-compassion. The test scale has been validated and can significantly predict mental health outcomes as measured by Beck *et al.* (1961) and Spielberger *et al.* (1983). High scores of self-compassion are strongly correlated with positive mental health outcomes; specifically, reductions in depression and anxiety (Neff, 2003b).

Depression Anxiety Stress Scales

The DASS was developed to facilitate defining, understanding, and measuring the ubiquitous and clinically significant emotional states usually described as depression, anxiety and stress (retrieved from www2.psy.unsw.edu.au/groups/dass/, October 28, 2014). It is a 42-item test that consists of a set of three self-report scales designed to measure the negative emotional states of depression, anxiety and stress. This instrument shows excellent internal consistency and high test-retest stability of scores (Page *et al.*, 2007).

Data collection and analysis

Participants were individually invited to a private room at the university counseling center to complete both the DASS and SCS with pen and paper. Each instrument was estimated to take between 5 and 10 min to complete. Participants took no longer than 20 min to complete both assessments. The DASS and SCS were completed at weeks 0 and 7 and paired sample t-tests were used to evaluate differences between pre and post intervention responses. Client responses were kept in a locked file cabinet in the MHP's office at the counseling center.

Intervention

Each week participants attended one 90 min group psychotherapy session incorporating the EAGALA model for 7 weeks in the livestock pavilion on the NDSU campus. The same counselor and equine professional who are both certified in the EAGALA model facilitated all seven sessions. Three horses were selected each week from a consistent herd of five equine partners. Weekly EAGALA sessions incorporated interventions modified from publications by DePrekel and Neznik (2015b) and DiGiacomo and Sierra Boys Ranch (n.d.) (Table 11.1) and are explained in greater detail below.

Table 11.1. Weekly intervention activities in group psychotherapy incorporating equines sessions

Session	Activity
1	Observing the Herd
2	Window of Tolerance
3	Collaborative Obstacle Build
4	Course of Life
5	Outside the Box
6	Forward Motion
7	Paint Your Story

Note: These intervention activities were modified from publications by DePrekel and Neznik (2015a and 2015b) and DiGiacomo and Sierra Boys Ranch (n.d.)

All sessions began with circling up in the arena to reflect on the week and past sessions (with the exception of week one) with the horses at liberty in the same space. Sessions were structured to provide experiences incorporating equines that would allow clients to experience and learn about the three major constructs of self-compassion (mindfulness, common humanity, and self-kindness) and additional anxiety management tools including: increased personal awareness (mindfulness), emotional regulation and distress tolerance (self-kindness), psychological flexibility, behavioral activation and universality (common humanity). Each session concluded with circling up and processing participants reflections, observations and insights and time for group members to say good-bye to their equine group members. The names and brief descriptions of each sessions were as follows:

1. Observing the Herd. The first session began with establishing group norms (i.e. attendance, confidentiality), as well as observing herd interactions and clients introducing themselves to the horses. Post-intervention process discussions included clients' sharing observations of the horses and themselves, frequently identifying mutual themes and

building a sense of common humanity within group members.

2. Window of Tolerance. The second session introduced the concept of Window of Tolerance (WOT) and identifying individual sensations that correlate with shifts between hyper and hypo arousal states when confronted with daily stress and anxiety. Clients moved horses around the arena to explore the idea of recognizing, understanding or expanding one's WOT. Post-intervention processing included discussions regarding how clients perceived stress in the equines and how both the clients and the equines mutually cared for one another as well as themselves illustrating self-kindness.
3. Collaborative Obstacle Build. This session engaged the group in a collaborative activity building on the concept of universality (something shared by all people) and common humanity (the idea that suffering is a shared human experience) (Neff, 2003b; Yalom and Leszcz, 2005). Participants were instructed to build obstacles that were representative of their collaborative anxiety story that they managed in their everyday college experience and to move their equine group members through the obstacles.
4. Course of Life focused on creating experiences for clients to increase personal mindfulness of their individual anxiety experiences. Individually, clients constructed their personal anxiety story utilizing available props and then partnered with an equine group member of their choice to walk through their story while sharing what each obstacle represented to their equine partner. Each participant could choose to share with the larger group if they wished.
5. Out of the Box created an opportunity for psychological flexibility, cognitive re-framing, and emotional regulation development. Each group member took a turn at being the 'leader' and instructed the other group members to move their equine partners into a boundary and keep them there. Only the acting group leader was allowed to speak and offer instructions. The exercise promoted psychological flexibility in allowing each client to 'lead' in a unique way and demonstrate alternate approaches to accomplishing a common task. Additionally, the Out of the Box exercise allowed for group members to engage in the role of 'leader' which many group members shared was unfamiliar and anxiety provoking.
6. Forward Motion targeted self-compassion's concepts of common humanity and mindfulness, in addition to anxiety management tools of behavioral activation. Group members were asked to consider what their anxiety story would look like as they completed the group and moved into their future. Utilizing available props, participants were asked as a group to create their future relationship with anxiety and then to partner with the equines and move into their future.
7. Paint Your Story. This final session allowed group members to self-select their equine partner and using non-toxic and washable finger paint to create their story for their equine partner to wear, which represented their shift from the first group session to the last group session.

Results

During the research period, a total of seven clients were recruited; however, one did not complete the study. Of the remaining participants, all six completed the DASS (n=6), but only five completed the SCS (n=5). There was a significant increase observed between pre (M = 2.15, SD = 0.45) and post-test scores (M = 2.80, SD = 0.41) in the common humanity subscale of the SCS; t(4) = −2.80, p=0.05 with an effect size of −1.50. No significant differences (p>0.05) were detected in the remaining subscales of the SCS (Table 11.2); however, increasing (desired) shifts were observed in self-kindness and total SCS subscales and decreasing (also desired) shifts were observed in self-judgement, isolation, overidentification. The mindfulness subscale remained equal from pre to post tests. No significant differences were observed in the DASS (Table 11.3); however, depression, anxiety, stress subscales did shift, but in the increasing (undesired) direction.

Discussion

The purpose of the present study was to determine whether a psychotherapy incorporating equines group would yield significant

Table 11.2. Results of Pre and Post-test Self-Compassion Scale Scores

SCS Subscales	Pre-test		Post-test		$t(4)$	p	Cohen's d
	M	SD	M	SD			
Self-kindness	2.24	0.95	2.72	1.05	−1.29	0.27	−0.48
Common humanity	2.15	0.45	2.80	0.41	−2.80	0.05	−1.50
Mindfulness	3.25	0.56	3.25	0.85	0	1	0
Self-judgement	4.12	0.18	4.01	0.58	0.37	0.73	0.27
Isolation	3.70	1.16	3.15	0.74	2.27	0.09	0.56
Overidentification	3.75	0.90	3.65	0.86	0.23	0.83	0.11
Total	2.37	0.63	2.67	0.58	−1.24	0.28	−0.50

Paired sample t-tests were used to determine differences in pre- and post-test scores of individuals participating in weekly 90-min psychotherapy incorporating equines group sessions for 7 weeks.

Table 11.3. Results of Pre and Post-test Depression Anxiety Stress Scales Scores

DASS Subscales	Pre-test		Post-test		$t(5)$	p	Cohen's d
	M	SD	M	SD			
Depression	16.33	14.00	17.33	14.62	−0.13	0.90	−0.07
Anxiety	17.33	11.50	19.67	8.89	−1.61	0.17	−0.23
Stress	21.33	9.93	24.67	11.78	−1.01	0.36	−0.31

Paired sample t-tests were used to determine differences in pre- and post-test scores of individuals participating in weekly 90-min psychotherapy incorporating equines group sessions for 7 weeks.

improvements in self-compassion and anxiety for participants as measured by the SCS and DASS, respectively. We found a significant improvement in only one subscale of the SCS, that of common humanity. Neff (2003a) defines common humanity as 'perceiving one's experiences as part of the larger human experience rather than seeing them as separating and isolating.' The idea of common humanity is tied into a sense of belonging (Tajfel and Turner, 1979), which has been shown to have protective factors against stressors (Dreisoerner *et al.*, 2020). Additionally, 'belonging' has been linked to improved student retention and success in higher education (Russell and Jarvis, 2019), important factors of sustainability for the learners and institutions.

Statistically significant shifts were not observed in 6 out of the 7 SCS subscales, however most subscale scores increased in desired directions except the mindfulness subscale. Research participants in the current study also shared during group processing that interacting with the horses lent them to be more mindful of what was occurring for them intrapersonally. Participants described an increased connection between cognitive awareness and how those cognitions related bodily sensations and experiences. The researchers posit the increased shift in participants' mindfulness functions to increase their ability to name, recognize and acknowledge feelings of anxiety. The increased personal awareness of their present states is an essential first step for implementing anxiety management skills. Furthermore, the psychotherapy incorporating equines group enabled those clients with some pre-existing fear or anxiety about horses to work through those feelings in a safe space.

Conversely, overall mean scores for the DASS indicated a slight, albeit insignificant, upward shift with an overall change of +1.0 from pre to post indicating higher levels of perceived depression, anxiety and stress for participants. The DASS subscale for anxiety showed the greatest shift with a variance of +2.3. This shift could be related to a number of confounding variables that were not considered such as client's increases in awareness of their intrapersonal states, state of being at the time of pre and post measures (beginning of the semester compared to the end), and the common occurrence in mental health treatment of decreased dysregulation as one becomes more emotionally attuned (Yalom and Leszcz, 2005; Van Der Kolk, 2014).

The lack of significant differences may also be due to a number of limitations within the study, the first being the duration of the pilot project. According to the American Psychological Association (2020), improvements for some individuals may occur in six to twelve therapy sessions, whereas Howard *et al.* (1986) found that 50% of psychotherapy clients improved after eight sessions, while 75% improved after six months. The intervention that we studied only had seven sessions and so was at the cusp of possibility for detectable change. The number of sessions for this pilot study was determined by logistical reasons (e.g. arena and facilitator schedules, academic calendar, and horse availability) as opposed to the optimal number of therapeutic sessions. While this is not ideal, it is indicative of real life and a consideration for this work in the academic setting.

Additional limitations include the fact that the DASS and SCS are both self-report instruments so the potential for bias exists. For example, a client may want to demonstrate improvement and so chooses the answer reflective of that phenomenon. Self-reporting also requires the ability to be introspective and accurately self-assess, which may be confounded by a mental health diagnosis. Another possibility is that the instruments we selected (DASS and SCS) were not sensitive enough to detect change or, conversely, that there simply was no detectable difference. It should also be noted that four of the six clients were participating in the group psychotherapy session as well as their individual counseling session, so essentially, they were receiving a 'double-dose' of treatment. We also did not collect any medication information from participants for this study, so this is another potentially confounding element. Additionally, we had a small sample size and participants self-selected for psychotherapy incorporating equines treatment. Finally, we had no control group in this study and so the improvements we did see in the common humanity subscale may not be reflective of the equine piece of the group psychotherapy session.

Attrition is often a significant challenge in administering group therapy in a college setting. Yalom and Leszcz (2005) reported that therapy groups of 8 sessions or less in university outpatient clinics have a drop-out rate of 39%. Because students with higher rates of social anxiety report more social impairment, attending groups can often become too internally distressing to persist (Nordstrom *et al.*, 2014). The aforementioned challenges contribute to disruptions in the therapeutic process due to poor attendance, late arrivals, or early termination of group. Horses have been shown to be a motivating factor for attendance and engagement in psychotherapy sessions incorporating equines (Rothe *et al.*, 2005; Trotter, 2012; Lentini and Knox, 2015). Multiple students expressed during our final session that they would not have persisted in group without the connection they felt to the horses. Additionally, there was an attrition rate of only 14% here (from 7 participants to 6). At a very minimum, clients must *attend* psychotherapy groups if they are to engage in treatment sessions and the horse may be an integral factor for consistent attendance. In addition to higher completion rates in the psychotherapy incorporating equines groups compared to non-psychotherapy incorporating equines college groups, we noticed that participants would stay after group to help put supplies away and socialize with group members. These brief social interactions created opportunities for clients to practice casual conversations and general social skills. Practicing social interactions in an emotionally safe environment is often an aspect of treating social anxiety (Yalom and Leszcz, 2005).

Given limited results related to self-compassion, we recommend that future research examine if psychotherapy that integrates both horses and the tenets of self-compassion (i.e. self-kindness, common humanity, mindfulness) can help women college students enhance their abilities to manage anxiety and become more resilient. We further recommend that the DASS, as a measure of anxiety, be replaced with the Counseling Center Assessment of Psychological Symptoms Instruments (CCAPS) (Center for Collegiate Mental Health (CCMH), 2018). The CCAPS was created to be used as an affordable and psychometrically strong clinical tool for college counseling centers. It is a multi-dimensional assessment that is updated regularly with peer-based norms drawn from large samples to provide reliable and valid measurements (CCMH, 2021). Due to the widespread use of the CCAPs among college counseling centers, it would be commonly known to consumers of the literature.

Incorporation of a robust qualitative component would also be essential (Mowbray *et al.*, 2003) for myriad of reasons. Chur-Hansen *et al.* (2010) highlighted the advantages of qualitative inquiry in reference to animal-assisted therapy because its open-ended nature allows themes that have not been considered important to be identified. They posited that identification of new themes is pivotal to further our understanding of 'mechanisms at work' (p. 142). A phenomenological inquiry would allow researchers to explore questions of motivation with regard to presence of equines, the relationship dynamics among equines and peer group members, the impact of those relationships on the tenets of self-compassion and anxiety, and insights to the intrapersonal experiences and shifts of clients as they persisted through the equine group experience.

Logical next steps include expanding the duration of the group to a more traditional timeframe of 8 to 12 weeks, as well as creating a facilitator manual of psychotherapy incorporating equines sessions. This manual would be designed to educate participants about the three key tenets of self-compassion (self-kindness, mindfulness, common humanity) in building participant resiliency in managing their anxiety. The aforementioned structure provided by a facilitator manual would aid in creating treatment integrity of the group experience and insulate against some mitigating variables that are inherent to an experiential facilitation model like EAGALA. Moreover, implementation of a fidelity measure currently being developed by Gergley (2020, Aug 6-9) that assesses adherence to the EAGALA model, as well as competency of the model's delivery, would greatly enhance the ability of researchers to replicate treatment interventions and test the efficacy and effectiveness of the model to treat anxiety in the college population.

Conclusion

While this is a pilot study with a small sample size, it is the first one we know of that incorporates equines as part of treatment options for students through their college counseling center in the United States. The findings suggest that psychotherapy incorporating equines for women college students with anxiety shows promise as a treatment intervention. As the mental health needs of college-aged students in the United States continues to increase and the demand for university counseling centers to provide services continues to rise, solutions to mitigate wait times and offer effective treatment options are imperative for the health and well-being of this population. While we found significant changes in only the common humanity subscale of the SCS, overall shifts in the means of both the SCS and the DASS were observed, and facilitators noted improved attendance for the group psychotherapy incorporating equines compared to office-based group psychotherapy at the NDSU Counseling Center. This pilot study suggests that a research design is warranted that is longer in duration, incorporates fidelity measures, and includes a qualitative component exploring the impacts of psychotherapy incorporating equines as it relates to self-compassion. Furthermore, the causal relationships and motivating factors related to attendance and attrition rates for psychotherapy incorporating equines compared to more traditional office-based psychotherapy, as well as groups incorporating other animals, such as dogs, should be investigated.

Author Note

The authors declare the following conflicts of interest: all authors have been trained as certified practitioners of the EAGALA model; the first and third authors as Equine Specialists and the second author as a Mental Health Practitioner. The authors received no financial support for the research, authorship, and/or publication of this article.

References

American Psychological Association (n.d.). Available at: https://www.apa-hai.org/wp-content/uploads/2021/01/Summary-of-Considerations-for-APA-Ethical-Standards-.pdf (accessed 16 April 2021).

American Psychological Association (2020). Available at: https://www.apa.org/topics/understanding-psychotherapy#:~:text=Some%20people%20begin%20to%20feel,conversation%20about%20what%20to%20do (accessed 9 November 2020).

American Psychiatric Association (2013) *Diagnostic and Statistical Manual of Mental Disorders*, 5th ed. Author, Washington, DC.

Animal Assisted Intervention International: Standards of Practice (2019). Available at: https://aai-int.org/wp-content/uploads/2019/02/AAII-Standards-of-Practice.pdf (accessed 16 April 2021).

Association of University and College Counseling Center Directors (AUCCCD) (2019) The Association for University and College Counseling Center Directors Annual Survey Reporting period: July 1, 2018 through June 30, 2019. Available at: https://www.aucccd.org/assets/documents/Survey/2019%20AUCCCD%20Survey-2020- 05-31-PUBLIC.pdf (accessed 9 November 2020).

Association of University and College Counseling Center Directors (AUCCCD) (2018) The Association for University and College Counseling Center Directors Annual Survey Reporting period: July 1, 2017 through June 30, 2018. Available at: https://www.aucccd.org/assets/documents/Survey/2018%20AUCCCD%20Survey-Public-June%2012-FINAL.pdf (accessed 17 October 2020).

Association of University and College Counseling Center Directors (AUCCCD) (2010) The Association for University and College Counseling Center Directors Annual Survey Reporting period: July 1, 2009 through June 30, 2010. Available at: http://files.cmcglobal.com/directors_survey_2009_nm.pdf (accessed 17 April 2021).

Bandelow, B. and Michaelis, S. (2015) Epidemiology of anxiety disorders in the 21st century. *Dialogues in Clinical Neuroscience* 17(3), 327–335.

Beck, A.T., Ward, C.H., Mendelson, M., Mock, J. and Erbaugh, J. (1961) An inventory for measuring depression. *Archives of General Psychiatry* 4, 561–571.

Bernhardsdottir, J. and Vilhjalmsson, R. (2013) Psychological distress among university female students and their need for mental health services. *Journal of Psychiatric and Mental Health Nursing* 20(8), 672–678. Available at: https://doi.org/10.1111/jpm.12002

Burton, L.E., Qeadan, F. and Burge, M.R. (2019) The efficacy of equine-assisted psychotherapy in veterans with posttraumatic stress disorder. *Journal of Integrative Medicine* 17(1), 14–19. Available at: https://doi.org/10.1016/j.joim.2018.11.001

CCMH (2021) Counseling Center Assessment of Psychological Symptoms (CCAPS) Instruments. Available at: https://ccmh.psu.edu/ccaps-34-62 (accessed 26 April 2021).

Center for Collegiate Mental Health (CCMH) (2018) CCAPS User Manual. University Park, PA.

Chur-Hansen, A., Stern, C. and Winefield, H. (2010) Gaps in the evidence about companion animals and human health: Some suggestions for progress. *International Journal of Evidence-Based Healthcare* 8, 140–146. Available at: https://doi.org/10. 10.1111/j.1744-1609.2010.00176.x

DePrekel, M. and Neznik, K. (2015a) *Equine Therapy Group*.

DePrekel, M. and Neznik, K. (2015b). *Equine Therapy Adult Group Workbook*.

DePrekel, M. (n.d.) *Equine Facilitated Group Psychotherapy School Curriculum*.

DiGiacomo, M. and Sierra Boys Ranch (n.d.) *EAGALA Model Sessions – How do we let the horses and clients decide the activity?*

Dreisoerner, A., Junker, N.M. and Dick, R. (2020) The relationship among the components of self-compassion: A pilot study using a compassionate writing intervention to enhance self-kindness. *Common Humanity, and Mindfulness. Journal of Happiness Studies*. Available at: https://doi.org/10.1007/s10902-019-00217-4

Eisenberg, D., Golberstein, E. and Hunt, J. (2009) Mental health and academic success in college. *Journal of Economic Analysis and Policy* 9(1), 1–40.

Equine Assisted Growth and Learning Association (EAGALA) (2019) *Fundamentals of EAGALA model practice: Equine Assisted Psychotherapy certification program* (9th ed.). Santaquin, UT.

Federation of Animal Science Societies (2010) *Guide for Care and Use of Agricultural Animals in Research and Teaching*, 3rd ed. Federation of Animal Science Societies, Champaign, IL.

Fine, A.H. (Ed.). (2015). *Handbook on animal-assisted therapy: Foundations and guidelines for animal-assisted interventions* (4th ed.). Elsevier Academic Press.

Fine, A.H. (Ed.). (2019). *Handbook on animal-assisted therapy: Foundations and guidelines for animal-assisted interventions* (5th ed.). Elsevier Academic Press.

Gergley, E.J. (2020, Aug 6-9) In: *EAGALA Treatment Integrity Measure [Conference presentation]. APA 2020 Convention*.

Howard, K.I., Kopta, S.M., Krause, M.S., and Orlinsky, D.E. (1986). The dose-effect relationship in psychotherapy. American Psychologist, 41(2), 159-64.

International Association of Human-Animal Interaction Organizations (IAHAIO) (2013) The IAHAIO definitions for animal assisted interventions and animal assisted activity and guidelines for wellness of animals involved. Available at: https://petpartners.org/wp-content/uploads/2015/07/8000IAHAIO-WHITE-PAPER-TASK-FORCE-FINAL-REPORT-070714.pdf (accessed 16 April 2021).

Kersten, G. and Thomas, L. (2004) Equine assisted psychotherapy and learning untraining manual. In: *EAGALA, Inc*, 3th ed. Santaquin, UT.

Kessler, R.C. and Greenberg, P.E. (2002) The economic burden of anxiety and stress disorders. In: Davis, K.L., Charney, D., Coyle, J.T. and Nemeroff, C. (eds) *Neuropsychopharmacology: The Fifth Generation of Progress*. Lippincott Williams & Wilkins, Philadelphia, pp. 981–992.

Klontz, B.T., Bivens, A., Leinart, D. and Klontz, T. (2007) The effectiveness of equine-assisted experiential therapy: Results of an open clinical trial. *Society and Animals* 15, 257–267.

Kohanev, L. (2015) *The power of the herd: A nonpredatory approach to social intelligence, leadership and innovation*. New World Library.

Kraft, D.P. (2011) One hundred years of college mental health. *Journal of College Mental Health* 59, 487–782.

Lentini, J.A. and Knox, M.S. (2015) Equine-facilitated psychotherapy with children and adolescents: An update and literature review. *Journal of Creativity in Mental Health* 10(3), 278–305. Available at: https://doi.org/10.1080/15401383.2015.1023916

McCullough, L., Risley-Curtiss, C. and Rorke, J. (2015) Equine facilitated psychotherapy: A pilot study of effect on posttraumatic stress symptoms in maltreated youth. *Journal of Infant*. Available at: http://doi.org/10.1080/15289168.2015.1021658

Morgan, B.A. (2017) Stress management for college students: An experiential multi-model approach. *Journal of Creativity in Mental Health* 12(3), 276–288. Available at: https://doi.org/10.1080/15401383.2016.1245642

Mowbray, C.T., Holter, M.C., Teague, G.B. and Bybee, D. (2003) Fidelity criteria: Development, measurement, and validation. *American Journal of Evaluation* 24(3), 315–340.

Neff, K.D. (2003a) Self-compassion: An alternative conceptualization of a healthy attitude toward oneself. *Self and Identity* 2(2), 85–101. Available at: https://doi.org/10.1080/15298860309032

Neff, K.D. (2003b) The development and validation of a scale to measure self-compassion. *Self Identity* 2, 223–250.

Nordberg, S.S., Hayes, J.A., McAleavey, A.A., Castonguay, L.G. and Locke, B.D. (2013) Treatment utilization on college campuses: Who seeks help for what? *Journal of College Counseling* 16(3), 258–274.

Nordstrom, A., Swenson, G.L. and Hiester, M. (2014) The effect of social anxiety and self-esteem on college adjustment, academics and retention. *Journal of College Counseling* 17(1), 48–63.

Notgrass, C.G. and Pettinelli, J.D. (2014) Equine assisted psychotherapy: The Equine assisted Growth and Learning Association's model of equine-based modalities. *Journal of Experiential Education* 38(2), 162–174. Available at: https://doi.org/10.1177/1053825914528472

Page, A.C., Hooke, G.R. and Morrison, D.L. (2007) Psychometric properties of the Depression Anxiety Stress Scales (DASS) in depressed clinical samples. *British Journal of Clinical Psychology* 46(3), 283–297.

Pascarella, E.T. and Terenzini, P.T. (2005) *How college affects students: a third decade of research*, Vol. 2. Jossey-Bass, San Francisco, CA.

PATH Intl (2011) Equine Specialist in Mental Health and Learning Student Manual. Available at: https://www.pathintl.org/images/pdf/resources/certifications/PATH-Intl-ESMHL-Student-Manual.pdf

Reetz, D., Krylowicz, B., Bershad, C., Lawrence, J. and Mistler, B. (2015) *Association for University and College Counseling Center Directors Annual Survey*. Association for University and College Counseling Center Directors (AUCCCD), Indianapolis, IN.

Rothe, E.Q., Jiménez Vega, B., Torres, R.M., Campos Soler, S.M. and Molina Pozas, R.M. (2005) From kids and horses: Equine facilitated psychotherapy for children. *International Journal of Clinical and Health Psychology* 5(2), 373–383.

Russell, L. and Jarvis, C. (2019) Student withdrawal, retention and their sense of belonging: Their experience in their words. *Research in Educational Administration & Leadership* 4(3), 494–525.

Salzer, M.S. (2012) A comparative study of campus experiences of college students with mental illnesses versus a general college sample. *Journal of American College Health* 60(1), 1–7. Available at: https://doi.org/10.1080/07448481.2011.552537

Spielberger, C.D., Gorsuch, R.L., Lushene, R., Vagg, P.R. and Jacobs, G.A. (1983) *Manual for the State-Trait Anxiety Inventory*. Consulting Psychologists Press, Palo Alto, CA.

Stewart, L.A., Chang, C.Y., Parker, L.K. and Grubbs, N. (2016) Animal-assisted therapy in counseling competencies.

Strahan, E.Y. (2003) The effects of social anxiety and social skills on academic performance. *Personality and Individual Differences* 34(2), 347–366. Available at: https://doi.org/10.1016/S0191-8869(02)00049-1

Tajfel, H. and Turner, J.C. (1979) An integrative theory of intergroup conflict. In: Austin, W.G. and Worchel, S. (eds) *The Social Psychology of Intergroup Relations*. Brooks/Cole, Monterey, pp. 33–37.

Trotter, K. (2012) *Harnessing the power of equine-assisted counseling: Adding animal assisted therapy to your practice*. Routledge, New York, NY.

Trotter, K.S., Chandler, C.K., Goodwin-Bond, D. and Casey, J. (2008) A comparative study of the efficacy of group equine assisted counseling with at-risk children and adolescents. *Journal of Creativity in Mental Health* 3(3), 254–284. Available at: https://doi.org/10.1080/15401380802356880

Van Der Kolk, B. (2014) *The Body Keeps the Score: Brain, Mind, and Body in the Healing of Trauma*. Penguin Books, New York, NY.

Werner, K., Jazaieri, H., Goldin, P., Ziv, M., Heimberg, R. *et al.* (2012) Self-compassion and social anxiety disorder. *Anxiety Stress Coping* 25(5), 543–558. Available at: https://doi.org/10.1080/10615806.2011.608842

Yalom, I. and Leszcz, M. (2005) *The theory and practice of group psychotherapy*, 5th ed. Basic Books, New York, NY.

Part 3: Adults

12 Human-Animal Interaction and Metaphor in Equine-Assisted Psychotherapy: Empirical Support for the EAGALA Model

Angela K. Fournier[1]*, Elizabeth A. Letson[2], Thomas D. Berry[3] and Emma Leigh Pasiuk[1]

[1]Department of Psychology, Bemidji State University; [2]Eagle Vista Ranch & Wellness Center; [3]Department of Psychology, Christopher Newport University

Abstract

Research was conducted on equine-assisted psychotherapy and learning (EAP), aimed at identifying and examining process variables. A total of 112 men and women participated in a correlational study, permitting the use of self-report data from EAP sessions within the model of the Equine-Assisted Growth and Learning Association (EAGALA). Human-animal interaction and metaphor were studied as process variables, while subjective distress and psychosocial learning were investigated as outcome variables. Participants reported a range of interactions between humans and horses, and the quantity of human-animal interaction was a significant predictor of reduced subjective distress. Metaphor data indicated horses most often represented family, friends, or feelings. Analyses suggest the relationship between human-animal interaction and psychosocial learning is mediated by the extent to which horses serve as metaphors for people and issues in the client's life, providing empirical support for EAGALA's notion that metaphor is critical to EAP outcomes. Findings are discussed as they relate to developing theory on the role of human-animal interaction in EAP and other animal-assisted therapies.

Keywords: Human-Animal Interaction, Equine-Assisted Psychotherapy, EAGALA Model, Metaphor

Introduction

Equine-assisted psychotherapy (EAP) is an experiential modality that involves partnering with horses for emotional growth and learning. EAP is typically a collaborative effort by a mental health professional and an equine specialist, working as a team alongside horses to attain client treatment goals (Equine Assisted Growth and Learning Association [EAGALA], 2015). The literature indicates that EAP has been applied frequently with youth and adult populations struggling with a variety of mental health disorders (e.g. Trotter *et al.*, 2008).

While EAP has been used to treat a multitude of psychological problems including emotional and behavioral disorders (Wilkie *et al.*, 2016), quantitative studies on its effectiveness

*Corresponding author: afournier@bemidjistate.edu

are scarce (Vidrine *et al.*, 2002). Many studies that have been conducted are limited by small sample size and a lack of control group comparisons (e.g. Trotter *et al.*, 2008; Bachi *et al.*, 2011). Furthermore, while there are some studies with strong design and promising findings (e.g. Pendry *et al.*, 2014), research to date has investigated EAP *outcomes* only. At the time of this writing, the investigation of EAP *processes* has been minimal. This research aims to elucidate underlying processes to further the development of theory explaining EAP.

Need for Empirical Evidence and Theory

Although EAP investigators have often hypothesized an underlying process, such as horses mirroring (e.g. Symington, 2012) or attuning (Knapp, 2013) to human emotion, there are no published studies directly observing these processes in EAP. Theoretical explanations of equine-related treatments are limited to writings on hippotherapy, also referred to as therapeutic riding. For example, Shumway-Cook and Woollacott (1995) speculate that hippotherapy outcomes can be explained by dynamic systems theory. In this regard, riding is thought to influence sensorimotor, cognitive, respiratory, behavioral, social, and psychological functioning (Casady and Nichols-Larsen, 2004). Kendall *et al.* (2014) reviewed the literature on therapeutic riding, developing hypotheses to test the psychological benefits of horses. They suggest researchers investigate three possible explanations for therapeutic-riding outcomes – (a) benefits are unrelated to the horses, (b) horses provide a positive context for the primary treatment to occur, or (c) horses have unique therapeutic qualities that bring about changes. Although Kendall *et al.* focused on riding, the EAP field would benefit from testing these hypotheses on the benefits of horses to psychotherapy.

In reviewing literature on the topic, Kakacek and Ottens (2008) discuss metaphoric communication as integral to the EAP process. Specifically, they state that metaphors are directed at (1) the horse's behavior, (2) props present in the arena/pasture (e.g. halter, gate), (3) lessons taken from solving a problem, and (4) extrapolation of lessons learned in the arena to life's challenges. These authors discuss horse characteristics as a second mechanism of change in EAP, suggesting characteristics unique to horses allow them to mirror human emotions.

Selby and Smith-Osborne (2013) conducted a systematic review of the literature on biopsychosocial interventions involving horses. They reviewed 103 studies, of which 14 met inclusion criteria. The studies addressed a variety of outcome measures, including self-esteem, locus of control, self-perception, loneliness, and depression. The authors concluded that equine-assisted interventions are promising but more rigorous research is needed. Outcome measures were not discussed in the context of an overall theory on equine-assisted treatments. Similarly, Anestis *et al.* (2014) conducted a systematic review on equine-related treatments, reviewing 14 empirical studies investigating either equine-related counseling or therapeutic horse riding. The authors found significant problems in the field, including (1) great variability in application of equine-related treatments, (2) an absence of theory explaining the change process, and (3) research limited by multiple threats to internal validity. Thus, they concluded there is a lack of empirical support for EAP and discourage its use until stronger empirical evidence is found. More recently, Lee *et al.* (2016) conducted a narrative synthesis of 24 EAP studies from 2005 to 2013. They determined EAP research is limited by nonexperimental research design, incomplete analysis of qualitative findings, and mixing of mounted and non-mounted therapies. They conclude by calling for theory development, particularly in understanding the horse's role in EAP. We address this gap in the literature, examining EAP within the EAGALA model (Equine Assisted Growth and Learning Association, 2015), as it is an established practice model which necessitates consistent application of prescribed methods. Furthermore, the model alludes to psychological mechanisms of change underlying practical techniques.

EAGALA Model

EAGALA is a nonprofit organization formed in 1999, providing credentialed training in EAP. Currently, EAGALA has a presence in 50 countries worldwide with over 4500 professionals working as certified practitioners in the delivery of EAP (Equine Assisted Growth and Learning Association, 2015). The four tenets of the EAGALA model are that sessions are (1) on the ground (no riding), (2) solution-oriented, (3) facilitated by a mental health professional-equine specialist team, and (4) guided by a code of ethics (Equine Assisted Growth and Learning Association, 2015). In addition to the four tenets, EAGALA utilizes clean language (Grove and Panzer, 1989) and symbolic modeling (Lawley and Tompkins, 2000) to better facilitate sessions based on client interpretation (Thomas and Lytle, 2016). Notgrass and Pettinelli (2015) provide a detailed description of the EAGALA model, arguing it is different from other modalities involving horses. The model follows the guiding principles of the Association for Experiential Education, and is similar to adventure therapy in its use of activities designed to create therapeutic opportunities and metaphor (Gass *et al.*, 2012).

EAP studies specifically addressing the EAGALA model are limited in quantity and scientific rigor. Notgrass (2011) conducted an online survey of EAP practitioners and found that of the 328 respondents, 84.8% followed the EAGALA training model. Schultz *et al.* (2007) investigated the effects of the EAGALA model on children who had experienced intrafamily violence. Global assessment of functioning (GAF) scores improved significantly from pre- to post-treatment. However, the study design was limited, as there was no control group. Nuremberg *et al.* (2014) conducted a randomized trial, comparing group treatment with horses (i.e. EAGALA model), dogs (i.e. traditional animal-assisted therapy), no animals, and treatment as usual on aggression in long-term psychiatric patients. Results showed participants receiving the EAGALA model had reduced violence-related incidents in the short term. Nuremberg *et al.* suggest EAGALA model outcomes were comparable to the more well-established canine-assisted psychotherapy, and in some cases may have even greater effects. However, their research did not gather process data nor speculate on why working with horses might benefit patients. Exploring variables such as interactions between humans and horses is needed to further our understanding of the EAGALA model and EAP.

Human-Animal Interaction

At the most basic level, EAP involves human-animal interaction (HAI) – clients interact with or are in the presence of horses. However, it is unknown which kinds of interactions occur between humans and horses and which interactions, if any, are related to treatment outcomes. Vitztum (2012) conducted a concept analysis of HAI in general, concluding it is studied as a construct with significant outcomes but without a clear understanding of its crucial ingredient(s). This is a limitation of EAP research as well. Behavioral scientists measure a variety of EAP outcomes, such as psychological symptoms (e.g. Klontz *et al.*, 2007) and physiological measures (e.g. Pendry *et al.*, 2014), and they may describe number of sessions and program objectives (e.g. Trotter *et al.*, 2008), but provide no description or analysis of the ways in which the humans and horses behave with one another. Instead, interaction is treated as a categorical variable; comparisons are made between clients who receive EAP and those who do not. In other words, EAGALA model and EAP research generally do not address particular behaviors from the horse or the client, nor do they address the quantity or degree of interaction between the two. Clients are given vague instructions to interpret on their own, to be consistent with the solution-oriented approach (Equine Assisted Growth and Learning Association, 2015), and sessions are described as having a wide range of interactions, from none at all (e.g. client sits in the pasture and scarcely looks at a horse) to a great deal (e.g. grooming, petting, talking to, haltering and leading a horse) (Thomas and Lytle, 2016). Case stories give the reader a general idea of what can happen between clients and horses, but do not detail which specific behaviors occurred or to what degree they occurred.

In response to the lack of theoretical understanding of HAI, and a suggested need for reliable instruments to examine the construct (Vitztum, 2012; Wilson and Netting, 2012), researchers developed the Human-Animal Interaction Scale (i.e. HAIS), which allows for systematic measurement of behavioral interactions between humans and non-human animals (Fournier *et al.*, 2016; Fournier *et al.*, 2017). The HAIS was evaluated among several samples and across a variety of animals, and was tested for both validity and reliability (Fournier *et al.*, 2016). One purpose of our research was to further our understanding of the EAGALA model and EAP sessions by using the HAIS.

Metaphor

As mentioned above, both EAP (e.g. Kakacek and Ottens, 2008) and the EAGALA model (Notgrass and Pettinelli, 2015) consider the use of metaphor, defined as 'a device for seeing something in terms of something else' (Burke, 1941, p.421), as a necessary component to therapy. A review of the literature shows that metaphor is often discussed as an important component of traditional psychotherapy. Witzum *et al.* (1988) discuss the use of metaphor in psychotherapy, stating it allows language to be 'examined as an index of the natural imagistic content of a patient's thinking' (p. 2). Otto (2000) encourages clinicians to maximize clients' engagement with and memory of content from sessions through the use of stories and metaphor. Abbatiello (2006) suggests metaphor helps to extend the boundaries of beliefs about thinking, introducing metaphors to the client in cognitive-behavioral therapy as a way of making cognitive distortions more salient.

Regarding therapies involving animals, O'Callaghan (2008) conducted a survey of animal-assisted therapy providers and found that sharing animal stories and metaphors was a technique used by the majority of respondents. Participants who reported using metaphors most commonly stated a purpose of developing insight and building rapport. Esbjörn (2006) surveyed practitioners offering equine-assisted interventions, including the EAGALA model, hippotherapy, and therapeutic riding. Respondents reported horses serving as metaphors for other aspects of clients' lives as the most important component of the therapy, in addition to providing physical connection and present focus. Similarly, Gilbert (2013) and Lujan (2012) surveyed EAP practitioners. Respondents described metaphor as a key component of EAP and providing metaphors for clients as an important perceived role for horses. Finally, Shultz (2005) stated that horses provide metaphors for important relationships and life challenges when providing a rationale for studying EAP with at-risk youth, yet these statements were not tested empirically. It should be noted that the last four studies are unpublished dissertations, which discussed metaphor as important to EAP but did not directly measure the construct.

With regard to the EAGALA model, metaphor is described as the 'foundation' of the model (Equine Assisted Growth and Learning Association, 2015, p. 63); it is assumed that metaphors allow a transformation from relating to horses to relating to deeper life issues (i.e. personal growth and learning). This is consistent with the principles of experiential education and adventure therapy. Hovelynck (1998) discusses the facilitator's role in experiential learning as creating a space for the learner's life metaphors to emerge, facilitating appreciation for the power of life metaphors, and aiding in the change of restrictive metaphors. Metaphors are described as serving both of these roles in the EAGALA model – introduction of topics/tasks and understanding and changing restrictive patterns in thought and behavior, with the latter taking precedence (Equine Assisted Growth and Learning Association, 2015; Thomas and Lytle, 2016). Symington (2012) used case examples to illustrate how the EAGALA model and metaphor influence client grief. Specifically, clients identified horses as key subjects in their grief process and worked through grief by working with the horses.

Taken together, the literature suggests metaphor may be a component of psychotherapy in both traditional and animal-assisted modalities; used as a tool for building rapport, remembering and connecting content across sessions, and developing insight. Within the field of EAP, investigators and practitioners have indicated the importance of metaphor use as a therapeutic variable necessary for beneficial client outcomes and progress. Additionally, researchers stipulate that horses in EAP sessions serve as metaphors for the client, family or friends, goals or aspirations, and challenges or losses. However, none of these descriptions have been supported with empirical evidence. To date, there are no studies involving direct observation and analysis of the function of metaphor in animal-assisted psychotherapy. Despite it being described as a key component of the EAGALA model and EAP, there have been no empirical investigations into metaphor presence or purpose in sessions. The current study serves as an initial attempt to identify the presence of metaphor in sessions following the EAGALA model and examine its relationship with other therapy-relevant variables.

Study Design & Hypotheses

Design

Although not yet tested empirically, EAGALA training materials suggest (1) clients engage with horses, (2) life metaphors emerge, and (3) clients gain insight to psychological issues (Equine Assisted Growth and Learning Association, 2015). This implies a mediation model, in which a mediating variable intervenes in the causal path between cause and effect variables (Baron and Kenny, 1986). In the case of EAP, EAGALA suggests metaphor is a mediating variable between HAI and personal growth and learning. Empirical evidence of this potential mediation is a first step in testing the validity of the EAGALA model as a theoretical and substantive approach to psychotherapy. A mediation analysis was conducted to test the mediating potential of metaphor. Thus, our research was conducted to understand EAP in general and the EAGALA model specifically. Using the EAGALA model, two post-test-only process variables – HAI and metaphor – were measured along with two repeated-measures variables – subjective distress and psychosocial learning. This mixed-methods study included both quantitative and qualitative data.

Hypotheses

First, based on research findings that EAP can result in improvements in mood and anxiety (e.g. Beetz *et al.*, 2012), it was hypothesized that HAI (i.e. interactions between humans and horses) would be a significant predictor of subjective distress, such that increased HAI would be associated with decreased distress after EAGALA model sessions. Second, little to no evidence has been collected regarding the relation between specific behaviors (within HAI) and a client's metaphor use. However, we hypothesized that increased HAI overall would be associated with increased metaphor use. Third, we expected that metaphor use would serve as a mediator in the relation between HAI and psychosocial learning.

Method

Participants & setting

Participants were 112 male and female adult clients who attended an EAGALA model session at a private practice in the Midwestern United States. Sessions were held in an arena with one or more live horses, a licensed mental health professional, and an equine specialist, both certified in the EAGALA model. Sessions consisted solely of groundwork (e.g. observing, grooming, leading horses); there were no horse-back riding or mounted activities. Participants were clients of the provider, seeking services for a variety of reasons. Specifically, 47.3% were seeking individual or family psychotherapy (n=53), 25% underwent group psychotherapy (n=28), and 27.7% received group experiential learning (n=31). The data described here are from the participating clients' first EAGALA-model session, occurring between June 2014 and July 2016.

Materials

Materials included self-report measures regularly used by the provider for assessment and program evaluation, including the Human-Animal Interaction Scale, Subjective Units of Distress Scale, and follow-up questions on Psychosocial Learning and Metaphor created for this study.

Human-Animal Interaction Scale (HAIS)

Originally used to measure HAI in the prison setting (Fournier *et al.*, 2007; Fournier, 2016), the HAIS is a 24-item paper-and-pencil self-report instrument completed by an individual after interacting with a non-human animal (e.g. animal-assisted therapy/activity session) (Fournier *et al.*, 2016; Fournier *et al.*, 2017). Respondents rate the extent to which they or the animal engaged in various behaviors on a 5-point Likert scale, ranging from 0, indicating 'not at all' to 4, indicating 'a great deal.' Items ask respondents to report on behaviors emitted by the *human* (e.g. pet the horse, talk to the horse) and behaviors emitted by the *animal* (e.g. make friendly sounds); the scale includes behaviors that may be perceived as *desirable* (e.g. play, groom) or *undesirable* (e.g. decline in interaction, behave aggressively). Higher scores indicate a greater quantity of desirable behavioral interaction. Research on the instrument suggests it has adequate reliability and validity (Fournier *et al.*, 2016).

Metaphor Self-Report

Created for this study, participants were administered an item on which to report their experience regarding metaphor. Respondents were asked to rate their level of agreement with the statement 'The horses represented something in my life,' on a scale from 0 ('not at all') to 10 ('completely'). After rating their agreement, participants wrote an explanation of their rating. From these participant explanations researchers identified and coded thematically similar metaphor use. The prompt was based on (1) EAGALA's suggestion that experiences in the session relate to life issues outside of the session, (2) literature on metaphor in general and metaphor in EAP (e.g. Burke, 1941; Esbjörn, 2006), and (3) observation and pilot testing at the private practice facility.

Subjective Units of Distress Scale (SUDS)

The SUDS is a Likert scale on which an individual rates their distress from 0 to 10 (Wolpe, 1990). The SUDS is established as a valid measure of distress when treating anxiety disorders (e.g. Kaplan *et al.*, 1995) and is often used to determine progress in therapy (e.g. Rothbaum *et al.*, 2005). In this study, participants rated current mood and anxiety, each on a 0–10 scale, where 0 indicated 'none' and 10 indicated 'worst ever.' Scores were combined for an overall score of subjective distress, ranging from 0 to 20.

Psychosocial learning self-report

Similar to the metaphor self-report, respondents were asked a question designed to assess psychosocial learning. Using the 0–10 scale, participants rated their agreement with the statement, 'I learned something about myself,' and wrote comments to explain their rating. These participant comments were examined for common learning themes and then coded for tabulation. This item was written to gather broad intrapersonal data, developed from observation and pilot testing, showing clients report learning a wide array of content about themselves (e.g. thought and behavior patterns) in EAGALA sessions.

Procedure[1]

Research procedures consisted of analysis of de-identified data from a convenience sample of EAGALA model clients. When consenting to services, clients of the provider could also consent to have data from self-report measures used in research at the aggregate level. Agreeing to the research was voluntary; clients were informed their decision would have no impact on their services. Data from clients who consented to the use of aggregate data for research were entered into a de-identified database, which was saved and stored separately from client treatment records. The research examined this de-identified data only, which included scores from the above self-report measures; diagnostic, treatment, and demographic information were excluded. Sessions included one or more horses from a herd of nine horses housed at the provider facility. The research did not involve any changes to horse care, nutrition, or management.

Clients were administered the SUDS scale before and after each session, rating both mood and anxiety on a 10-point scale. After the session, clients reported interactions with the horses on the HAIS, and also completed the Metaphor and Psychosocial Learning self-reports. These administrations were part of regular practice by the provider and their use was not dependent on research participation. Clients had the right to decline the measures. The data presented here are from adult clients who attended a session, completed the self-report measures, and consented to their data being used in research.

Results

During the research period, services were provided to a total of 120 clients. Of those, 1.7% (n=2) declined to complete the self-report measures and 5.0% (n=6) of those who completed measures declined to have their scores used in research. Thus, the total sample was 112, including 76.8% women (n=86) and 23.2% men (n=26). This gender ratio is consistent with the demographics of this specific facility. Although participants may have attended more than one session, for consistency their participation in this study was limited to self-report data from the first session only.

Human-animal interaction

Descriptives

Table 12.1 provides mean scores on the HAIS, including individual items, subtotals, and the overall total. Participants reported the full range of behaviors, with all items endorsed by at least one participant. The table also provides the percentage of participants endorsing each of the 0–4 ratings, and the percentage of participants who rated the item one or higher. Taken together, the information in Table 12.1 reflects the variability between items, with some behaviors rated much higher (e.g. Watch, Spend time near) than others (e.g. Kiss, Take pictures). While there are no cut-off scores on the HAIS, higher scores reflect a greater quantity of positive behavioral interaction between the human and horse(s). Total scores, including behaviors emitted by humans and horses, ranged from −1 to 63, with a mean of 25.6 (SD = 12.4), a median of 24, and a mode of 32. A negative score indicates there were higher ratings on undesirable behaviors (e.g. Decline interaction, Make unfriendly sounds) than desirable behaviors (e.g. Pet, Groom, Offer food). There was just one negative score (i.e. −1), from a participant who reported some approach behaviors from the horses but declined or avoided any interaction with the horses.

In addition to total scores, the HAIS provides a human behavior subtotal and horse behavior subtotal. Similar to the total score, there was great variability within these subtotals. Human behavior scores ranged from −2 to 45 (M = 18.78, SD = 8.63) and horse behavior scores ranged from −2 to 22 (M = 6.94, SD = 4.01). Watching the horse(s) was the highest rated human behavior (M = 3.13, SD = 0.89) and Initiating interaction with the human(s) was the highest rated horse behavior (M = 2.44, SD = 0.81). For both humans and horses, desirable behaviors were rated much higher than undesirable behaviors. Ratings on Aggressive behavior and Declining or avoiding interaction were among the lowest for both subscales.

Subjective Distress

Participants rated their distress before and after each session. Paired-samples t-tests indicated distress decreased from pre-test (M = 7.42, SD = 3.75) to post-test (M = 3.58, SD = 3.55), t (92) = 9.39, p<0.001. To determine whether there was a relationship between subjective distress and interactions with the horses, Pearson r correlations were calculated between SUDS scores and HAIS scores. Table 12.2 shows the correlation matrix for calculations between HAI and other variables. There were negative correlations between post-test distress and human behavior, r (92) = −0.443, p<0.001, horse behavior, r (91) = −0.294, p=0.005, and total HAI, r (92) = −0.405, p<0.001. These correlations were not present at pre-test, p's > 0.05. To test the hypothesis that HAI would predict EAP outcomes, a simple linear regression was calculated. A significant regression equation was found for total HAI to predict distress, F (1, 90) = 17.64, p<0.001, R = 0.164, explaining 15.5% of the variance.

Table 12.1. Self-Reported Interactions between Humans and Horses (n=112)

		Percent Reporting (%)						
	HAIS Item	Mean (SD)	0	1	2	3	4	>1 (%)
Human Behavior toward Horse(s)	Watch	3.11 (0.89)	0.0	1.8	27.7	25.9	42.0	100.0
	Spend time near**	3.13 (1.08)	4.0	2.0	21.2	22.2	50.5	98.2
	Pet**	2.82 (1.06)	1.8	10.0	25.5	30.0	32.7	96.0
	Talk to**	2.24 (1.26)	7.3	22.7	34.5	10.0	25.5	93.7
	Play with**	1.56 (1.14)	21.3	28.9	31.3	10.2	8.3	78.7
	Hold*	0.84 (1.10)	53.2	27.3	9.3	4.6	5.6	63.6
	Hug*	1.14 (1.36)	47.3	20.0	14.5	8.2	10.0	52.7
	Kiss	0.55 (0.91)	56.8	30.2	7.4	2.8	2.8	47.3
	Groom*	1.53 (1.46)	36.4	17.3	18.2	13.6	14.5	46.8
	Offer food to	0.60 (0.91)	61.3	25.8	8.3	0.9	3.7	43.2
	Try tricks/training*	0.96 (1.25)	52.7	18.2	15.5	7.3	6.4	38.7
	Take pictures	0.27 (0.59)	71.3	24.1	2.8	0.9	0.9	31.3
	Decline interaction**	0.31 (0.59)	73.0	23.3	1.9	0.9	0.9	28.7
	Behave aggressively	0.06 (0.17)	68.7	31.3	0.0	0.0	0.0	27.0
Horse Behavior toward Human	**Initiate interaction****	2.44 (0.81)	1.9	6.5	43.5	36.1	12.0	98.1
	Friendly sounds*	1.75 (0.94)	13.1	19.4	52.6	10.3	4.7	93.5
	Accept food	0.34 (0.72)	67.1	27.4	0.9	3.7	0.9	86.9
	Play*	1.13 (1.02)	31.8	37.1	21.80	5.6	3.7	86.9
	Obey tricks/training	1.26 (1.03)	29.0	30.0	30.8	6.5	3.7	71.0
	Sniff	2.11 (1.03)	6.5	17.6	44.4	21.3	10.2	68.2
	Lick	0.36 (0.72)	72.8	20.7	3.7	0.9	1.9	39.8
	Decline interaction	0.36 (0.72)	13.1	39.4	37.2	8.4	1.9	32.9
	Unfriendly sounds	0.44 (0.64)	60.2	31.5	13.0	6.5	1.9	37.2
	Mess or inconvenience	0.22 (0.49)	82.8	14.4	0.9	1.9	0.0	33.9
	Behave aggressively	0.27 (0.48)	76.1	21.1	1.9	0.9	0.0	18.2

Note: Items in bold correlate significantly with post-test distress. *$p<0.01$, **$p<0.001$
Responses on the HAIS can range from 0, indicating 'not at all' to 4, indicating 'a great deal'.

To examine the predictive power of human behavior vs horse behavior, a multiple regression model was calculated on distress, with human and horse behavior entered as predictors. Several analyses were carried out to determine whether required assumptions were met. An analysis of standard residuals identified three participants as outliers; the participants were removed from the analysis. Tests of collinearity indicated that multicollinearity was not a concern (human behavior tolerance = 0.52, VIF = 1.93; horse behavior tolerance = 0.52, VIF = 1.93). The Durban-Watson test indicated the data met the assumption of independent errors (1.92), and the histogram and P-P Plot indicated the data met the assumption of normally distributed residuals. The scatterplot of standardized predicted values showed that the data met the assumptions of homogeneity of variance and linearity. Finally, the data met the assumption of non-zero variances (human behavior = 73.68; horse behavior = 16.73; distress = 9.13). Using the enter method, it was found that human and horse behavior

Table 12.2. Correlation Matrix on HAI, Distress, Metaphor, and Psychosocial Learning (n=112)

Variables	*1.*	*2.*	*3.*	*4.*	*5.*	*6.*	*7.*	*8.*	*9.*
1. HAI	-								
2. Human HAI	0.96**	-							
3. Animal HAI	0.86**	0.69**	-						
4. Anxiety (Pre)	0.04	0.03	−0.01	-					
5. Anxiety (Post)	−0.33**	−0.39**	−0.15	0.38**	-				
6. Mood (Pre)	−0.14	−0.10	−0.12	0.40**	0.09	-			
7. Mood (Post)	−0.33**	−0.33**	−0.31**	0.28**	0.66**	0.38**	-		
8. Metaphor	0.34**	0.36**	0.22*	0.03	−0.33**	−0.05	−0.36**	-	
9. Psychosocial Learning	0.41**	0.39**	0.29**	−0.03	−0.38**	−0.16	−0.43**	0.80**	-

*$p < 0.01$, **$p < 0.001$

explained a significant amount of the variance in distress scores, F (2, 83) = 7.75, p=0.001. The analysis showed that human behavior was a significant predictor of post-test distress (β=−0.39, t (83) = −2.82, p=0.006), but horse behavior was not (β=−0.01, t (83) = −0.03, p=0.98).

Psychosocial Learning

Participants rated the extent to which they agreed with a statement that they had learned something about themselves, and then were invited to write comments explaining their answer. A total of 90 participants rated their agreement. Ratings ranged from 0 to 10, with a mean of 5.74 (SD= 3.49). Pearson r correlations were calculated between scores on the HAIS and psychosocial learning ratings, shown in Table 12.2. There were significant correlations between learning and human behavior, r (81) = 0.39, p<0.001; horse behavior, r (88) = 0.29, p=0.006; and total HAI r (81) = 0.41, p<0.001. Simple linear regression resulted in a significant equation, where HAI explained 15.9% of the variance in psychosocial learning, F (1, 76) = 15.51, p<0.001, R=0.41. Similar to analyses on subjective distress, a multiple regression analysis was calculated, assessing human behavior and horse behavior as predictors of learning. The data met all necessary assumptions for the analysis. Using the enter method, it was found that the regression equation was significant, F (2, 73) = 7.23, p=0.001, R=0.41, explaining 14.2% of the variance in learning. Similar to the findings on distress, results showed that human behavior was a significant predictor of psychosocial learning (β=0.32, t (75) = 2.32, p=0.023), but horse behavior was not (β=−0.11, t (75) = 0.81, p=0.42).

When asked to explain their ratings on the psychosocial learning self-report, 64.4% of the 90 participants (n=58) provided written comments. These comments were analyzed for themes following the constant comparative method (Brunner, 2004). To increase the reliability of the findings, the researchers inductively drew themes, then discussed the themes until reaching agreement concerning the importance of each theme and its conclusion. Once a distinct number of themes was identified, researchers rated the written responses, categorizing each response as fitting one of the themes. All the comments were rated by two researchers who rated independently of one another; percent agreement between raters was 88%. Participant responses fell within six themes. Table 12.3 provides the list of themes that emerged, in order of frequency, as well as an example comment for each theme. Although no predictions were made, one-way ANOVAs were calculated to explore whether HAI or subjective distress differed across the learning themes. There were no significant differences, p's > 0.05.

Comparisons were made between participants who did (n=58) and did not (n=32) write comments explaining their ratings. A one-way ANOVA showed that participants who wrote comments about learning rated their agreement with the statement on learning higher ($M=6.35$, $SD=3.10$) than those who did not comment ($M=4.89$, $SD=3.92$), $F(1, 84)=3.72$, $p=0.05$. The presence of written comments was not dependent on any other variables (i.e. HAI, subjective distress, metaphor), p's >0.05.

Metaphor

Participants rated the extent to which the horses represented something in their life, then wrote comments explaining their ratings. The ratings and comments were summarized and tested for associations with HAI and psychosocial learning.

Descriptives

Of the 112 participants, 86.6% (n=97) rated the extent to which the horse(s) represented something in their life. The ratings for these horse metaphors ranged from 0 to 10, ($M=6.02$, $SD=3.10$). When asked to explain their ratings on the metaphor self-report, 59.8% of the 97 participants (n=58) provided written comments. Horses represented a feeling or state, family or friends, a relationship or relationship construct (e.g. trust), challenges or barriers, life lessons, and themselves. A small number of respondents (n=4) provided a concrete description of the session rather than describing a metaphor. Referring again to Table 12.3, the most common metaphor was horses representing a feeling or state, with 28.1% of participants reporting in that category. To explore whether horse metaphor was related to other variables, one-way ANOVAs were calculated, comparing posttest distress, HAI, and psychosocial learning across the five metaphor themes. There were no significant differences found, p's >0.05.

Similar to the data on psychosocial learning, comparisons were made between participants who did and did not write comments explaining their ratings on the metaphor self-report. A one-way ANOVA showed that participants who wrote comments about horses as metaphors rated their agreement with the statement on horse metaphor higher ($M=7.29$, $SD=2.39$) than those who did not comment ($M=4.88$, $SD=3.67$), $F(1, 90) = 14.54$, $p<0.001$. The presence of comments was not dependent on any other variables (i.e. HAI, subjective distress, metaphor), p's >0.05.

Mediation

To test the hypothesis that increased HAI would be associated with increased metaphor use,

Table 12.3. Themes and Examples from Qualitative Data (n=58)

Theme	%	n	Example
Psychosocial Learning			
Strengths/Weaknesses	34.0	18	*'I am strong, and can be ok'*
Relationship Awareness	28.3	15	*'I struggle meeting new people; trust takes time'*
Emotional Awareness	17.0	9	*'I'm terrified'*
Acknowledgement/Acceptance	9.4	5	*'I can overcome anything'*
Goals/Interests	7.5	4	*'I need to learn to let go'*
Horse Metaphor			
Feeling or State	28.1	16	*'that horse is my anxiety'*
Family or Friends	19.3	11	*'the people I need to trust'*
Relationship Constructs	14.0	8	*'interactions with others, communication, trust'*
Lesson	12.3	7	*'pushing leads to resistance'*
Self	10.5	6	*'it's me in my vulnerable stage'*
Challenge	8.8	5	*'the white horse is struggles in my life'*

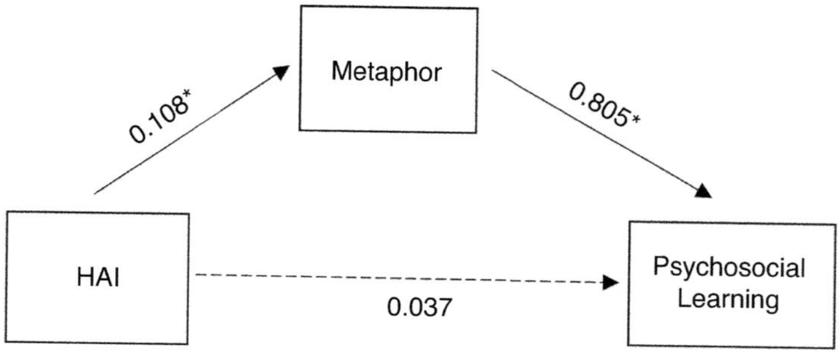

Fig. 12.1. Standardized coefficients showing metaphor mediating the relationship between HAI and psychosocial learning. Solid lines indicate significant paths; dotted lines indicate nonsignificant paths. *p<0.001

correlations were calculated between HAIS scores and metaphor ratings. As shown in Table 12.2, there was a positive correlation between HAI and metaphor; this was true for human behavior, horse behavior, and total HAI. A mediation analysis was conducted, examining the relationship between HAI, metaphor, and psychosocial learning. In Step 1 of the mediation model, the regression of HAI on learning, ignoring the mediator, was significant, $b=0.120$, $t\ (77) = 3.91$, $p<0.001$. Step two showed that the regression of HAI on the mediator, horse metaphor, was also significant, $b=0.108$, $t\ (77) = 3.66$, $p<0.001$. Step 3 of the mediation process showed that the mediator (horse metaphor), controlling for HAI, was significant, $b=0.805$, $t\ (77) = 9.66$, $p<0.001$. Step 4 of the analysis revealed that, controlling for the mediator (horse metaphor), HAI was not a significant predictor of learning, $b=0.037$, $t\ (77) = 1.62$, $p=0.109$. A Sobel test was conducted and found full mediation in the model ($z=3.41$, $p<0.001$). Thus, it was found that horse metaphor mediated the relationship between HAI and psychosocial learning. Fig. 12.1 illustrates the mediation, each path labeled with regression coefficients.

Discussion

The purpose of the present study was to further our understanding of EAP, providing empirical evidence of two process variables – HAI and metaphor – and their effects on EAP outcome. More specifically, the EAGALA model was examined, testing the assumption that metaphor is a key component of the process. HAI was measured, providing a first look at the actual behaviors that occur in psychotherapy with horses. Clients reported a range of behaviors occurring between them and horses, with variability between respondents. While each item on the HAIS was endorsed by at least one participant, some behaviors were rated much higher than others. This variability is consistent with case stories depicting a variety of interactions during EAGALA model sessions (Thomas and Lytle, 2016) and provides a more objective description of EAP interactions (Fournier *et al.*, 2016). For both humans and horses, desirable behaviors were much more prevalent than undesirable behaviors. Regression analysis showed that the overall quantity of reported interactions was correlated with decreased subjective distress (i.e. mood and anxiety states).

Data supported the hypothesis that HAIS scores would predict subjective distress, suggesting the quantity of behavioral interactions between humans and horses has an impact on the outcomes of EAGALA model-based treatment. The absence of a correlation between pre-test distress and HAI, coupled with the correlation with post-test scores, warrants further experimental research testing a potential cause and effect relationship. Regression analyses indicated that HAI was a significant predictor of subjective distress and psychosocial learning. In both cases, when entering both human behavior and horse behavior into the equation, human behavior was a significant predictor but horse behavior was not. This pattern should be explored further, as it is important for an understanding of the EAGALA model and EAP, and may reflect nuances in the effects of HAI and therapeutic outcomes of animal-assisted therapies.

Metaphor was explored as a second process variable, as it is described as a critical component in the EAGALA model (Equine Assisted Growth and Learning Association, 2015; Thomas and Lytle, 2016). To date, the literature discusses the use of metaphor in psychotherapy (e.g. Witzum *et al.*, 1988; Schoo, 2009) and the prevalence of animal metaphors in society (e.g. Palmatier, 1995). Empirical investigations of metaphor have focused on their use to describe humans, categorized as positive or negative (e.g. Goatly, 2006; Sommer and Sommer, 2011). Writings on metaphor in psychotherapy discuss its function in facilitating the therapeutic process (e.g. Abbatiello, 2006), but have not measured their salience or meaning for the client. Metaphor was measured here via client ratings and reports on whether the horses in the session represented something in their life. The data confirmed that metaphors do emerge during sessions. Although literary investigations into animal metaphors tend to find that metaphors describing humans as animal or animal-like are negative or pejorative (Goatly, 2006; Sommer and Sommer, 2011), participant comments on horse metaphor show the opposite. When horses were reported as representing humans (i.e. family or friends), comparing the person to a horse was not a way of disparaging the target. Instead, statements on horses representing humans illustrated the target's relationship to the client or role in their life. The horse provided a vehicle for understanding and interacting with the target in a kind of emotional simulation.

Clients saw horses as a variety of representations, including a feeling or state, family or friends, relationships, lessons, or challenges. This is consistent with previous research, which found horses as metaphors for relationships, family members, and challenges (Shultz, 2005; Esbjörn, 2006; Symington, 2012). Analyses indicated the relationship between HAI and psychosocial learning was mediated by metaphor, providing the first empirical evidence supporting EAGALA's assertion that metaphor is a key variable in the EAGALA model (e.g. Equine Assisted Growth and Learning Association, 2015; Thomas and Lytle, 2016). Combined with results on subjective distress, the findings suggest HAI may have direct effects on EAP outcomes (i.e. decreased distress, improved mood and anxiety), as well as indirect effects via influencing the therapy's mechanisms of change (e.g. metaphor).

There were no differences in psychosocial learning between the different metaphor themes. For example, people who identified horses as representing their family and friends did not report any greater or lesser learning than those for whom a horse represented a feeling. However, metaphor ratings (i.e. the degree to which the horse represented something in their life) were strongly related to learning, fully mediating the relationship between HAI and learning. This is consistent with EAGALA's suggestion that it's the client's interpretation of what happens in the arena, rather than what any observer may perceive to have happened, that leads to personal growth and learning (Equine Assisted Growth and Learning Association, 2015), and suggests there is not a specific kind or quantity of human-horse interaction necessary for specified outcomes. Rather, interacting with the horses provides an experience within which therapeutic issues arise and can be addressed.

Limitations

Results must be considered within the context of several limitations. Findings on metaphor and learning relied in part on written comments, which were not completed by all participants. Respondents self-selected whether to write comments, leaving this data vulnerable to participant bias. Analyses suggest the decision to write a comment or not was related to their agreement with the statement, and perhaps to the degree to which they actually experienced metaphor and psychosocial learning. For both variables, participants who commented reported higher ratings than those who did not. It is possible that an individual who does not perceive to have learned something about themselves would then have little if anything to write about. Similarly, if the horses did not represent something in their life, they would have less to write about than someone for whom the horses did serve as a metaphor. However, ratings spanned the full range (0–10) for both participants who did and did not comment; this was true for metaphor and learning. Although the response rate was substantial, (59.8% for metaphor, 64.4% for learning), methods to increase the number of participants providing written comments would improve future studies, confirming whether metaphors emerge during session and are related to psychosocial learning for all clients.

A second limitation lies in the way metaphor data were elicited – via directive statements. Clients were asked whether the horses represented something in their life, but unsolicited reports of metaphor may be even more important to study. EAGALA training materials state that anything in the pasture or arena (e.g. landscape, gate, rope, props) can serve as a vehicle for connecting to the client's internal world (Equine Assisted Growth and Learning Association, 2015). Thus, asking directly about the horses may have focused a client's attention away from a more important metaphor for them (e.g. the gate to the pasture as entrance into the unknown). This may explain why there were no differences in psychosocial learning between different horse representations. Perhaps the metaphor that really impacted their process was not detected by the methods used here. Future investigations might benefit from exploring metaphors beyond the horses, in a nondirective manner.

Results are limited by the correlational nature of the study. While analyses indicate significant relationships between variables, experimental methods are needed to confirm the causal pathways depicted in Fig. 12.1. In addition, this research should be extended through replication using alternative measures of distress, metaphor, and learning. There are a variety of instruments available to measure mood and anxiety states, such as the profile of mood states (McNair *et al.*, 1971) or the positive and negative affect schedule (Watson *et al.*, 1988), but metaphor measurement is more challenging. The measures of metaphor and learning used here are new instruments not yet validated or normed. They were used because of a lack of appropriate existing measures. While there are tests of one's ability to create metaphor from verbal stimuli (e.g. Primi *et al.*, 2006) and taxonomies for categorizing metaphors to determine author style (e.g. Crisp *et al.*, 2002), there are no standardized instruments or methods to measure metaphor as was done here. Refinement and validation of the metaphor self-report may be valuable for further study of metaphor as a process variable in EAP and other forms of psychotherapy.

These data are from participants' first EAGALA session only. Although this was intentional, to ensure consistency, it is unknown how the findings would compare if taken from multiple sessions. Important questions to be addressed in future research are whether the same metaphor is addressed throughout a multiple-session treatment regimen and whether the relationship between HAI, metaphor, and learning is affected by length of treatment. Finally, this research was limited to a relatively small convenience sample and thus replication with a larger, random sample is recommended.

Conclusion

Equine-assisted psychotherapy (EAP) programs are increasing in number (McConnell, 2010) and scientists are calling for better research on outcomes as well as theoretical explanations for those outcomes (e.g. Anestis *et al.*, 2014). The findings examined here provide substance for development of theory, connecting process and outcome variables within the EAGALA model. Further research is needed to replicate and extend these findings. Should the process variables of HAI and metaphor withstand further scrutiny, their utility in explaining other animal-assisted therapies must be explored. Indirect effects of HAI occurred through metaphor, a purportedly key mechanism of the EAGALA model. It is important to gather empirical evidence of other EAP methods and determine whether HAI functions similarly, providing direct and indirect effects via mechanisms of change specific to that method.

More broadly, the field of HAI is in need of theory to explain findings and guide more rigorous research (Wilson and Barker, 2003). It is important to know if direct and indirect effects of

HAI are unique to the EAGALA model and EAP or whether the same principles apply to psychotherapy with other species. It is crucial that we determine whether it is the technique used (e.g. EAGALA model, PATH), the animal involved (e.g. horses, dogs), some interaction between these variables, or some other variable(s) altogether that lead to positive outcomes. The research on EAP and other animal-assisted therapies is largely applied in the field, evaluating the outcomes of various programs involving horses and other animals. EAP research would benefit from more basic research, wherein variables are randomized and tested on healthy individuals first, then extended to clinical populations as appropriate.

Note

[1] All procedures were reviewed and approved by an institutional review board, in accordance with the U.S. Department of Health and Human Services – Federal Policy for the Protection of Human Subjects, 45 Code of Federal Regulations Part 46. Procedures were also reviewed and approved by the institutional animal care and use committee, in accordance with the animal welfare act.

References

Abbatiello, G. (2006) Cognitive-behavioral therapy and metaphor. *Perspectives in Psychiatric Care* 42(3), 208–210.

Anestis, M.D., Anestis, J.C., Zawilinski, L.L., Hopkins, T.A. and Lilienfeld, S.O. (2014) Equine-related treatments for mental disorders lack empirical support: A systematic review of empirical investigations. *Journal of Clinical Psychology* 70(12), 1115–1132.

Bachi, K., Terkel, J. and Teichman, M. (2011) Equine-facilitated psychotherapy for at-risk adolescents: The influence on self-image, self-control and trust. *Clinical Child Psychology and Psychiatry* 17(2), 298–312.

Baron, R.M. and Kenny, D.A. (1986) The moderator– mediator variable distinction in social psychological research: Conceptual, strategic, and statistical considerations. *Journal of Personality and Social Psychology* 51(6), 1173–1182.

Beetz, A., Uvnäs-Moberg, K., Julius, H. and Kotrschal, K. (2012) Psychosocial and psychophysiological effects of human-animal interactions: The possible role of oxytocin. *Frontiers in Psychology* 3, 1–15.

Brunner, J. (2004) Life as narrative. *Social Research* 71, 691–711.

Burke, K. (1941) Four master tropes. *The Kenyon Review* 3(4), 421–438.

Casady, R.L. and Nichols-Larsen, D.S. (2004) The effect of hippotherapy on ten children with cerebral palsy. *Pediatric Physical Therapy* 16(3), 165–172.

Crisp, P., Heywood, J. and Steen, G. (2002) Metaphor identification and analysis, classification and quantification. *Language and Literature* 11(1), 55–69.

Equine Assisted Growth and Learning Association (2015) Fundamentals of the EAGALA model: practice untraining manual, 8th edn. EAGALA, Santaquin, UT.

Esbjörn, R.J. (2006) *When horses heal: A qualitative inquiry into equine facilitated psychotherapy* (Unpublished doctoral dissertation). Institute of Transpersonal Psychology.

Fournier, A.K., Geller, E.S. and Fortney, E.V. (2007) Human-animal interaction in a prison setting: Impact on criminal behavior, treatment progress, and social skills. *Behavior and Social Issues* 16(1), 89–105.

Fournier, A.K. (2016) PenPals: An examination of human-animal interaction as an outlet for healthy masculinity in prison. In: Blazina, C. and Kogan, L. (eds) *Men and Their Dogs: A New Understanding of Man's Best Friend*. Springer, Switzerland, pp. 175–194.

Fournier, A.K., Berry, T.D., Letson, E. and Chanen, R. (2016) The human-animal interaction scale: Development and evaluation. *Anthrozoös* 29(3), 455–457.

Fournier, A.K., Letson, E. and Berry, T.D. (2017) *HAIS: Human-Animal Interaction Scale and manual*. Angela Fournier, Bemidji, MN.

Gass, M.A., Gillis, H.L. and Russell, K.C. (2012) *Adventure therapy: Theory, research, and practice*. Routledge, New York.

Gilbert, R. (2013) *Equine assisted psychotherapy: Provides perspectives and beliefs* (Unpublished doctoral dissertation). Available from ProQuest Dissertation and Theses database (UMI No. 3599176).

Goatly, A. (2006) Humans, animals, and metaphors. *Society & Animals* 14(1), 15–37.

Grove, D.J. and Panzer, B.I. (1989) *Resolving traumatic memories. Metaphors and symbols in psychotherapy*. Irvington, New York.

Hovelynck, J. (1998) Facilitating experiential learning as a process of metaphor development. *The Journal of Experiential Education* 21(1), 6–13.

Kakacek, S.L. and Ottens, A.J. (2008) An arena for success: Exploring equine-assisted psychotherapy. *Michigan Journal of Counseling* 35(1), 14–23.

Kaplan, D.M., Smith, T. and Coons, J. (1995) A validity study of the subjective unit of discomfort (SUD) score. *Measurement and Evaluation in Counseling and Development* 27(4), 195–199.

Kendall, E., Maujean, A., Pepping, C.A. and Wright, J.J. (2014) Hypotheses about the psychological benefits of horses. *Explore: The Journal of Science and Healing* 10(2), 81–87.

Klontz, B.T., Bivens, A., Leinart, D. and Klontz, T. (2007) The effectiveness of equine-assisted experiential therapy: Results of an open clinical trial. *Society & Animals* 15(3), 257–267.

Knapp, S. (2013) *More than a mirror: Horses, humans & therapeutic practices*. Horse Sense of the Carolinas, Inc.

Lawley, J. and Tompkins, P. (2000) *Metaphors in mind: Transformation through symbolic modeling*. Highgate, London.

Lee, P.T., Dakin, E. and McLure, M. (2016) Narrative synthesis of equine-assisted psychotherapy literature: Current knowledge and future research directions. *Health & Social Care in the Community* 24(3), 225–246.

Lujan, K. (2012) *The challenges and discoveries in using equine assisted psychotherapy approaches by counseling practitioners in the southwest* (Unpublished doctoral dissertation). University of New Mexico.

McConnell, P.J. (2010) *National survey on equine assisted therapy: An exploratory study of current practitioners and programs* (Unpublished doctoral dissertation). Walden University.

McNair, D.M., Lorr, M. and Droppleman, L.F. (1971) *Profile of mood states manual*. Multi-Health Systems, Tonawnda, NY.

Notgrass, C.G. (2011) *Equine-assisted psychotherapy: A study of U.S. practitioners trained by the Equine-Assisted Growth and Learning Association* (Unpublished doctoral dissertation). St. Louis University: St. Louis, MO.

Notgrass, C.G. and Pettinelli, J.D. (2015) Equine-assisted psychotherapy: The Equine-Assisted Growth and Learning Association's model overview of equine-based modalities. *Journal of Experiential Education* 38(2), 162–174.

Nuremberg, J.R., Schleifer, S.J., Shaffer, T.M., Yekllin, M., Desai, P.J. *et al.* (2014) Animal-assisted therapy with chronic psychiatric inpatients: Equine-assisted psychotherapy and aggressive behavior. *Psychiatric Services in Advance, October* 1, 1–7.

O'Callaghan, D.M. (2008) *Exploratory study of animal assisted therapy interventions used by mental health professionals* (Unpublished doctoral dissertation). University of North Texas.

Otto, M.W. (2000) Stories and metaphors in cognitive-behavior therapy. *Cognitive and Behavioral Practice* 7, 166–172.

Palmatier, R.A. (1995) *Speaking of animals: A dictionary of animal metaphors*. Greenwood Press, Westport, CT.

Pendry, P., Smith, A.N. and Roeter, S.M. (2014) Randomized trial examines effects of equine facilitated learning on adolescents' basal cortisol levels. *Human-Animal Interaction Bulletin* 2(1), 80–95.

Primi, R., Miguel, F.K., Cruz, M.B.Z., Couto, G., Barros, D.P. *et al.* (2006) *Teste de Criação de Metáforas (Formas A, B e C)*. LabAPE, Universidad São Francisco, Itatiba.

Rothbaum, B.O., Astin, M.C. and Marsteller, F. (2005) Prolonged exposure versus eye movement desensitization and reprocessing (EMDR) for PTSD rape victims. *Journal of Traumatic Stress* 18(6), 607–616.

Schoo, A. (2009) Counseling: Using metaphorical narratives. *International Journal of Reality Therapy* 28(2), 9–14.

Schultz, P.N., Remick-Barlow, G.A. and Robbins, L. (2007) Equine-assisted psychotherapy: A mental health promotion/intervention modality for children who have experienced intra-family violence. *Health and Social Care in the Community* 15(3), 265–271.

Selby, A. and Smith-Osborne, A. (2013) A systematic review of complimentary and adjunct therapies and interventions involving equines. *Health Psychology* 32(4), 418–432.

Shultz, B.N. (2005) *The effects of equine-assisted psychotherapy on the psychosocial functioning of at-risk adolescents ages 12-18* (Unpublished doctoral dissertation). M.A. Denver Seminary, Denver CO.

Shumway-Cook, A. and Woollacott, M.H. (1995) *Motor control: Theory and practical applications*, 2nd Ed. Lippincott Williams & Wilkins, Baltimore, MD.

Sommer, R. and Sommer, B.A. (2011) Zoomorphy: Animal metaphors for human personality. *Anthrozoös* 24(3), 237–248.

Symington, A. (2012) Grief and horses: Putting the pieces together. *Journal of Creativity in Mental Health* 7(2), 165–174.

Thomas, L. and Lytle, M. (2016) *Transforming therapy through horses: Case stories teaching the EAGALA model in action*. Equine Assisted Growth and Learning Association, Santaquin, UT.

Trotter, K.S., Chandler, C.K., Goodwin-Bond, D. and Casey, J. (2008) A comparative study of the efficacy of group equine assisted counseling with at-risk children and adolescents. *Journal of Creativity in Mental Health* 3(3), 254–284.

Vidrine, M., Owen-Smith, P. and Faulkner, P. (2002) Equine-facilitated group psychotherapy: Applications for therapeutic vaulting. *Issues in Mental Health Nursing* 23(6), 587–603.

Vitztum, C. (2012) Human–animal interaction: A concept analysis. *International Journal of Nursing Knowledge* 24(1), 30–36.

Watson, D., Clark, L.A. and Tellegan, A. (1988) Development and validation of brief measures of positive and negative affect: The PANAS scales. *Journal of Personality and Social Psychology* 54(6), 1063–1070.

Wilkie, K.D., Germain, S. and Theule, J. (2016) Evaluating the efficacy of equine therapy among at-risk youth: A meta-analysis. *Anthrozoös* 29(3), 377–393.

Wilson, C.C. and Barker, S.B. (2003) Challenges in designing human-animal interaction research. *American Behavioral Scientist* 47(1), 16–28.

Wilson, C.C. and Netting, F.E. (2012) The status of instrument development in the human– animal interaction field. *Anthrozoös* 25, 11–55.

Witzum, E., Hart, O. and Friedman, B. (1988) The use of metaphors in psychotherapy. *Journal of Contemporary Psychotherapy* 18(4), 270–290.

Wolpe, J. (1990) *The practice of behavior therapy*. Pergamon, Oxford.

13 Investigating Human Experiences and Perceptions of Unmounted Activities Commonly Practiced in Equine-Assisted Services

Emma Leigh Pasiuk[1]*, Christine Rudd[1], Emily Webberson[1], Nichole C. Anderson[2], Nathan Hall[1], Anastasia C. Stellato[1] and Katy Schroeder[3]
[1]Department of Animal and Food Sciences, Texas Tech University, 2500 Broadway, Lubbock, TX, 79409, USA; [2]School of Veterinary Medicine, Texas Tech University, 7671 Evans Drive, Amarillo, TX, 79106, USA; [3]Department of Counselor Education, University of Iowa, N350 Lindquist Center North, Iowa City, IA 52242, USA

Abstract
Equine-assisted services (EAS) are known to be beneficial for participants, though factors that influence their in-session experiences of horse-human interactions (HHI) remain underexplored. A mixed-methods, repeated measures design was used to investigate relationships between unmounted activities and human mood state, their activity enjoyment, and their perceptions of the horse they interacted with. We aimed to determine if participants' (n = 56) reported mood states reflected their behavior toward the EAS horses (n = 14) across activity conditions. Horses and humans completed three activity conditions: grooming, leading the horse, and leading the horse around a maze. To assess participants' mood states throughout the interaction, we measured participants' mood states before the interaction began and after each activity. A post-interaction survey was used to assess participants' enjoyment of each activity and the influence of certain horse characteristics on their experience. All interaction activities were filmed for analysis of human behavior. Results from two logistic regression models suggest that activity type was not significantly associated with the arousal (p = 0.09) or valence (p = 0.70) of participants' reported mood states; however, a trend in the post hoc comparisons indicated participants were more likely to report an activated mood state before the interaction began compared to after the first activity (i.e., post-groom) (OR: 3.15, CI: 7.79−1.28, p = 0.013). Most participants (n = 46/54) reported pleasant moods throughout the interactions, although qualitative results suggest participants experienced a wider range of affective states. Participant behavior supported their reported pleasant mood as most behaviors across conditions were affiliative toward the horse. These results identify factors that influence human experiences when interacting with horses that are incorporated into EAS.

Keywords: horse-human interactions, human behavior, human mood state

*Corresponding author: epasiuk@ttu.edu

Introduction

Equine-assisted services (EAS) encompass a variety of horse-human interactions (HHI), including adaptive riding, educational services, and speech, physical, and occupational therapy. EAS also includes mental health treatment inclusive of mounted and unmounted equine activities (Wood, 2021). These activities can vary from exercises such as herd observation, to more physical interactions, such as grooming and riding. The mental health and equine professionals facilitating these services intentionally select activities to assist clients in achieving their treatment or educational objectives. The horses incorporated into mental health and educational services are considered members of the facilitation team and hold valuable roles, such as the role of a teacher, metaphor or symbol, and a relational partner (Lee and Makela, 2015; Eagala, 2019; Ekholm-Fry, 2021).

Previous research has established that horse riding (i.e. mounted activities) provides several well-documented benefits to EAS participants, and literature suggests unmounted activities have benefits for specific client populations as well (Usadi and Levine, 2017; Eagala, 2018; Waun, 2020). Several studies have investigated the impacts of mental health treatments that incorporate horses using unmounted activities (e.g. grooming, leading, herd observation). The intended outcomes of such activities include but are not limited to, improving the client's knowledge of horse behavior and communication as well as developing relational skills through interactions with horses (Burgon, 2003; Cerino *et al.*, 2011). Research exploring the incorporation of horses in groundwork activities found a range of beneficial human outcomes including decreased tension (Hama *et al.*, 1996), increased sense of belonging (Frame, 2006), empathy (Burgon, 2011), confidence (Burgon, 2011; Gilbert, 2013), awareness of internal states (Vincent *et al.*, 2021), and connection with the horse (Frame, 2006). A recent study exploring the effects of groundwork EAS activities on individuals with chronic mental illness found the longest-lasting effect for participants was the ability to be calm (Hemingway *et al.*, 2019).

There is research that explores human mood states and behavior when interacting with other species (Turner *et al.*, 2003; Dell *et al.*, 2019). HAI researchers often measure the frequency and duration of affiliative behaviors, like initiating physical contact, laughing, and smiling, to assess the quality of human-animal relationships and bonds (Fournier *et al.*, 2016; Arrazola and Merkies, 2020; Pelyva *et al.*, 2020; Hartmann *et al.*, 2021). For example, colleagues Turner and Rieger (2001) investigated if human-reported mood correlated with subsequent interactions between singly living people and their cats; their findings indicated that participants who reported feeling negative mood states, like numbness or anger, initiated interactions with their cats significantly less compared to participants who reported feeling a positive mood. Findings from Riddoch *et al.* (2022) identified behaviors like physical affection, proximity, and attunement when exploring dog owner perceptions of behaviors that facilitate human-dog bonding. These results suggest that positive internal states during HAI may be more likely accompanied with affiliative behavior, and this has been echoed in HAI research that includes horses (Fournier *et al.*, 2021). Exploration of the relationship between human mood state and behavior during HAI is imperative, especially when interacting with horses, who are large prey species that are hypersensitive to communication. However, little research exists that explores the intersection between human mood and behaviors that occur when people interact with horses.

A single unmounted HHI can take place in a variety of environments (e.g. paddock, indoor arena, outdoor arena) and include multiple activity types (e.g. herd observation, grooming, leading). These multifaceted interactions make it difficult for researchers to measure and identify the influence of each factor. For example, one recent study that explored the impact of horse grooming on women veterans' mindfulness skills (Vincent *et al.*, 2021) indicated a significant pre-post intervention increase in women's self-identified mindfulness. The researchers reported significant findings; however, they did not investigate what experiential factors during the grooming activity contributed to participants' mindfulness changes and how participants perception of the horse they interacted with influenced their outcomes. Similarly, Atherton *et al.* (2020) investigated the impact of a 6-week intervention incorporating horses on the psychosocial and behavioral outcomes of adolescents experiencing addiction. Although findings indicated significant increases in adolescents' self-reported prosocial behaviors, such as increased 'supporting' behaviors, the pre-post design could not determine which session(s) and activity(i.e.s) influenced participants' positive behavioral outcomes.

To begin filling these knowledge gaps, it is necessary to identify the mechanisms of change and the role species-specific variables may have in mental health and educational services incorporating horses (Scopa *et al.*, 2019). In a recent study, Fournier *et al.* (2021) set out to test the anodyne-agent model. This model proposes that animals incorporated into therapeutic environments are effective as they serve as an anodyne (i.e. the animal enhances positive emotions and decreases negative emotions), and/or serve as a stand-in or agent (e.g. attachment figure, metaphor, symbol, social mediator; Fournier *et al.*, 2021). Findings supported the anodyne hypothesis as participants in both animal conditions (i.e. horse and dog) had significantly higher reports of positive mood compared to those in the no-animal control condition (Fournier *et al.*, 2021). Qualitative findings indicated a species-specific difference evidenced by a trend in participants' reporting that dogs may serve more as an anodyne (e.g. reduction in distress) while horses more often serve as an agent (e.g. social supporter) during the group learning session (Fournier *et al.*, 2021). Findings from Fournier *et al.* (2021) provide preliminary evidence suggesting participant perception and experience during animal interactions may be impacted by species-specific variables.

The current literature in mental health and educational services incorporating horses has not yet established clear relationships between horse characteristics (e.g. sounds, movements, coat color) and different activity types (e.g. grooming vs leading), which limits our understanding of how these variables might influence humans' experience during HHI. In addition, although research has largely demonstrated positive experiences during HHI, it has yet to be explored whether specific groundwork activities change participant mood state. The current study aims to explore how HHI-related factors influence a human's experience during activities commonly employed in mental health and educational services by assessing participants' mood, behavior, and activity enjoyment during unmounted, structured HHIs. Our research objectives were to (1) identify if horse characteristics influenced human experience during structured grooming, leading, and a maze activity, (2) describe participant experience and determine their level of enjoyment of structured HHI activities, (3) determine if participant-reported mood state varies across activity and time during structured HHI activities, 4) identify human behavior and physiological responses during structured HHI activities, (5) explore the relationship between participant-reported mood states and the behaviors displayed during the structured HHI activities. We hypothesized that there would be significant differences in participants' ratings of the influence each horse characteristic had on their overall experience during HHI activities, and that participant-reported mood states would differ across activities.

Methods

This report comes from a larger study that utilized mixed methods, repeated measures designed to investigate horse and human experiences during structured HHI activities commonly practiced in EAS. This investigation was approved by the Institutional Review Board (IRB2021-101) and by the Institutional Animal Care and Use Committee (#21039–04).

Participants

Human

Adult participants (n=60) were recruited from a university community via hard copy recruitment flyers posted in various buildings on campus and the university's weekly email announcement system. Individuals were excluded from participation if they were pregnant, non-ambulatory, or experienced severe allergic reactions to the equine environment.

Horse

Horses (n=16, 5 mares, 11 geldings) from a PATH Intl. Premier Accredited Center were selected to complete this study. Horses were included if they were in good physical health (e.g. no lameness) and if participating in the current study did not exceed their maximum workload for the day (i.e. 6 working hours). Two Quarter horses were removed from the study, resulting in a final sample of 14 horses. Removal was due to behavioral concerns for one horse and scheduling conflicts for the other horse. Horses ranged from 6 to 27 years of age (M = 14.3 ± 6.1 years) and varied in the length of time they worked in the facility's EAS program (M = 2.44 ± 1.87 years). The height of the therapy horses ranged from 13.3 to 15.3 hands (M = 14.46 ± 0.67 hands; 1 hand = 4 inches). Six different breeds were represented, including, 8 Quarter horses, 2 Haflingers, 1 Fjord, 1 draft cross, 1 Arabian cross, and 1 Pony of the Americas (POA).

Procedures

Human

Data were collected from April 2021–August 2021. Prior to the HHI activities, participants were brought to a private office to complete informed consent paperwork and release documents for the facility. Each participant went through three phases of testing which were pre-interaction and baseline measures, the HHI activities, and post-interaction measures. During the pre-interaction phase participants were fitted with a wrist sensor and were asked to sit quietly in the room by themselves for 5 min to collect baseline heart rate (HR). Participants were then asked to complete a pre-interaction survey that collected demographic information (e.g. sex, education level, age) and then received a safety demonstration covering equine body language (i.e. ear position, tail movement, posture, and blind spots), how to safely approach the horse, and methods for safely moving from one side of the horse to the other. Participants were then led to a 100 × 200 ft indoor arena to begin the three, 7 min HHI activity conditions (i.e. grooming, leading, and maze activity). The arena was equipped with four cameras to record human behavior during all conditions. Participants' mood states were recorded via an iPad (Apple, Cupertino, CA) questionnaire before the interaction and after each activity condition for a total of four mood state measures per participant. Both the horse and participant received a 7 min rest period between each HHI activity. After the final activity, participants were given the option to say goodbye to the horse (e.g. extra petting, talking, hugging) before reporting their final mood state. A researcher then led the participants back to the private office where they completed the post-interaction survey. The HR wrist sensor was then taken off and participants were thanked for their time and the experiment was concluded.

Horse

The horses completed 6, 7-min phases which began with a baseline rest phase. Horses were then led into the research arena and completed three activity phases with a 7-min resting phase following the completion of each activity. Physiological measurements (i.e. eye temperature and heart rate variability) were recorded during the baseline rest phase in a familiar tie area; subsequently followed by a grooming, leading, and maze activity inside the research arena. The familiar tie area (i.e. area of rest phases) and the arena were equipped with four cameras to record the horse's behavior during all phases.

HHI activities

Grooming activity

Before the activity began, a researcher recited a script that described basic grooming methods and tools. During the 7-min grooming activity, the participants had a choice of using three different brushes: a curry comb, a soft brush, and a mane and tail brush. If they decided to use the mane and tail brush, they were asked to not brush the horse's tail to maintain participant safety and horse comfort levels.

Leading activity

Prior to the leading task, a researcher recited a script that introduced basic horse leading which covered methods of stopping, turning, walking on, lead rope tension, and the appropriate way to hold the lead rope. All participants were then prompted by the researcher to practice stops and direction changes at predetermined times throughout the 7-min activity.

Maze activity

The activity began with participants practicing their leading skills with a lap around the maze in which they were to practice two stops before beginning the pattern. The activity involved leading the horse over and around poles and cones in a predetermined pattern (Fig. 13.1). Participants were directed during their first round through the maze, and if they had time left over, they were asked to complete the maze again. If participants finished a second lap through the maze before the 7 min were completed, they were asked to complete one more lap along the maze's perimeter.

Measures

Pre-interaction survey

Participants completed a 10-item questionnaire to collect demographic information, prior horse experience, and comfort level around horses. Surveys and questionnaires were administered in Qualtrics® (Qualtrics, Provo, UT) via Apple iPad (Apple, Cupertino, CA, OS version 14.4.2), with paper copies available in the event of an interruption in internet service.

Pick-A-Mood

The Pick-A-Mood (PAM, Fig. 13.2) is a one-item, pictorial self-report measure created to assess mood state (Desmet *et al.*, 2016a). The PAM contains three sets of cartoon characters, a male, a female, and a robot, each character set displays eight mood expressions (i.e. tense, irritated, sad, bored, calm, relaxed, cheerful, excited), and an additional mood 'neutral', for participants to choose from. The PAM demonstrates appropriate construct validity (75–96% correct identification of mood states) for reporting mood states and has been employed in various studies assessing human mood states (Bjorn *et al.*, 2013; Desmet *et al.*, 2016b; Zhuang and Gadiraju, 2019). The robot cartoon set was used for this study, as its character does not indicate sex, age, or race. The Pick-A-Mood was administered before the first activity as well as following each of the three HHI activities.

Post-interaction survey

After the final activity, participants were led back to the private office to complete a 13-item post-interaction survey. Six Likert-type questions assessed how much (i.e. not at all – the entire time) certain horse characteristics (i.e. size, coat color, sounds, movements, and behaviors) influenced their interactions with the horse. Three Likert-type questions assessed if participants enjoyed (i.e. strongly disagree – strongly agree) each HHI activity. The survey concluded with three open-ended questions to qualitatively assess participant experience (i.e. what horse characteristics and behaviors influenced their enjoyment/lack of enjoyment of the activities, the most memorable part(s) of their experience, and anything else they would like to share about their horse interaction).

Human behaviour

All HHI activities were video recorded using GoPro Hero9 Black action cameras (GoPro, San Mateo, California) and AKASO EK7000 action cameras (AKASO, Frederick, Maryland), mounted in each quadrant of the arena capturing four different angles. To examine whether participants' reported mood states align with participant behaviors during the activity conditions, participant behaviors were assessed using an ethogram (Table 13.1) by two independent observers using Noldus Observer XT 16 (Noldus, Wageningen, the Netherlands). The ethogram was modified from previous

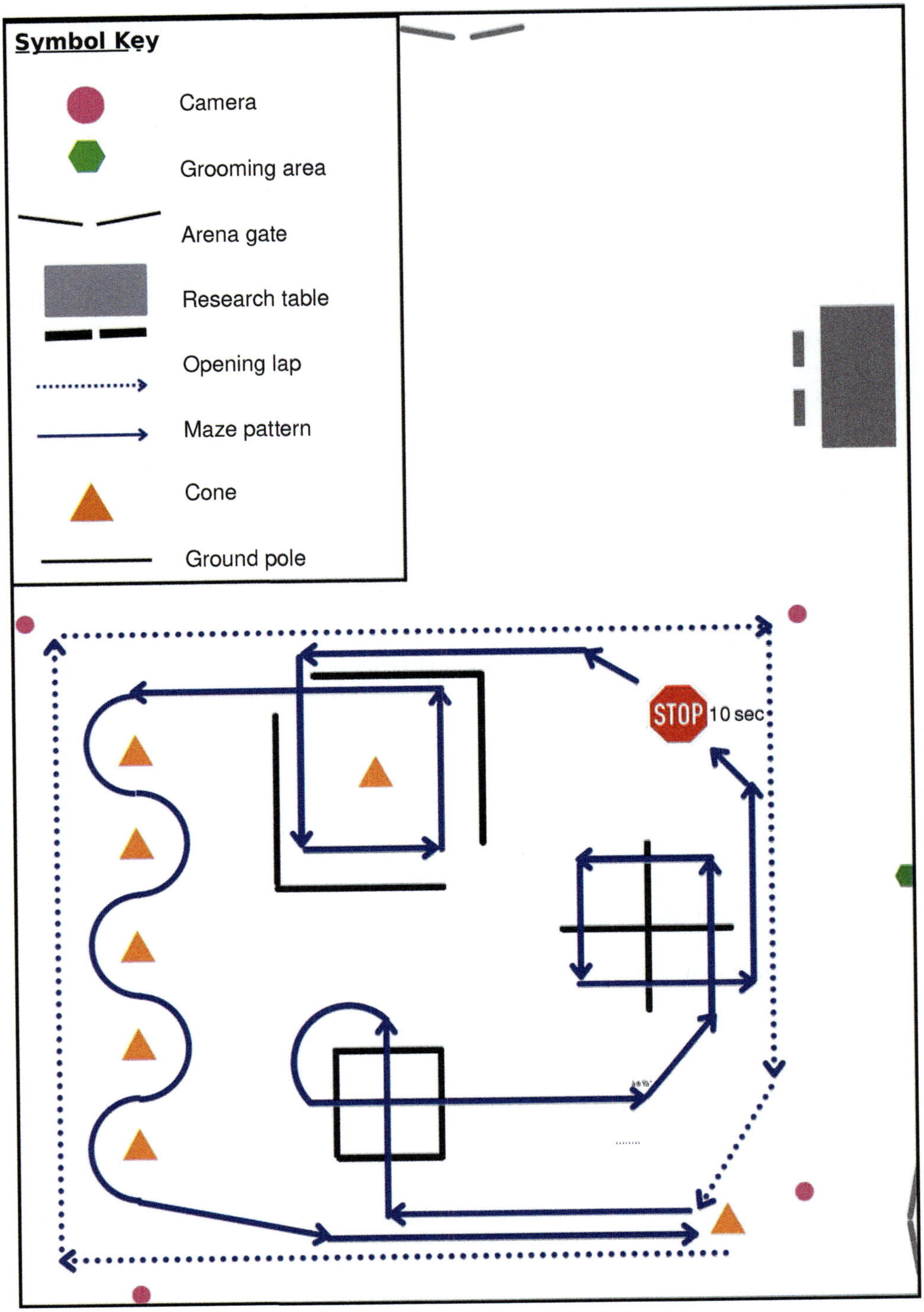

Fig. 13.1. Diagram displaying the maze activity pattern. Participants (*n*=56) led the horse through one practice lap indicated by the dotted arrows in a clockwise direction around the pattern. Once they made it to the cone in the lower right corner, they began the course in the direction of the solid arrows, beginning with the closed square of poles and ending with the serpentine cones, before returning to the start cone.

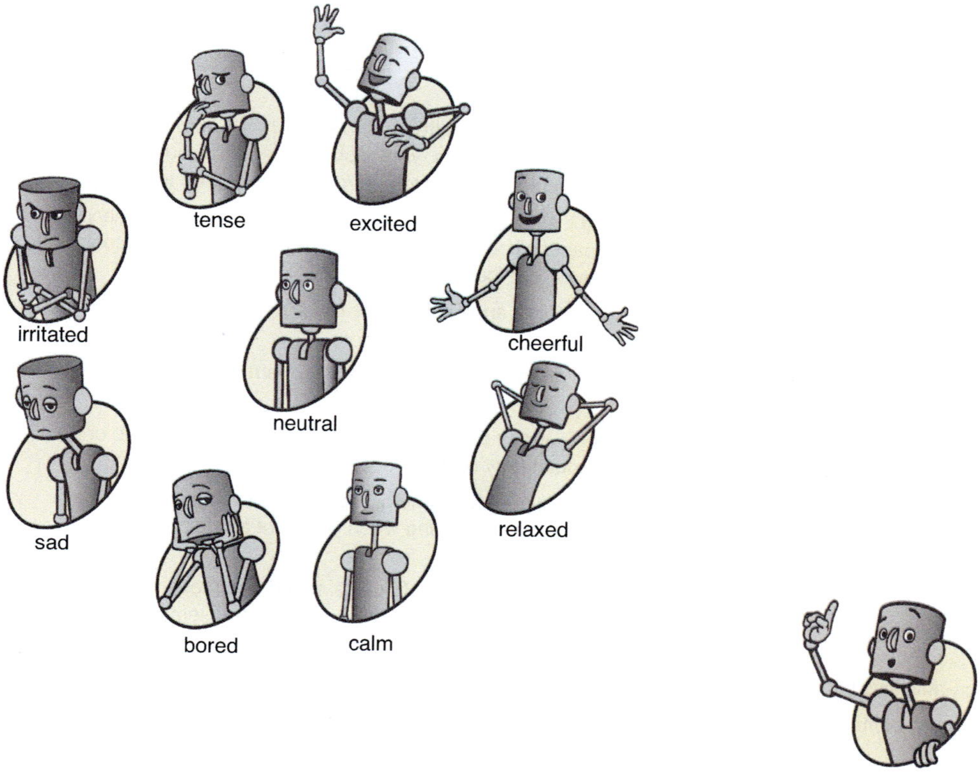

Fig. 13.2. Pick-A-Mood (PAM) Visual, including valence [pleasant (excited, cheerful, relaxed, calm), unpleasant (tense, irritated, sad, bored)] and arousal [activated (irritated, tense, excited, cheerful), deactivated (sad, bored, calm, relaxed)].

HAI research that developed and evaluated a behavior scale to explore social behaviors and affiliation (Horvath *et al.*, 2008; Feldman, 2011; Feldman, 2012; Fournier *et al.*, 2016), such as smiling, laughing, initiating physical contact, accepting physical contact, hugging, and petting/stroking/touching the animal. Each behavior was therefore categorized as an affiliative or non-affiliative behavior. The area of the horse's body the participants groomed was also recorded. Continuous sampling was used to score the frequency of all behaviors performed during HHI activities. For the behaviors 'petting/patting/stroking the horse's body,' 'front half of the horse,' and 'back half of the horse,' the participant had to be displaying the behavior for more than 4 s. This was to differentiate from the 'initiate contact' behavior which included any initiated contact under 4 s. Additionally, in the behavioral definition of 'avoiding contact,' we defined contact initiated by the horse as the horse turning their head towards the participant. Finally, for the behavior 'push away,' it was not scored when the participants were intentionally asking the horses to turn away from them in the context of directing the horse during handling, as this was a different behavior than trying to avoid contact during the leading and maze activities.

To establish interrater reliability (IRR), 10% of the data set (six participant sessions) was randomly selected. Of the six sessions, two sessions were used for ethogram training and were rated by both observers to establish IRR with a moderate Kappa of $k=0.75$. The remaining four sessions were then scored by both observers and IRR was assessed once more, resulting in a Kappa of $k=0.83$. With IRR established the remaining data was split in half for each observer to rate.

Table 13.1. Ethogram for human behaviors during the grooming, leading, and maze HHI activities.

Behavior	Definition
Affiliative behaviors	
Initiating physical contact	Any steps toward and/or attempts to extend their arm/hand toward the horse to initiate contact, where physical contact may or may not be achieved. Includes any attempts/contact that is under 4 s.
Accepts physical contact initiated by horse	The horse initiates contact by extending and orienting their head in the participant's direction, and the human accepts initiated contact by reaching toward or touching the horse.[a]
Hugging	Wraps one or both arms around the horse's body in an affectionate manner.[a]
Petting/Patting/Stroking the horse's body	Places hand on the horse's body, moving from one point to another.[a, b]
Touching or reaching for the horse's face	Extend their hand or tool to the horse's face to groom, greet, or pet. Physical contact may or may not be achieved.[a]
Smiling	Lips are pulled up in the corners, eyes may be squinting, and teeth may be showing. Only coding when you have a clear visual of the participant smiling.
Laughing	The human is eliciting a spontaneous sound and movement of the face that is the instinctive expression of lively amusement or being content.
Non-affiliative behaviors	
Avoiding contact	The human steps away or retracts their arm/hand from the horse to avoid contact initiated by the horse. Contact initiated by the horse is defined as the horse turning its head toward the participant.[a]
Startle	Flinching, quickly turning or pulling away, or abrupt stiffness in their body posture/language.
Push away	A limb, body, or lead rope is used to block physical contact or increase the distance between participant and horse.
Areas of grooming	
Front half of the horse	Grooming tool used to brush the horse's body from their withers and elbow forward for more than 4 s. All grooming that occurs in front of the horse's heart rate monitor.
Back half of the horse	Grooming tool used to groom the horse's body from their withers and elbow backwards for more than 4 s. All grooming that occurs behind the horse's heart rate monitor.
No grooming	Participant has stopped grooming.

[a]Adapted from Fournier *et al.* (2016).
[b]Adapted from Horvath *et al.* (2008).

Human heart rate

Human heart rate was collected using the Empatica E4 (Empatica Srl, Milan, Italy) a wearable wireless wristband. After completing the consent form and the pre-interaction survey, the E4 was fitted to each participant's non-dominant wrist to wear during the duration of the equine activities.

Data analysis

Quantitative data

All statistical analyses were computed in Jamovi (Version 2.3, Sydney, Australia). Two mixed logistic regression models were developed to examine the associations between interaction conditions (pre-interaction, post-grooming, post-leading, post-maze) and the following participant-reported mood states: valence (pleasant vs unpleasant), and arousal (activated vs deactivated), with participant included as a repeated measure. Friedman's test was used to assess differences in participant enjoyment across activities and to assess differences between how participants reported being influenced by horse characteristics, a repeated measures ANOVA was used. Descriptive analyses were run to assess the frequency and percentage of total behaviors (affiliative vs non-affiliative) observed during each activity condition. The Empatica E4 HR wrist sensor data could not be analyzed due to the number of movement artifacts. It is possible the electrodes did not maintain consistent contact with the participants' skin; therefore, the data was unusable.

Qualitative data

Two open-ended post-interaction survey responses were analyzed using inductive qualitative content analysis (Mayring, 2014), and pattern analysis was used for the third question in the survey. Inductive qualitative analysis employs *in vivo* coding to identify subcategories and categories from the data (Hsieh and Shannon, 2005). For all questions, the data was prepared in a Microsoft Excel spreadsheet (Version 16.0, Redman, Washington) for coding by organizing responses into meaning units, each representing a principal idea. For example, a participant's response could contain several ideas, each coded as a distinct meaning unit. This process was followed by a round of initial coding for each open-ended question. Peer debriefing with a second member of the research team was employed during the recursive rounds of coding to discuss code labels and definitions. A coding scheme was then developed for each question to assign code labels to meaning units. A third member of the research team was trained in the coding scheme for each question and served as a second coder to independently review and assign codes to the meaning units in each question. To establish intercoder agreement, both coders independently coded 10% of each question's meaning units and assessed the percentage of agreement which produced 90, 67, and 72% agreement (McAlister, 2017). Coder differences were discussed, and data was re-coded to resolve discrepancies. This occurred in several rounds until agreement was achieved. Following this process, meaning units were sorted into subcategories and categories to describe the data as-a-whole.

Results

Sixty adult participants (50 females, 10 males) agreed to participate in the study (M = 35.2 years of age ±14.7 years). Data from four participants were removed as they interacted with the two horses that were excused from the study. As such, our analysis only included 56 participants who interacted with the 14 horses remaining in the study. Participant demographics and horse experiences are described in Table 13.2.

Quantitative data

Human mood state

Our hypothesis that participant-reported mood states would be different across activity conditions was not supported. Descriptive statistics indicate most participants reported pleasant mood states across all activities (pre-interaction: n=45, 83%; after grooming: n=50, 91%; after leading: n=47, 87%; after maze: n=54, 94%, Fig. 13.3). The number of participants who reported a deactivated mood state increased two-fold from pre-interaction (n=12, 22%) to post-grooming (n=25, 45%). Results from the regression models suggest that activity type was not significantly associated with reports of arousal (p=0.09) and valence (p=0.70). When exploring post hoc comparisons of human-reported mood within each individual activity condition, participants were more likely to report an activated mood state before the interaction activities began compared to after grooming (OR: 3.15, CI: 7.79–1.28, p=0.013, Fig. 13.4).

Activity enjoyment

Friedman's test showed a significant difference between participants' enjoyment and the activity conditions (X^2 (2, n=56) = 8.33, p=0.02).

Table 13.2. Participant demographic results.

Variables	Values	*N*=56	Percentage (%)
Ethnicity			
	Asian American	5	9.0
	African American	5	9.0
	Hispanic/Latino	9	16.0
	White	34	61.0
	Multiple Ethnicities	3	5.0
University affiliation			
	Undergraduate Student	13	23.0
	Graduate Student	11	20.0
	Faculty	6	11.0
	Staff	20	36.0
	Combination of Affiliations	1	2.0
	No Affiliation Reported	1	2.0
	Other	4	7.0
Previous horse experience			
	Less than 6 months	33	59.0
	6 months–1 year	2	4.0
	2–5 years	2	4.0
	6–10 years	7	13.0
	11 + years	12	21.0
Comfort level around horses			
	Extremely Uncomfortable	1	2.0
	Uncomfortable	2	4.0
	Neutral	14	25.0
	Comfortable	18	32.0
	Extremely Comfortable	21	38.0

Durbin-Conover pairwise comparisons indicated that participants reported enjoying the maze activity more than the leading (p=0.02) and the grooming activity (p<0.01).

Horse characteristics

Our hypothesis that horse characteristics would influence participants' experience during the structured HHI activities was supported. A repeated measures ANOVA indicated a significant difference in the way participants reported being influenced by horse characteristics ($F(4) = 31.6$, $p<0.001$). Paired sample t-tests revealed that the horse's coat color had significantly less influence than other horse characteristics like size ($t(55) = -3.56$, $p<0.001$), behaviors ($t(55) = -8.46$, $p<0.001$), or movements ($t(55) = -7.95$, $p<0.001$). Also, the horse's size had significantly less influence than the horse's behaviors ($t(55) = -5.43$, $p = < 0.001$) and movements ($t(55) = -4.86$, $p = < 0.001$). No other significant differences were detected between horse characteristics e.g. behavior and movement ($t(55) = 0.00$, p=1.00); size and sounds ($t(55) = 1.40$, p=0.15); coat color and sounds ($t(55) = -1.53$, p=0.13); Fig. 13.5.

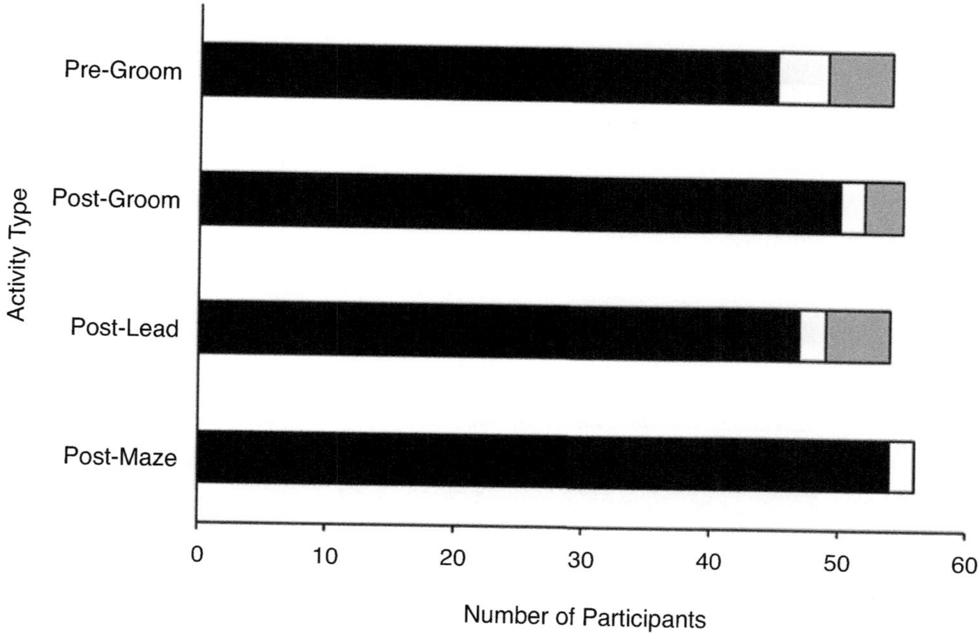

Fig. 13.3. Number of participants (N=56) who reported their valence mood state (i.e. pleasant, unpleasant, neutral) before the interaction began (i.e. pre-interaction, n=54) and after each structured HHI activity (grooming, n=55, leading, n=54, maze, n=56). Pleasant mood (black bars), unpleasant mood (gray), neutral mood (white).

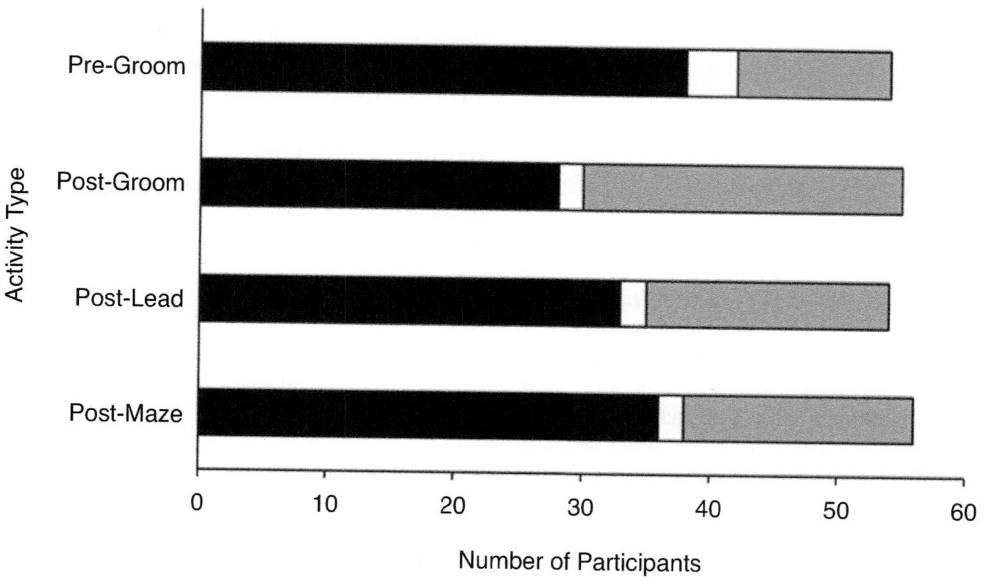

Fig. 13.4. Number of participants (N=56) who reported their arousal mood state (i.e. activated, deactivated, or neutral) before the interaction began (i.e. pre-interaction, n=54) and after each structured HHI activity (grooming, n=55, leading, n=54, maze, n=56). Activated mood (black bars), deactivated mood (gray), neutral mood (white).

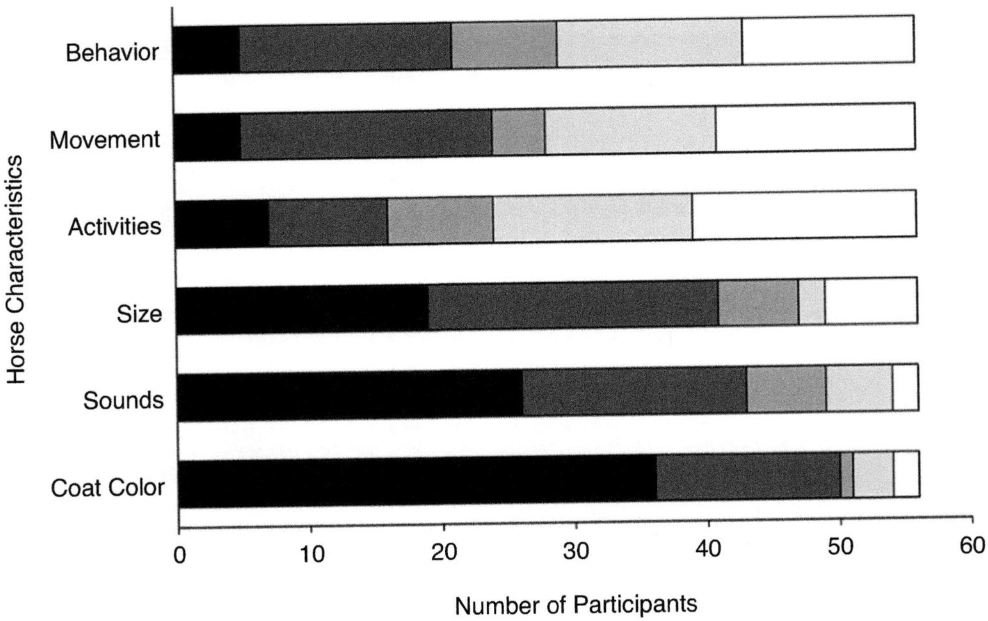

Fig. 13.5. Number of participants (N=56) reporting how influential certain horse characteristics were to their experience during structured HHI activities. The level of influence was based on the following Likert type responses: not at all (black bars), some of the time (dark gray), half of the time (light gray), most of the time (off-white), the entire time (pure white bars).

Human behaviour

Participants groomed the front half of the horse (34%, 327/967) more frequently than the back half of the horse (27%, 264/967). Affiliative behaviors represented 97% (570/588) of behaviors observed in the grooming, 85% (215/254) of behaviors in the leading and 90% (146/162) of behaviors in the maze activity. Initiating contact (21%, 125/588), touching/reaching toward the horse's face without a grooming tool (121/588, 21%), and accepting contact from the horse (18%, 104/588) were the most frequent affiliative behaviors in the grooming condition. Initiating contact (leading: 27%, 69/254, maze: 25%, 40/162) and smiling (leading: 28%, 71/254, maze: 25%, 41/162) were the most common affiliative behaviors in the leading and maze activities. A small proportion (3%, 18/588) of behaviors during grooming were non-affiliative. A larger portion of the observed behaviors in the leading (15%, 39/254) and maze activities (10%, 16/162) were non-affiliative. The most frequently performed non-affiliative behavior in the grooming condition was avoiding contact (2%, 14/588). The most frequently performed non-affiliative behaviors in the leading and maze activities were pushing away (leading: 10%, 25/254, maze: 4%, 6/162) and avoiding contact (leading: 4%, 11/254, maze: 4%, 7/162).

Qualitative data

Question 1

All participants (n=56) answered the first open-ended question in the post-interaction survey which asked, 'What characteristics and behaviors of the horse do you think might have influenced your enjoyment, or lack of enjoyment, of the activities?' The qualitative content analysis produced a total of 101 meaning units and three major categories which were: *Affiliative Experiences, Disengaged Experiences,* and *Unexpected Experiences.*

Affiliative Experiences: This category was comprised of 64 meaning units and describes the factors that participants believed contributed to their positive experiences during the HHI such as strengthening their bond and relationship with the horse. Within this category emerged three subcategories: *Horse Traits, Horse Behaviors,* and *Human Perceptions.*

The first subcategory, *Horse Traits,* was comprised of 21 meaning units and described the calm demeanor and other personality traits of the horse that impacted participants' enjoyment of the interaction. An example of a response in this subcategory was 'Enjoyed the activities because the horse was calm, if he was nervous/scared I wouldn't have acted the same.' Another subcategory, *Horse Behaviors,* described the types of horse behaviors including sounds and movements that contributed to participants enjoying the interaction and consisted of 33 meaning units. An example of a response in this subcategory was 'Enjoyed horse turning toward to rub against me even if it was to scratch.' The last subcategory, *Human Perceptions* described responses in which participants' report perceiving the horse as having a positive experience or enjoying the interaction. This subcategory consisted of 10 meaning units, and an example of a response was '[The horse], he seemed to enjoy the maze which increased my enjoyment.'

Disengaged Experiences: The second major category represented 15 meaning units that describe factors that participants indicated as anxiety-provoking or led to negative experiences during the HHI such as hypervigilance, distraction, and tension. Within this category, two subcategories emerged which were *Horse Behaviors* and *Human Perceptions of Horse's Negative Experience.*

The first subcategory, *Horse Behaviors,* described the horse's non-compliance and disengaged behaviors as well as sounds and movements the horse made that contributed to participants' lack of enjoyment. This subcategory consisted of 9 meaning units; an example of a response was 'Horse disinterested in being touched and grooming, didn't sniff, look, or move toward me and had slow pace.' The second subcategory, *Human Perceptions of Horse's Negative Experience,* was comprised of 6 meaning units and described responses in which participants commented on feeling like they were doing something wrong and perceived the horse as not enjoying the interaction. An example of a response was 'Difficult to read [Horse] behaviors because no prior grooming or leading experience. When leading he pulled away and I worried I did something to aggravate him.'

Unexpected Experiences: The last major category had 16 meaning units and described factors that participants viewed as unexpected and led to mixed or conflicted feelings during the interaction such as enjoying the activities while being sad that the horse did not. An example of a response was 'Initial movement during grooming intimidating, as time passes understanding their behaviors becomes intriguing and almost entertaining.'

Question 2

All participants (n=56) answered the second open-ended question in the post-interaction survey which asked, 'What was the most memorable part(s) of your experience with the horse today?' The qualitative content analysis produced a total of 79 meaning units and two major categories *Memorable Features of the Activities* and *Memorable Horse Interactions.*

Memorable Features of the Activities: The first category had 14 meaning units. Participants commonly noted the novelty of the activities with the horses as memorable, some reported this experience being their first time interacting with a horse. One participant wrote, 'The entire experience was something I've never had before and pretty much all parts of the experience were a surprise and that includes the behavior of the horse and stuff like that.' Other participants reported the activities, such as grooming, brought a sense of calmness to their interactions with one participant explaining 'Grooming: the horse stayed calm and I was surprised how calming it was for me.' Finally, some participants noted only the activities as the most memorable factor of their experience. An example of a response in this category was 'Leading the horse through obstacle.'

Memorable Horse Interactions: This category had 66 meaning units and describes horse-focused memorable events, rather than an activity focus. This includes responses about building a connection with the horse, or the behavior of the horse that participants reported made their experience memorable. Some responses in this category focused on the participants and horse sharing or increasing their bond and connection with one another or noting that they both enjoyed the activity. This could be responses that mention communication, trust, or an increase in confidence and comfort. An example of a response in this code was:

I really liked to walk [the horse], especially through the maze and over the obstacles. I imagine there is some component of trust that needs to be established for the horse to feel comfortable walking over the obstacles and following the walker. It made me happy and feel relaxed to think that maybe [the horse] felt a connection with me, and it caused me to feel a fondness and connection with him.

Other participants described the horse's physical or personality traits as being most memorable. Attributes include participants mentioning the horse being responsive or accepting. An example of a response in this code was: '...[The horse] made the maze activity memorable because I was unsure of my next phase of the course and stopped quickly and she did not respond in kind. She was patient while I figured out my next step...' The category also described responses that focus on the horse's behavior as being the most memorable. This could be meaning units that describe being unsure of a particular behavior, enjoying a behavior, noticing a behavior that stuck with them, or describing the absence of a specific behavior. This also includes compliance, non-compliance, or mixed compliance in which the participant writes about the horse doing a particular behavior such as pawing or perking ears. An example of a response in this category was, '...[The horse] seems more willing to go with me. Again, the perked ears were surprisingly delightful and felt like a reward. I was also happy that he seemed to pick up his pace doing the course...'

Question 3

Eighty-nine percent (89%) of participants (n=50) answered the last open-ended question in the post-interaction survey which asked, 'Is there anything else you would like to share about your experience with the horse today?' The qualitative content analysis produced a total of 71 meaning units which were coded based on patterns in the data. Patterns were defined as multiple meaning units that repeated the same central idea and incorporated similar keywords (e.g. 7 meaning units described that this was participants' *first time* or *first experience* with a horse). Six patterns were identified: *Novel Experience, Experiment & Procedural Feedback, Encouraged Future Horse Interactions, Appreciated & Enjoyed Opportunity, Increased Positive feelings/ Decreased Negative Feelings,* and *Horse Attributes.* Responses included comments regarding the research design as one participant reported, 'Having handled horses before this activity was really slow in pace. Maybe try something a little faster pace. Sometimes 7 min is a little to short to do an activity.' Other respondents discussed that their experience during the study encouraged them to pursue future horse interactions, one respondent said, 'I think I will be actively seeking an opportunity to ride and get good at it.' Participants also reported that their interaction with the horse increased positive feelings or decreased negative feelings, one participant disclosed, '...the stress of chronic pain and all the stress from COVID, went away for a little bit.' Other common responses to this question repeated similar sentiments to the first open-ended question discussing attributes of the horse they interacted with, one participant recalled, 'The first thing I noticed about [Horse] when I walked into the arena was his size. He is quite short, so I assumed he would be a docile horse.' The remaining responses touched upon the enjoyment and novelty of interacting with horses and the participants' appreciation.

Discussion

Results of our study suggest that participants experienced pleasant mood states during HHI activities, demonstrated affiliative behaviors, and that their experience was influenced by a horse's physical and behavioral characteristics. Though activity type was not significantly associated with the arousal or valence of participants' mood states, qualitative content analysis reveals a breadth of affective states experienced during activities. It is possible that the relationship between mood states and activity type was not captured due to the lack of variation in reported mood states, with most participants reporting pleasant mood states throughout the interaction. Our findings of positive mood states during HAI are supported by previous literature, as dog and cat interactions have been found to heighten positive moods (Turner *et al.*, 2003; Dell *et al.*, 2019). Post hoc comparisons of each activity type indicated that participants' arousal changed from activated (i.e. cheerful or excited) before the interaction began (i.e. pre-interaction) to deactivated (i.e. relaxed or calm) after the first activity (i.e. post-grooming). It is possible that the increased arousal reported before the interaction began was due to anticipation or other feelings (e.g. excitement, fear), which diminished once the interaction activities began. On the other hand, the change in arousal echos previous research that measured the effect of stroking a horse on humans' heart rate and self-reported levels of arousal (Hama *et al.*, 1996; Hemingway *et al.*, 2019). For example, Hama *et al.* (1996) found that participants' heart rate eventually decreased to baseline levels when they engaged in stroking horses for 90 s. Further, results indicated that participants' reports of tension decreased following the act of stroking a horse.

Hemingway *et al.* (2019) conducted a similar study exploring the impacts of embodied nature-based equine-assisted interventions on young people with chronic mental illness. Participants in this intervention took part in similar unmounted activity conditions employed in the current study (i.e. grooming and leading the horse over objects), and the results indicated calmness was the most consistent outcome among participants. Thus, our findings may suggest that grooming activities yield calming effects for humans. Participant reports from the post-interaction survey support this trend as many participants who reported the grooming activity as their favorite activity described the experience as 'calming' or 'relaxing.' Further, during code labeling, participants specified the grooming activity in half of the meaning units that comprised the code 'Calmness During Activities' as memorable factors of the interaction. Participants' behavioral data reflect their reports of positive grooming experiences, as almost all the observed behaviors in this activity were affiliative. This information is beneficial for practitioners as certain activities may support specific client populations and their therapeutic goals more than others. As the most performed behaviors were initiating contact, touching the horse's face, and accepting contact from the horse, grooming could be an activity employed to cultivate deeper emotional connection through physical touch. This idea is supported by previous research that suggests a grooming activity can create and enhance the connection between horses and humans (Meola, 2017; Atherton *et al.*, 2020; Acri *et al.*, 2021; Lassell *et al.*, 2021).

In the current study, almost all the participants agreed they enjoyed all the groundwork activity conditions. This aligns with previous research reporting the physical and mental benefits associated with mounted activities, like therapeutic riding, that often incorporate certain groundwork activities like grooming (Kaiser *et al.*, 2004; Cerino *et al.*, 2011). Participants who reported the grooming as their favorite activity explained that '*...you can learn so much about them during that time*' and '*starting with grooming was a way to build trust with the horse.*' Participants who reported leading as their favorite activity condition explained that they 'gained confidence' as they learned to lead the horse and enjoyed how 'responsive' the horse was to follow their direction. Further, participants reported enjoying the maze activity significantly more than the leading and grooming activity. It is possible participants felt they had built confidence and skill through the first two activities of grooming and leading, resulting in the maze activity being the most enjoyable. This is echoed in the qualitative data as participants who reported the maze as their favorite activity expressed how they '*...appreciated his trust in me to get us through...*' and that some participants '*...felt more confident and comfortable*' by the end of the maze activity. It may also be that the level of challenge and mastering of skills needed to complete the maze activity led participants to enjoy the maze interaction activity the most. Research suggests that HHI may provide a rich environment for participants to master various skills, which is one mechanism that can cultivate self-efficacy and self-confidence (Bandura, 1977; Hauge *et al.*, 2015; Battisti *et al.*, 2017). Our results highlight the human experience during groundwork activities and suggest future research in the field of EAS may benefit from further exploring the effects of activity type on different client populations' experiences of HHIs.

Results from the current study indicate that certain horse characteristics influence participants' experience during an HHI; however, there was a lack of agreement as to how influential characteristics were across participants. Although most participants agreed that the horse's coat color did not influence their experience at all, there was greater variation in the reported influence of other horse characteristics (i.e. size, sounds, behaviors, and movements) suggesting the impact of certain horse characteristics may vary depending on the individual and context. Further, participants' open-ended responses suggest certain characteristics can enhance their interaction with a horse. One participant described how the horse's size influenced their experience reporting, '*...I have almost no experience with horses and I couldn't believe that all of a sudden I could lead this giant animal and he seemed so comfortable...*' Another participant shared, '[The horse's] *small size made me feel comfortable.*' It is possible that the size of the animal they are interacting with emphasizes the processes present in sessions more than previously realized, such as skill building and cultivating a comfortable environment for interactions. Previous literature suggests horses' large size could assist veterans with post-traumatic stress disorder in regulating their emotions as those who may be quick to anger or experience outbursts often modify their behavior around an animal of such size (MacLean, 2011; Waun, 2020). Results from the current study can guide EAS facilitation teams by enhancing their understanding of clients' individual preferences by exploring which horse characteristics feel most salient to them. Thus, incorporating horses whose characteristics may emphasize the successes of the clients may aid in reaching their educational or treatment goals. Future research may consider selecting specific horse characteristics identified in the literature (i.e. size, movements, vocalizations) to determine the influence such characteristics may have on human behavior, mood, and activity enjoyment during HHIs.

There were several limitations in the current study that may have influenced the study findings. First, the activity conditions were not randomized, therefore we cannot ascertain which activity was driving the trend in reported mood state arousal. However, the protocol for this study was designed to ensure the safety of both the horses and humans involved, with little to no previous horse experience required for humans to participate. As seen in Althobaiti *et al.* (2019), structured, non-randomized procedures are often used in HHI research to allow the horse and human to become comfortable and familiar and more likely to experience a safe interaction. As such, the HHI activities were standardized and developed to gradually get the participant familiar with the horse and the activities involved (e.g. participants began with a grooming activity to become familiar with the horse, then learned the skills to lead before practicing leading the horse through a maze activity). To better establish causal links between specific HHI activities and human experiences, future research could include randomization designs. Further, prior to reporting their last mood state, participants were allowed to thank their horses at the end of the last activity. This time included extra time for petting or hugging; therefore, it is possible that participants' final reported mood state was influenced by this final interaction. In addition, as behavior coding and qualitative data analysis happened in the same timeframe, it is possible that analysis of the qualitative data influenced the behavioral coding and vice versa. To mitigate this bias, second coders were employed for behavioral and qualitative content analysis.

Conclusion

In this study, we investigated how certain horse characteristics and groundwork activities commonly practiced in EAS influence participants' self-reported mood state, enjoyment, and behavior during an HHI. Mood states and behavioral data indicate participants had pleasant interactions with the EAS horses. Additionally, results showed participants enjoyed the maze activity significantly more than the grooming and leading activity. Qualitative data analysis provided context and detail regarding which HHI components contributed to participants' overall experiences. Future research would benefit from employing mixed methods and randomized designs to explore how various unmounted equine activities influence human experiences during HHI.

Acknowledgments

We thank those who participated in the study and the equine-assisted service centers for access to the facility and their horses to complete this research.

Funding

The authors have no funders to declare.

Conflict of Interest

The authors declare no conflicts of interest.

Author Contributions

ELP, CR, NA, and KS conceptualized the data; ELP, CR, NA, KS, and NH carried out methodology; formal analysis was carried out by ELP, CR, EW, NH, KS, and ACS; CR, ACS, and KS performed the writing – reviewing and editing; ELP carried out writing – original draft, reviewing, and editing; ACS and KS supervised the study.

Ethics Statement

The authors confirm that the research meets any required ethical guidelines, including adherence to the legal requirements of the study country.

Data Availability

The data are not publicly available due to ethical restrictions.

References

Acri, M., Morrissey, M., Peth-Pierce, R., Seibel, L., Seag, D. *et al.* (2021) An equine-assisted therapy for youth with mild to moderate anxiety: Manual development and fidelity. *Journal of Child and Family Studies* 30(10), 2461–2467. DOI: 10.1007/s10826-021-02011-4.

Althobaiti, T., Katsigiannis, S., West, D. and Ramzan, N. (2019) Examining horse-human interaction by means of affect recognition via physiological signals. *IEEE Access* 7. DOI: 10.1109/ACCESS.2019.2922037.

Arrazola, A. and Merkies, K. (2020) Effect of human attachment style on horse behaviors and physiology during equine-assisted activities – A pilot study. *Animals* 10(1156), 1–13. DOI: 10.3390/ani10071156.

Atherton, L., Meola, C. and Pritchard, K. (2020) Innovative equine facilitated psychotherapy intervention for adolescents addiction treatment: A pilot study. *International Journal of High Risk Behaviors and Addiction* 9(3), 1–7. DOI: 10.5812/IJHRBA.103877.

Bandura, A. (1977) Self-efficacy: Toward a unifying theory of behavioral change. *Psychological Review* 84(2), 191.

Battisti, H.E., Battisti, F.L. and McAbee, R. (2017) Does long-term equine assisted learning have an effect on childhood weight management? *Journal of Obesity Weight Loss Therapy* 7, 288. DOI: 10.4172/2165-7904.1000338.

Bjorn, V., Moor, K., Heegaard, P. and Instefjord, A. (2013) Paper Presentation. *Investigating quality of experience in the context of adaptive video streaming: Findings from an Experimental User Study* [Paper Presentation]. Norwegian Informatics Conference, Uppsala, Sweden.

Burgon, H. (2003) Case studies of adults receiving horse-riding therapy. *Anthrozoos* 16(30), 263–276.

Burgon, H.L. (2011) Queen of the world': Experiences of 'at-risk' young people participating in equine-assisted learning/therapy. *Journal of Social Work Practice* 25(2), 165–183.

Cerino, S., Cirulli, F., Chiarotti, F. and Seripa, S. (2011) Non conventional psychiatric rehabilitation in schizophrenia using therapeutic riding: The FISE multicentre Pindar project. *Annali Dell'Istituto Superiore Di Sanita* 47(4), 409–414. DOI: 10.4415/ANN_11_04_13.

Dell, C., Chalmers, D., Stobbe, M., Rohr, B. and Husband, A. (2019) Animal-assisted therapy in a Canadian psychiatric prison. *International Journal of Prisoner Health* 15(3), 209–231. DOI: 10.1108/IJPH-04-2018-0020.

Desmet, P.M.A., Vastenburg, M.H. and Romero, N.H. (2016a) Pick-A-Mood Manual. Available at: www.diopd.org (accessed 8 July 2021).

Desmet, P.M.A., Vastenburg, M.H., Romero, N. and Pieter, D.M.A. (2016b) Mood measurement with Pick-A-Mood: Review of current methods and design of a pictorial self-report scale. *Journal of Design Research* 14(3), 241–279.

EAGALA (2018) Equine Assisted Growth and Learning Association: Eagala Military Services. Available at: https://www.eagala.org/military (accessed 24 April 2023).

EAGALA (2019) Equine Assisted Growth and Learning Association. Available at: https://www.eagala.org/ (accessed 9 March 2022).

Ekholm-Fry, N. (2021) Conceptualization of psychotherapy incorporating equine interactions in the United States. *Human-Animal Interaction Bulletin* 9(3), 94–114.

Esposito, N.W. and Fournier, A. (2023) *How Change Happens in Equine-Assisted Interventions: A Theory of Horses, Humans, and Psychotherapy*, 1st edn. Routledge. DOI: 10.4324/9781003088011.

Feldman, R. (2012) Oxytocin and social affiliation in humans. *Hormones and Behavior* 61(3), 380–391. DOI: 10.1016/j.yhbeh.2012.01.008.

Feldman, R. (2011) Maternal touch and the developing infant. In: Hertenstein, M. and Weiss, S. (eds) *Handbook of Touch*. Springer, New York, pp. 373–407.

Fournier, A.K., Berry, T.D., Letson, E. and Chanen, R. (2016) The human-animal interaction scale: Development and evaluation. *Anthrozoos* 29(3), 455–467.

Fournier, A.K., Letson, E.A., Berry, T.D., Laitala, J., Proels, L. *et al.* (2021) The anodyne-agent model: A framework for conceptualizing the animal's role in animal-assisted intervention. *Human-Animal Interaction Bulletin* 9(1), 55–75.

Frame, D. (2006) *Practices of Therapists Using Equine Facilitated/Assisted Psychotherapy in the Treatment of Adolescents Diagnosed with Depression: A Qualitative Study*. ProQuest Dissertation Publishing Ann Arbor, MI.

Gilbert, R. (2013) *Equine Assisted Psychotherapy: Prover Perspectives and Beliefs*. ProQuest Dissertation Publishing Ann Arbor, MI.

Hama, H., Yogo, M. and Matsuyama, Y. (1996) Effects of stroking horses on both humans' and horses' heart rate responses. *Japanese Psychological Research* 38(2), 66–73.

Hartmann, E., Rehn, T., Christensen, J., Nielsen, P. and McGreevy, P. (2021) From the horse's perspective: Investigating attachment behavior and the effect of training method of fear reactions and ease of handling—A pilot study. *Animals* 11(2), 1–17. DOI: 10.3390/ani11020457.

Hauge, H., Kvalem, I.L., Enders-Slegers, M.J., Berget, B. and Braastad, B.O. (2015) Persistence during tasks with horses in relation to social support, general selfefficacy and self-esteem in adolescents. *Anthrozoos* 28(2), 333–347. DOI: 10.1080/08927936.2015.11435406.

Hemingway, A., Carter, S., Callaway, A., Kavanagh, E. and Ellis, S. (2019) An exploration of the mechanisms of action of an equine-assisted intervention. *Animals* 9(6), 1–22. DOI: 10.3390/ani9060303.

Horvath, Z., Doka, A. and Miklosi, A. (2008) Affiliative and disciplinary behavior of human handlers during play with their dog affects cortisol concentrations in opposite directions. *Hormones and Behavior* 54(1), 107–114.

Hsieh, H. and Shannon, S. (2005) Three approaches to qualitative content analysis. *Qualitative Health Research* 15(9), 1277–1288. DOI: 10.1177/1049732305276687.

Kaiser, L., Spence, L., Lavergne, A. and Bosch, K. (2004) Can a week of therapeutic riding make a difference? – A pilot study. *Anthrozoos* 17(1), 63–72. DOI: 10.2752/089279304786991918.

Lassell, R., Fields, B., Busselman, S., Hempel, T. and Wood, W. (2021) A logic model of a dementia-specific program of equine-assisted activities. *Human Animal Interaction Bulletin* 9(2), 1–33.

Lee, P. and Makela, C. (2015) Horses' roles in equine-assisted psychotherapy: Perspectives of mental health practitioners. *Journal of Psychology and Behavioral Science* 3(1). DOI: 10.15640/jpbs.v3n1a9.

MacLean, B. (2011) Equine-assisted therapy. *Journal of Rehabilitation Research and Development* 48(7), ix–xii.

Mayring, P. (2014) *Qualitative Content Analysis: Theoretical Foundation, Basic Procedures, and Software Solution*. Klagenfurt. Available at: https://nbn-resolving.org/urn:nbn:de:0168-ssoar-395173

McAlister, A.M. (2017) *Qualitative Coding: An Approach to Assess Inter-Rater Reliability*. American Society for Engineering Education.

Meola, C. (2017) The effects of an equine-assisted learning supervision intervention on counselors'-in-training performance anxiety, counseling self-efficacy, and supervisory working alliance (Publication No.10760206). [Doctoral Dissertation East Carolina University], ProQuest.

Pelyva, I., Kresak, R., Szovak, E. and Toth, A. (2020) How equine-assisted activities affect the prosocial behavior of adolescents. *International Journal of Environmental Research and Public Health* 17(8), 1–13. DOI: 10.3390/ijerph17082967.

Riddoch, K.A., Hawkins, R.D. and Cross, E.S. (2022) Exploring behaviors perceived as important for human-dog bonding and their translation to a robotic platform. *PLOS One* 17(9), 1–22. DOI: 10.1371/journal.pone.0274353.

Scopa, C., Contalbrigo, L., Greco, A., Lanata, A., Pasquale Scilingo, E. *et al.* (2019) Emotional transfer in human-horse interaction: New perspectives on equine-assisted interventions. *Animals* 9(1030), 1–21. DOI: 10.3390/ani9121030.

Turner, D.C. and Rieger, G. (2001) Singly living people and their cats: A study of human mood and subsequent behavior. *Anthrozoös* 14, 38–46.

Turner, D.C., Rieger, G. and Gygax, L. (2003) Spouses and cats and their effects on human mood. *Athrozoos* 16(3), 213–228. DOI: 10.2752/089279303786992143.

Usadi, E.J. and Levine, S.A. (2017) Why we don't ride: Equine assisted psychotherapy, military veterans and moral injury. *Journal of Trauma & Treatment* 6(2), 1–5. DOI: 10.4172/2167-1222.1000374.

Vincent, A., Ballard, I. and Farkas, K.J. (2021) Mind full or mindful? A cohort study of equine-facilitated therapy for women veterans. *Journal of Creativity in Mental Health*. DOI: 10.1080/15401383.2021.

Waun, D. (2020) Exploring the effectiveness of equine assisted learning on service members diagnosed with post-traumatic stress disorder and/or traumatic brain injury (Publication No.27830015) [Doctoral Dissertation Trevecca Nazarene University] ProQuest.

Wood, W. (2021) Building the science of diverse equine-assisted services from the ground up. *Human-Animal Interaction Bulletin* 9(3), 1–4.

Zhuang, M. and Gadiraju, U. (2019) In what mood are you today? An analysis of crowd workers *mood,* performance and engagement [Conference Presentation]. In: *11th ACM Conference on Web Sciences*, Boston, MA. DOI: 10.1145/32925223326010.

14 "Before, he fought every day with the horse and with me": Reducing Violence in a Guatemalan Community Through a Horse-Handling Program

Judith L. Gibbons[1*], Catherine A. Cunningham[2], Leslie Paiz[2], Katelyn E. Poelker[1] and Marco Antonio Montufar Cardenas[3]

[1]Saint Louis University; [2]La Ronda Equestrian Club; [3]Fundación Equinos Sanos Para El Pueblo

Abstract

Community and family violence are endemic in Guatemala. We evaluated the effectiveness of a horse-handling program to reduce violent attitudes and aggressive behavior. Eighteen community members who worked with horses in their daily lives (16 men, 2 women, ages 15 to 58) participated in four weekly sessions of embodied experiences with horses. The program taught Monty Roberts' Join-Up®, a method of non-violent handling, as well as desensitization of horses to feared objects. Compared to the pretest, on a posttest participants were less likely to endorse violent attitudes with respect to wife-beating, corporal punishment of children, and physical punishment of horses. Their horses also were less reactive to the owner's approach. Female relatives reported that following the program participants provided better care for horses and were calmer and less aggressive with other people. These findings imply that an equine-facilitated program shows promise for transforming communities in which violence is prevalent.

Keywords: Equine-human interactions, attitudes towards violence, embodied experience, Guatemala

Introduction

The most perilous risk for children and adolescents in Guatemala is immersion in a climate of violence, as a victim, perpetrator, or community member (UNICEF, 2012). Guatemala ranks second among countries of the world in rates of violent deaths of children and adolescents, ages newborn to 20 (UNICEF, 2014). Children are also vulnerable to physical abuse, an offense that is greatly underreported, either for fear of retribution or a belief that severe physical punishment is appropriate for disciplining children (UNICEF, 2009). According to one survey, 25% of Guatemalan adults reported knowing a child being physically abused, and of those known

*Corresponding author: gibbonsjl@slu.edu

cases, less than one third had been reported to authorities (UNICEF, 2009). Gang violence also plagues the country with an estimated 8000 to 14,000 members (Guatemala Human Rights Commission (GHRC), 2012; United States Department of State, 2014). For youth who struggle to find employment, gang membership is a tempting option, with initiations beginning as early as age 14 (Guatemala Human Rights Commission (GHRC), 2012). In a recent ethnographic study, Guatemalan adolescents claimed, 'Violence is who we are' (Bellino, 2014, p. 149).

Violence can be endemic; domestic violence, intimate partner violence, and child physical abuse are interrelated and intergenerational (Renner and Slack, 2006). Attempts by the Guatemalan government and by US AID to reduce violence in Guatemala have been to a large extent unsuccessful (Office of Inspector United States Department of State, 2013; Office of Inspector General, U.S. Agency for International Development, 2014). Research in general (originating mostly from North America and Europe) on interventions aimed at reducing intra-familiar and intimate partner violence has failed to identify an empirically-supported effective intervention (Gondolf, 2011) and some have called for more science to support the commonly administered programs (e.g. Dutton and Corvo, 2006). Moreover, cultural sensitivity is essential for any intervention (Carroll *et al.*, 2014), and there is a scarcity of evidence about violence prevention and attenuation in low- and middle-income countries, including Guatemala. A PsycINFO search revealed no articles that included the key words violence prevention, violence reduction, or violence intervention and Guatemala; however, movement and sport therapy have been reported to be useful in improving self-esteem and a coherent worldview in Guatemalan women who had been victims of violence (Ley and Rato Barrio, 2011; Ley and Rato Barrio, 2012). It is clear that more efforts are needed to develop programs addressing the problem of violence in Guatemala, and to evaluate their effectiveness.

A promising approach for reducing violence is human-animal interaction (HAI), especially interaction with horses. Equine-assisted interventions have been applied to treat a wide range of psychological and physical problems as well as to promote human development and well-being (e.g. Masters, 2010; Selby and Smith-Osborne, 2013). Horses have been 'engaged as change agents' (Selby and Smith-Osborne, 2013, p. 419) with varying success, to promote behavioral changes in combat veterans (Lanning and Krenek, 2013), children who have experienced family violence (Schultz *et al.*, 2007), and at-risk youth (Burgon, 2011). Horses may be particularly well-suited for these applications because people can become attached to horses (DeAraugo *et al.*, 2014), and horses are sensitive detectors of nonverbal behavior (Vidrine *et al.*, 2002; Hallberg, 2008). What 'horses bring to the therapeutic environment [is] cooperation, patience, willingness, receptiveness, and, after millennia of domestication, an orientation toward people' (Selby and Smith-Osborne, 2013, p. 418).

Research evaluating the effectiveness of equine-partnered experiences on anger, aggression, emotional regulation, or violence, is scarce, primarily descriptive, and/or poorly controlled (Selby and Smith-Osborne, 2013; Anestis *et al.*, 2014). In addition, the intervention itself, the interaction with horses, has varied from grooming and ground work to riding and vaulting. Finally, in many studies the equine part of the program was considered adjunctive or complementary to other counseling or psychological interventions, while in other studies equine-based programs have been implemented as stand-alone. The names and acronyms of programs facilitated by horses have also varied substantially, and have included equine-facilitated learning (EFL), equine-guided education (EGE), equine-assisted activities (EAA), equine-facilitated psychotherapy (EFP), equine-assisted experiential therapy (EAET), equine-facilitated body and emotion-oriented psychotherapy (EBEP), equine-assisted therapies (EAT), equine-assisted activities and therapies (EAAT), relational equine-partnered counseling (REPC), equine-assisted counseling (EAC), equine-facilitated mental health and education services (EFMH/ES), and hippotherapy. The use of diverse nomenclature and acronyms has further clouded the understanding of the efficacy of employing horses as change agents. Except when referring to others' interventions, we will use the general term, equine-facilitated learning (EFL).

The rationale for and detailed description of an equine-assisted program were provided by Johansen *et al.* (2014). The manual-based program emphasizes emotion recognition and nonverbal communication and the proposed mechanisms of change rest in attachment, nonverbal communication, and positive emotions. The authors reported that patients who had not responded to other forms of psychotherapy showed improved awareness of their own emotions and anger reduction, although no systematic data were presented (Johansen *et al.*, 2014).

In a study of 63 children (ages 4 through 16) who had experienced intra-familiar violence, a significant increase in well-being as measured by the Children's Global Assessment of Functioning (GAF) Scale was revealed following 19 equine-assisted psychotherapy sessions that included approaching, grooming, lifting hooves, and engaging in other ground work with horses (Schultz *et al.*, 2007). Children with a history of physical abuse and neglect showed the greatest improvement.

In another study, the behaviors of chronic psychiatric patients who had shown violent behavior in the hospital were evaluated before and after various interventions (Nurenberg *et al.*, 2015). Patients were randomly assigned to one of four treatment groups: equine-assisted psychotherapy, canine-assisted psychotherapy, life skills training, or regular hospital care. The group that received equine-assisted therapy showed significant decreases in aggressive behavior post-treatment. However, the random assignment to groups had resulted in an equine-assisted group with higher aggression scores on the pretest. Therefore the findings could have reflected only a statistical artifact, regression toward the mean.

Externalizing behaviors, such as noncompliance, hyperactivity, and aggression were significantly reduced in children and adolescents (ages 8 to 17) following a series of interactions with horses as reported by their parents (Kemp *et al.*, 2014). The activities with horses included 'backing up a horse, asking a horse to yield his hind or front quarters, desensitizing a horse, asking a horse to circle around a person, jump over obstacles either on a loose rope or at liberty' (Kemp *et al.*, 2014, p. 561). The pre-post design incorporated three evaluation times: Time 1 – before starting intervention, Time 2 following 6 to 7 weeks of psychotherapy but before equine intervention, and Time three after equine intervention. Changes were reported between Time two and Time 3. There were also decreases in depression and anxiety, and (only in adolescents) trauma symptoms.

Therapeutic horsemanship, involving both grooming and riding, was associated with reductions in antisocial behavior and aggression of child and adolescent boys over 22 months of participation (Pearson, 1997). A retrospective review of charts for the seriously emotionally disturbed youth revealed that reported aggressive incidents decreased in all groups, independent of the frequency of riding. Overall, boys who rode more often and were more engaged in the riding program received fewer reports of aggressive incidents.

In a study by Trotter *et al.* (2008), at-risk adolescents experienced a 12-week equine assisted counseling program (126 participants) or a classroom-based life skills program (control group of 38 participants). There were pre- and posttest measures of both self-reports and parents' reports. Parents of students in the equine-assisted program reported significant reductions in their children's aggressive behavior; that was not the case for the control group.

Other literature provides some indications of decreased anger and violence following experiences with horses. In a review of studies of equine-facilitated interventions with children and adolescents, Lentini and Knox (2009) pointed to several case studies of reduced aggression following equine interventions, but also one report of increased aggression. In a pilot study, following a 5 day riding camp, children and adolescents (ages 7 to 17) reported less anger (Kaiser *et al.*, 2004). Similarly, Kaye Gehrke (2009) has provided an anecdotal account of teaching leadership through experiences with horses. She emphasized that people learn that anger and aggression are not effective with horses, and they are then able to apply that knowledge to their own leadership activities. Several well-controlled studies have demonstrated benefits of equine-facilitated learning, but the measures were not specific to anger, aggression, or emotion regulation. Instead, they showed improved social competence and decreased psychological distress (Klontz *et al.*, 2007; Pendry and Roeter, 2013; Pendry *et al.*,

2014). Moreover, cortisol levels, an indicator of stress and arousal, were reduced in children receiving an equine-facilitated intervention (Pendry *et al.*, 2014). Although not measured directly, aggression, anger, and emotion regulation were likely to have been affected in those studies. On the other hand, controlled studies designed to improve adolescents' self-image, self-control, trust, self-esteem, empathy, locus of control, depression, and loneliness have failed to reveal significant changes after EFL (Ewing *et al.*, 2007; Bachi, 2012).

Despite the widespread use of equine-assisted therapies or interventions, there is a gap between evidence and practice (Bachi, 2012). The multiple equine-based programs that are used to alleviate psychological distress, change behavior, or foster human development have limited empirical support (Selby and Smith-Osborne, 2013; Anestis *et al.*, 2014). For equine-assisted therapies or interventions to be effective and sustainable, more sound research is needed on the specific practices, target groups, and contexts that are effective (Selby and Smith-Osborne, 2013).

Monty Roberts and Join-Up®

Handling and training according to Monty Roberts' Join-Up® are achieved through nonviolent communication to build a relationship of trust between the horse and the handler (About Join-Up®, 2015; Roberts, 2002a; Roberts, 2002b). Through the Join-Up® exercise, the horse sees the handler as the leader of the herd and comes to respect the handler without use of physical punishment ('About Join-Up®', 2015; Roberts, 2002a; Roberts, 2002b). Roberts' Join-Up® has been shown to benefit horses as a less stressful and more effective training approach (Fowler *et al.*, 2012). A systematic study (Fowler *et al.*, 2012) matched horses on temperament and randomly assigned them to two groups. One group was trained according to traditional methods in the United Kingdom and the other trained by the Monty Roberts horsemanship technique, a nonviolent method. Upon being saddled and accepting a rider for the first time, the maximum and average heart rates were higher for horses trained by traditional methods, as compared to those trained by the Monty Roberts technique. Finally, horses trained by the Monty Roberts technique showed better performance on a series of walking, trotting, and cantering exercises. These results suggest that not only is the Monty Roberts technique more effective in training horses, but also that horses handled through these techniques show less anxiety and fear as indicated by a diminished heart rate response to initial saddling and riding.

The Join-Up® approach differs greatly from the conventional horse training methods employed in Guatemala. The traditional way of handling equines in Guatemala involves coercion, violence, and punishment. Typically horses are first thrown to the ground, and then loaded with a heavy burden that they are forced to endure for long periods of time. Additionally, handlers often beat them to make them walk faster. These violent practices can cause suffering and psychological damage to working equines, who cannot be rehabilitated (M. R. Sapón Pellecer, Director General, Fundación Equinos Sanos para el Pueblo-ESAP, personal communication, May 7, 2015). Both the punitive horse-handling and the high rates of violence against women may be linked to the ideology of machismo, a hypermasculinity that prescribes aggression and domination (McClaurin, 1996; Velzboer-Salcedo and Novick, 2000; Menjívar, 2011).

Embodied Cognition and Learning

Recent research in learning and cognition has begun to emphasize the importance of the learner's active participation in acquiring knowledge and skills. The horse-handling program is consistent with current developments in the field of embodied cognition, the claim that 'cognitive processes are deeply rooted in the body's interactions with the world' (Wilson, 2002, p. 625). With respect to learning, engagement with one's environment can facilitate learning, thereby stressing the importance of the active role learners play in understanding the world around them (Kontra *et al.*, 2012). Research has supported this idea; performing an action facilitates learning more effectively than simply watching someone else perform the action (Kontra *et al.*, 2012).

When applying the principles of embodied cognition to a community setting, the experiential learning paradigm (Kolb, 1984) is informative. Experiential learning promotes that 'learning is a continuous process grounded in experience' (Kolb, 1984, p. 27), involving 'transactions between the person [the learner] and the environment' (Kolb, 1984, p. 34). The Join-Up®-inspired program is an application of this framework, as learning is situated in doing and participating. In the current study, participants interacted with horses, using body language that facilitated cooperation and calmness. That is, they did more than passively watch demonstrations or videos or listen to lectures. Rather, participants directly experienced the principles they were exposed to during the course of the program and were able to practice those skills.

As previously explained, the goal of the horse-handling program was to promote non-violent interactions in both human-human and human-equine interactions. Research with female Guatemalan victims of violence supports the efficacy of experiential learning through movement and sport therapy in a related domain (Ley and Rato Barrio, 2011). Additional research suggests that movement and sport-centered therapy are culturally appropriate, thus contributing to the overall effectiveness of the approach (Ley and Rato Barrio, 2012). Results indicated after this active therapeutic experience, women reported better psychological health. This suggests experiential learning is useful both in a Guatemalan population and in addressing violence-related concerns. Experiential learning may have another advantage as well. Because people in general can experience and enjoy interacting with horses, it is less susceptible to the stigma associated with talk therapy (Brandt, 2013). Although prior literature in Guatemala has focused on the utility of experiential learning with the victims of violence (Ley and Rato Barrio, 2011; Ley and Rato Barrio, 2012), the current study proposes it be used as a way to reduce violence among potential perpetrators and other community members.

Current Study

The current study employed a Join-Up®-inspired curriculum in a rural Guatemalan community, Las Cumbres (a pseudonym), a community in the state of Chimaltenango. Because work with horses is a central part of community life in Las Cumbres, a training program with equines was implemented as the vehicle to promote themes of non-violence. Over the course of the multi-week intervention, two groups were trained sequentially to work with horses using the Join-Up®-inspired approach.

The primary purpose of the program was to improve the handling and care of the participants' horses and the purpose of the research was to assess changes in participants' attitudes and behaviors toward their horses as well as the transfer and application of the non-violent horse-handling to participants' interactions with others. In other words, the goal of the translational research was to promote horse-handling and family interactions that were friendly, humane, and nonaggressive and that avoid coercion or physical punishment. Multiple measures and sources including evaluation videos with the horses, interviews with participants and family members, and attitudes toward violence questionnaires were used to provide a diverse and in-depth assessment.

In order to assess the effectiveness of the program, a series of hypotheses were developed to answer the study's central question, 'Can violent and aggressive behavior and attitudes toward horses and people be reduced through a Join-Up®-inspired horse-handling program?'

H1. Horses will be less reactive to handling after their owners receive the program, as compared to horses' responses before the program began.

H2. Attitude measures will show less endorsement of the use of aggression, punishment, and coercion with wives, children, and horses after the program.

H3. Female relatives of participants will report changes in participants' behaviors that are consistent with the goals of the program.

Methods

Participants

Community

The community of Las Cumbres with a population of 2300 is located in the state of Chimaltenango in highland Guatemala. Although the larger state includes both Spanish and Kaqchikel speakers ('Idioma [Language] (N.d), Las Cumbres itself is entirely Spanish speaking. The community faces many challenges including a high illiteracy rate and widespread poverty. The economy is based on agriculture, with corn, beans, and other vegetables among the major crops ('Economía [Economy], n.d.).

Program participants

Participants were drawn from individual volunteers from Las Cumbres who worked with horses in their daily lives. Although 21 participants started the program, three dropped out before the program was completed, leaving a final sample of 18. Participants (n=18) were mostly male (n=16) and included both adolescents (n=10, M *age* = 15.70, SD = 0.82) and adults (n=8, M *age* = 36.13, SD = 13.45), ranging in age from 15-58 years. The two female participants were adolescents. For the purpose of this description, adults were defined as those 18 and over at the start of the study, and thus able to provide informed consent for their participation. Adolescents were those under 18 at the study's start, requiring parental consent, as well as their own assent, for participation.

Focus group participants

While program participants were mostly male community members, focus group participants represented the community's female voice; the focus group comprised 12 female relatives of participants, including wives (n=2), mothers (n=6), sisters (n=4), a daughter (n=1), and an aunt (n=1). Some of the focus group members had multiple relationships to horse program participants. For example, one focus group member was both the wife of an adult participant and the mother of an adolescent participant; a second focus group member was the sister of one participant and the daughter of a second participant. Focus group members ranged in age from 18-55 years (M *age* = 38.15, SD = 32).

Measures

Pre and posttest measures

Human-horse evaluations

Evaluations of horse-human interactions were conducted with the horse of each participant at pretest and posttest; the procedure was based on the Burn *et al.* (2010), was video-recorded, and was coded independently by two raters.

Stranger evaluation

Horses were first evaluated on their reaction to being approached by a stranger who, in all cases, was highly familiar with horses, but unfamiliar to that particular horse. The first rating was a rating of alertness prior to being approached. Three codes were possible: apathetic/depressed, alert, or irritable/nervous (e.g. ears back, tail swishing, stomping). Horses were then rated on their reaction to approach by the stranger and to the stranger touching their chin. Both responses were scored using one of six possible codes: (1) friendly, (2) no reaction, (3) moves head away, (4) moves body away, (5) irritable (i.e. ears back, tail swishing, stomping), or (6) aggressive (tries to kick or bite with ears back, rears).

Owner evaluation

Horses were then approached by the owners. Videos were coded on a series of five actions adapted from Wulf *et al.* (2013): the owner's approach, lifting the front right leg of the horse, touching the ears, putting on an unfamiliar halter, and introducing a saddle pad or blanket to the horse's back. All actions were scored using one of the six codes described above (friendly, no reaction, moves head away, moves body away, irritable, or aggressive).

Inter-rater reliability

Two raters coded each video independently. Weighted Kappa, for ordinal data, was calculated as an inter-rater reliability statistic for each of the eight behaviors described above (three from the stranger video and five from the owner video). Scores were combined across pretest and posttest to yield a single inter-rater statistic for each behavior. Interpretations of Kappa are based on Landis and Koch (1977). The average Kappa was 0.67 (substantial agreement). The lowest Kappa was 0.32 for the load (fair agreement), followed by ears and halter at 0.59 (moderate agreement), 0.62 for the lift, 0.65 for the owner approach, and 0.73 for the stranger approach, with the last three suggesting substantial agreement. Finally, the chin at 0.82 and horse alertness at 1.00 suggest almost perfect agreement and perfect agreement, respectively. Scores that differed for the two raters were then discussed and the consensus score used for analysis.

Participant interviews

Participants were interviewed at pretest and again at posttest about their past and present experiences with horses, as well as what they hoped to gain or had gained from the program. They also provided basic demographic data at this time. Due to the type of information collected, the interviews were structured. During interviews, the violence measure (described below) was read aloud to participants.

Attitudes toward violence scales

A series of three established questionnaires were combined to form an overall measure of attitudes towards violence. The composite measure consisted of 23 items and assessed attitudes toward violence with horses (an adaptation of the Beliefs about Punishment Scale; Corral-Verdugo *et al.*, 1995), children (a selection of items from the Corporal Punishment of Children subscales of the Velicer Attitudes Toward Violence Scale; Velicer *et al.*, 1989, as revised by Anderson *et al.*, 2006), and wives (a selection of items from the Inventory of Beliefs about Wife-Beating-Short Form; Saunders *et al.*, 1987, as synthesized by Ley and Rato Barrio, 2012). All items and responses were translated from English into Spanish by a bilingual member of the research team, with the translation checked and verified by two other bilingual members of the team and a third outside expert. Efforts were made, not only to translate accurately, but also to use Guatemalan Spanish, including local expressions that would be easily understood by the rural community members. During interviews, participants were first asked whether they agreed or disagreed with an item. If they agreed, participants were then asked if they agreed a lot (*mucho*) or just a little (*poco*). If they disagreed, participants were asked if they disagreed a lot or just a little. These sequential responses are useful in testing adolescents and persons from different cultural backgrounds who might not be familiar with the typical Likert scales (Wichstrøm, 1995; Rohner and Khaleque, 2005). Responses were then converted to a 4-point Likert scale with 1 (*strongly disagree*) to 4 (*strongly agree*). Higher scores indicated stronger endorsement of violent attitudes. Internal reliability as measured by Cronbach's alpha was adequate at both pretest ($\alpha = 0.80$) and posttest ($\alpha = 0.74$) for the composite measure.

Physical punishment of horses (adaptation of the Beliefs of Punishment Scale; Corral-Verdugo et al., 1995). Six items were adapted from Corral Verdugo *et al.*'s scale to capture the acceptability of violent treatment of horses during training. Sample items included, 'Punishment is the best way to correct the behavior of a horse' and 'Hitting the horse will teach it to behave well.'

Corporal Punishment of Children subscale of the Velicer Attitudes toward Violence Scale (Velicer et al., 1989, as revised by Anderson et al., 2006). Six items were chosen from the Children subscale of the Velicer Attitudes toward Violence Scale to ascertain participants' attitudes toward the role of violence in child discipline. Sample items included, 'Children should be beaten when they have a tantrum' and 'Disobedience in children should be punished physically.'

Inventory of Beliefs about Wife Beating-Short Form (Saunders et al., 1987, as synthesized by Ley and Rato Barrio, 2012). To assess participants' attitudes about wife beating, they answered 11 questions on the acceptability of husbands physically abusing their wives. Items included, 'When a husband mistreats his wife, it is the wife's fault' and 'A little violence from time to time from a husband toward his wife maintains the marriage.' Four items were reverse-scored.

Join-Up®-inspired curriculum

The Join-Up®-inspired curriculum, based on Monty Roberts' program, teaches non-violent communication between humans and equines through the Join-Up® experience ('About Join-Up®', 2015; Roberts, 2002a; Roberts, 2002b). According to Roberts (2002a, b) Join-Up® uses the horse's inherent methods of nonverbal communication and herd behavior. Humans learn to display calm and confident nonverbal behavior that facilitates the horse's acceptance of their leadership.

Participants in the current study participated in a 4-week session to experience Roberts' method of horse-handling. The first three sessions took place at the stables owned and maintained by a member of the research team. Thus, participants initially completed the exercises with unfamiliar horses. The final session and a closing ceremony took place in the participants' home community, Las Cumbres, where they worked with their own horses.

Week 1. During the first week, participants watched a horse-human Join-Up® demonstration. Then they practiced a human-to-human Join-Up® exercise in which participants learned about necessary body position and body language. Finally, they performed the Join-Up® exercise with a horse.

Week 2. The second session focused on emotional regulation and control of arousal when working with a horse. Participants first listened to a presentation and demonstration and then practiced those skills by grooming a horse. Desensitization, introducing scary objects to the horse in a progressive and non-coercive manner, was also demonstrated and practiced.

Week 3. The third session continued with training on desensitization and introduction of the Dually training halter. The Dually training halter, designed by Roberts, is a patented schooling halter that uses pressure and release to shape the horses' behavior. Participants viewed another desensitization demonstration and then practiced desensitization themselves and use of the halter with horses. In addition, they continued to practice the Join-Up®. As a token of appreciation of their participation in

Fig. 14.1. Join-Up® between a participant and his horse.

the program and as encouragement to continue practicing the grooming exercises, participants were given grooming brushes at the close of the session.

Week 4. In the fourth and final week, participants worked on the Join-Up® exercise with their own horses (See Fig. 14.1). They also completed posttest interviews and were recognized in a graduation ceremony where they were each given a certificate and a Dually halter.

Procedure

The community representative to a local horse-welfare non-governmental organization (NGO) offered his community, Las Cumbres, to participate in the horse-handling program. The research protocol was reviewed and approved by the Saint Louis University IRB, a local AD hoc ethics committee that addressed the research participation of people and horses, and the Consejo Comunitario de Desarrollo (COCODE), the community governing body. Participant volunteers over 18 years of age were read the consent form aloud; those under 18 were read the assent form, and their parents the consent form. The confidentiality of the interview was emphasized and permission was obtained to photograph and videotape the human-horse evaluations.

At pretest, participants and their horses were evaluated once with a stranger and the second time with the owner. The pretest took place immediately before the start of the program. Pretest evaluations with both the stranger ($M = 49$ s, $SD = 8$) and owner ($M = 158$ s, $SD = 130$) were brief. Participants also responded to questions regarding violent attitudes towards horses, children, and wives. Those questions were read aloud to participants during the interview in which they also provided demographic information and their background and experiences working with horses. As described above, the program lasted a total of 4 weeks. At posttest, horses were evaluated again with a stranger and the owner using the same protocol. Participants were also re-interviewed at this time. Posttest evaluations and interviews occurred immediately after the end of the program. Again, evaluations with the stranger ($M = 55$ s, $SD = 16$) and owner ($M = 147$ s, $SD = 30$) were short.

In addition, female family members of the participants were invited to participate in a focus group. Wives, mothers, and sisters shared and explained any changes they had noticed in their family member's behavior or attitudes since participating in the program. They were asked specifically about any changes in the participants' behavior toward horses and family members. The focus group was conducted in a classroom in the local school 1 week after the end of the program and lasted 32 min. The focus group was audio recorded and transcribed. The dialogue was coded according to thematic analysis (see Braun and Clark, 2006). Three members of the research team independently categorized statements extracted from the focus group transcript; all identified the primary three themes, and a small number of disagreements with respect to categorization were reconciled by discussion.

Study design

Originally, the study was designed as a randomized cross-over design and participants were randomly assigned to one of two groups. With this approach, participants are tested at three times: at pretest, after the first group receives the intervention, and finally, after both groups have received the intervention. Because of missing data, the design was simplified to a pretest/posttest analysis.

Results

The mixed-method design yielded both quantitative and qualitative data. The questionnaire data on attitudes toward violence were limited to the participants themselves, while the participant interviews and focus groups with wives and mothers provided other perspectives. In addition, the videos of stranger and owner evaluations with the horses provide a direct observation of the human-equine relations that were a major focal point of the project.

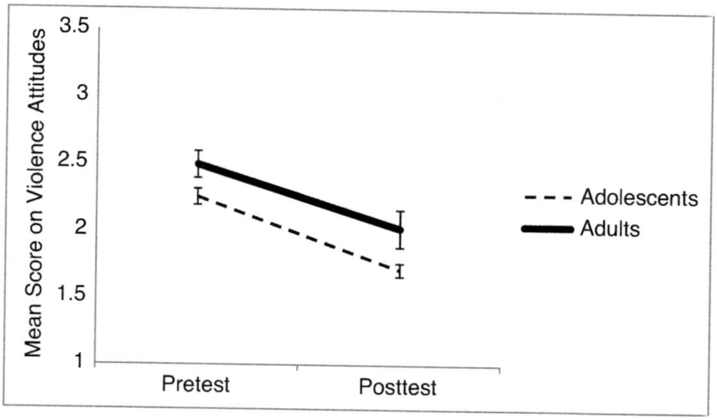

Fig. 14.2. Mean scores on the measure of violent attitudes pre- and posttest for adolescents and adults. The error bars represent SEM.

Attitudes toward violence scales

A 2 (Time: Pretest vs Posttest) x 2 (Age Group: Adolescent vs Adult) mixed model ANOVA was conducted to explore differences in attitudes toward violence between adolescents (15 through 17) and adults (those ages 18 and older). Time was analyzed as the within-subjects factor and adult vs adolescent status was the between-subjects variable. To assess sphericity, the Greenhouse-Geisser estimate was used; it indicated sphericity was not violated (V = 1.00). The main effect of time was significant with posttest scores lower than pretest scores, $F(1, 16) = 56.58$, $p<0.001$, partial $\eta^2 = 0.80$ indicating a decreased endorsement of violent attitudes after the program's conclusion. The main effect of age group was also significant, as adolescent participants were lower on endorsement of violent attitudes at pre- and posttest than adult participants, $F(1, 16) = 5.98$, $p=0.024$, partial $\eta^2=0.27$ (Fig. 14.2). The interaction between time and age group was not significant.

The means (depicted in Fig. 14.2) indicated that on the pretest most participants endorsed violent and coercive ways to punish horses and children and to treat wives. On the posttest the average response was to reject such attitudes. For example on the pretest, 50% of participants agreed that 'Punishing a child physically when he deserves it will make him a stronger, more responsible adult' compared to only 33.3% at posttest.

Human-horse evaluations

Wilcoxin signed-rank tests for ordinal data were used to assess potential differences between pretest and posttest scores. Because all horses were rated as alert unanimously by both coders at pretest and posttest, this variable was not analyzed further. Only the owner approach decreased significantly between pretest and posttest, $Z=-2.04$, $p=0.04$, $r=0.34$ (medium effect). Despite not reaching traditional levels of significance, the other six behaviors were trending toward decreased reactivity.

Posttest interviews

Participants' posttest interviews revealed what they had learned from the program. Fifty percent indicated that they had learned to treat horses kindly and not to mistreat them. Remaining participants gave less specific answers. For example, one said, 'I learned more about training horses.' When asked what they enjoyed most about the program, 47% said the Join-Up® when participants experienced communicating with a horse without physical force. The remaining responses indicated participants enjoyed everything about the program (18%) and gained self-confidence (12%).

Table 14.1. Focus group Responses

	Theme	Example
What changes have you seen in participants?	Better care and treatment of horses (27 comments)	'They work the horses less.' 'He is different with his horse. He used to hit his horse a lot and no more. He is less angry.' 'They would rather go without food themselves than not feed the animals.'
	The participants themselves changed (21 comments)	'He is more calm. Always smiling.' 'They fight less.' 'He scolded me a lot and now does less.'
	Direct application of work with horses to people (six comments)	'Before he mistreated me and the mare and now he doesn't.' 'Now he is more loving with the family and the animals.' 'When he has a conflict in the street, he tries to use what you taught him.'

Focus groups with wives, mothers, and sisters

Following the horse-handling program, the focus group with some female relatives of the participants provided an alternate perspective about the program's impact. The three themes that emerged in response to the question, 'What changes have you observed in participants?' were participants' more responsible and caring treatment of horses (27 comments), changes in participants themselves (21 comments), and a correspondence between the work with horses and changed interactions with people (six comments). Examples of the themes can be found in Table 14.1.

Discussion

Violence may be an everyday part of life for many youth in Guatemalan communities. One way that change might come about is through learning to handle horses in a gentle and non-coercive manner. In this study, an equine-facilitated program inspired by Monty Roberts' Join-Up® resulted in less endorsement of violence by participants, reports of better care of horses, less anger and aggression toward family members, and calmer responses from their own horses. A strength of the study was the use of three sources of information – the participants' self-reports, their female relatives' views, and horses' reactions to handling. The findings are consistent with other indications that horse-handling interventions may lessen anger expression, aggression, and violence (Schultz *et al.*, 2007; Trotter *et al.*, 2008; Kemp *et al.*, 2014; Nurenberg *et al.*, 2015).

This study is unique in that the program was endorsed by the community, bolstered by community involvement, and involved participants who already worked with horses in their daily lives. Because the participants learned better handling of their own working equines, with whom they formed a willing partnership, they were likely to realize economic benefits as well as psychological ones (Fundación Fundación Equinos Sanos Para El Pueblo (ESAP), 2010).

Like other equine-based programs, we saw the advantages of horses as talented facilitators of learning. Horses are social, sensitive to nonverbal cues, and authentic in their responses. In order to perform an effective Join-Up®, as well as desensitize the horse to a frightening object, participants had to learn to display calm and confident nonverbal behavior. Thus, a likely mechanism for the changes observed in this study is improved emotion regulation, especially changing cognitions and modulating responses (Gross, 1998). This explanation is consistent with a number of previous studies that implied improvements in emotion regulation following an equine-assisted intervention (Pearson, 1997; Trotter *et al.*, 2008; Kemp *et al.*, 2014). Future studies should specifically evaluate changes in the ability to regulate emotions among those receiving EFL programs.

There was evidence that horses, as well as their human companions, benefitted from the program. Family members reported that participants groomed, fed, and generally showed increased attention to their horses following the program. Moreover, on the posttest the horses themselves reacted more calmly to the approach of the owner.

Evidence suggests modifying simple, daily actions can have a lasting impact on future behavior and the current study was structured around similar logic. The participants in this study had daily routine interactions with their working equines, providing opportunities to practice what they were learning. The value of effecting change through changing daily habits is well documented in the health and nutrition literature. For example, simple dietary and activity modifications are associated with positive health outcomes (e.g. Golan *et al.*, 1998).

In our study the program was effective for both youth and adults. Although young people were less likely than adults to endorse violent attitudes on either the pre- or posttest, the attitude changes occurred in both age groups. While equine-facilitated learning may be useful for people of all ages, youth may be at a critical juncture for negotiating their futures. Adolescence is an important stage for development of interpersonal relationships; entry into gangs or engaging in violence as an adolescent is associated with a bleak future (Dmitrieva *et al.*, 2014; Gilman *et al.*, 2014). The ability to envision a future without violence may aid in the construction of a different worldview, one that replaces physical punishment of children and wife battering with positive parenting and collaborative domestic relations. Moreover, adolescents have the opportunity to carry their acquired skills and knowledge forth to new generations.

Despite its strengths, this study has many limitations. The design could not be executed as planned, and the pre-post design retained many threats to internal validity. Because the community members were actively engaged in buying and selling horses, only 10 of the 18 participants completed the posttest with the same horse as their pretest. In addition, the design did not allow us to address the essential elements for change. Were community involvement, positive interpersonal relationships with the research team, increased emotional regulation and perceived control, economic benefits, or the specific non-violent leadership inspired by Join-Up® necessary and/or sufficient to result in the observed changes? Participants reported that the Join-Up® experience itself was powerful and life-changing, but further research is needed to reveal the specific outcomes of that experience, and how it transformed attitudes and behavior.

Another issue is that of sustainability. The program was resource-heavy, requiring a great deal of time, money, and other resources. Transportation to and from the rural community was difficult for the research team and for participants. NGO's interested in implementing an equine-facilitated learning program would be hard-pressed to invest the necessary resources. Shorter and more sustainable models should be tested to evaluate whether they would result in similar outcomes as the model tested here.

In sum, this research adds to the accumulating evidence that equine-facilitated interventions can positively transform individuals and communities. However, many questions remain before programs can be considered to be empirically grounded and sustainable. Future research is needed to identify the critical elements of the program and to specify the population for which and the contexts in which an equine handling program is effective in reducing violent attitudes and behavior.

Acknowledgements

The authors would like to thank Join-Up® International for providing Dually halters for the participants, Lisa Willoughby and Jillon Vander Wal for statistical consultation, Santiago Albert for the photograph of a Join-Up®, and The Brooke partner program Equinos Sanos Para El Pueblo (Healthy Horses for the People) and its director Mario Sapón for introducing the research team to the community.

References

About Join-Up® (2015). Available at: http://www.montyroberts.com/ab_about_monty/ju_about/

Anderson, C., Benjamin Jr, A., Wood, P. and Bonacci, A. (2006) Development and testing of the Velicer Attitudes Toward Violence Scale: Evidence for a four-factor model. *Aggressive Behavior* 32, 122–136.

Anestis, M., Anestis, J., Zawilinski, L., Hopkins, T. and Lilienfeld, S. (2014) Equine-related treatments for mental disorders lack empirical support: A systematic review of empirical investigations. *Journal of Clinical Psychology* 70, 1115–1132. DOI: 10.1002/jclp.22113.

Bachi, K. (2012) Equine-facilitated psychotherapy: The gap between practice and knowledge. *Society & Animals* 20, 364–380. DOI: 10.1163/15685306-12341242.

Bachi, K., Terkel, J. and Teichman, M. (2012) Equine-facilitated psychotherapy for at-risk adolescents: The influence on self-image, self-control and trust. *Clinical Child Psychology and Psychiatry* 17, 298–312. DOI: 10.1177/1359104511404177.

Bellino, M. (2014) Educating for human rights consciousness. Special Issue: The social construction of human rights, *Listening: A Journal of Communication Ethics, Religion, and Culture* 136–157.

Brandt, C. (2013) Equine-facilitated psychotherapy as a complementary treatment intervention. *The Practitioner Scholar: Journal of Counseling and Professional Psychology* 20, 23–42.

Braun, V. and Clark, V. (2006) Using thematic analysis in psychology. *Qualitative Research in Psychology* 3, 77–101.

Burgon, H. (2011) 'Queen of the world': Experiences of 'at-risk' young people participating in equine-assisted learning/therapy. *Journal of Social Work Practice* 25, 165–183. DOI: 10.1080/02650533.2011.561304.

Burn, C., Denison, H. and Whay, T. (2010) Relationships between behaviour and health in working horses, donkeys, and mules in developing countries. *Applied Animal Behavior Science* 126, 109–118. DOI: 10.1016/j.applanim.2010.06.007.

Carroll, L., Perez, M. and Taylor, R. (2014) *The evidence for violence prevention across the lifespan and around the world*. The National Academies Press, Washington, DC.

Corral-Verdugo, V., Frías-Armenta, M., Romero, M. and Muñoz, A. (1995) Validity of a scale measuring beliefs regarding the "positive" effects of punishing children: A study of Mexican mothers. *Child Abuse & Neglect* 19, 669–679. DOI: 10.1016/0145-2134(95)00025-4.

DeAraugo, J., McLean, A., McLaren, S., Caspar, G., McLean, M. *et al.* (2014) Training methodologies differ with the attachment of humans to horse. *Journal of Veterinary Behavior: Clinical Applications and Research* 9, 235–241. DOI: 10.1016/j.jveb.2014.05.001.

Dmitrieva, J., Gibson, L., Steinberg, L., Piquero, A. and Fagan, J. (2014) Predictors and consequences of gang membership: Comparing gang members, gang leaders, and non–gang-affiliated adjudicated youth. *Journal of Research on Adolescence* 24, 220–234.

Dutton, D. and Corvo, K. (2006) Transforming a flawed policy: A call to revive psychology and science in domestic violence and research. *Aggression and Violent Behavior* 11, 457–483. DOI: 10.1016/j.avb.2006.01.007.

Economía [Economy] (n.d.). Available at: http://chimaltenango.webcindario.com/index_archivos/Page2147.htm

Ewing, C., MacDonald, P., Taylor, M. and Bowers, M. (2007) Equine-facilitated learning for youths with severe emotional disorders: A quantitative and qualitative study. *Child and Youth Care Forum* 36, 59–72. DOI: 10.1007/s10566-006-9031-x.

Fowler, V., Kennedy, M. and Marlin, D. (2012) A comparison of the Monty Roberts technique with a conventional UK technique for initial training of riding horses. *Anthrozoös* 25, 301–321. DOI: 10.2752/175303712X13403555186217.

Fundación Equinos Sanos Para El Pueblo (ESAP) (2010) Economic valuation on the impact of the working equine in the Peten and Chimaltenango communities. Available at: http://esap.org.gt/wp-content/uploads/2012/02/ESAP-Economic-Valuation-of-the-working-equine.pdf

Gilman, A., Hill, K. and Hawkins, J. (2014) Long-term consequence of gang membership for adult functioning. *American Journal of Public Health* 104, 938–945. DOI: 10.2105/AJPH.2013.301821.

Golan, M., Fainaru, M. and Weizman, A. (1998) Food consumption and cardiovascular risk factors in weight-loss program with the parents vs. children as the sole agents of change. *International Journal of Obesity* 22, 1217–1224. DOI: 10.1038/sj.ijo.0800749.

Gondolf, E. (2011) Lessons from a successful and failed random assignment testing batterer program innovations. *Journal of Experimental Criminology* 6, 355–376. DOI: 10.1007/s11292-010-9104-6.

Gross, J. (1998) The emerging field of emotion regulation: An integrative review. *Review of General Psychology* 2, 271–299.

Guatemala Human Rights Commission (GHRC) (2012) *Fact sheet: Gangs in Guatemala*. Available at: http://www.ghrc-usa.org/wp-content/uploads/2011/12/GangFactSheet.pdf

Hallberg, L. (2008) *Walking the way of the horse: Exploring the power of the horse-human relationship*. iUniverse, Bloomington, IN.

Idioma [Language] (N.d). Available at: http://chimaltenango.webcindario.com/index_archivos/Page1102.htm

Johansen, S., Arfwedson Wang, C., Binder, P. and Malt, U. (2014) Equine-facilitated body and emotion-oriented psychotherapy designed for adolescents and adults not responding to mainstream treatment: A structured program. *Journal of Psychotherapy Intervention* 24, 323–335. DOI: 10.1037/a0038139.

Kaiser, L., Spence, L., Lavergne, A. and Vanden Bosch, K. (2004) Can a week of therapeutic riding make a difference?—A pilot study. *Anthrozoö s* 17, 63–72. DOI: 10.2752/089279304786991918.

Kaye Gehrke, E. (2009) Developing coherent leadership in partnership with horses—A new approach to leadership training. *Journal of Research in Innovative Teaching* 2, 222–233.

Kemp, K., Signal, T., Botros, H., Taylor, M. and Prenticea, K. (2014) Equine facilitated therapy with children and adolescents who have been sexually abused: A program evaluation study. *Journal of Child and Family Studies* 23, 558–566. DOI: 10.1007/s10826-013-9718-1.

Klontz, B., Bivens, A., Leinart, D. and Klontz, T. (2007) The effectiveness of equine-assisted experiential therapy: Results of an open clinical trial. *Society and Animals* 15, 257–267. DOI: 10.1163/156853007X217195.

Kolb, D. (1984) *Experiential learning: Experience as the source of learning and development*. Prentice Hall, Englewood Cliffs, NJ.

Kontra, C., Golden-Meadow, S. and Beilock, S. (2012) Embodied learning across the lifespan. *Topics in Cognitive Science* 4, 731–739. DOI: 10.1111/j.1756-8765.2012.01221.x.

Landis, J. and Koch, G. (1977) An application of hierarchical Kappa-type statistics in the assessment of majority agreement among multiple observers. *Biometrics* 33, 363–374.

Lanning, B. and Krenek, N. (2013) Examining effects of equine-assisted activities to help combat veterans improve quality of life. *Journal of Rehabilitation Research & Development* 50, xv–xiii. DOI: 10.1682/JRRD.2013.07.0159.

Lentini, J. and Knox, M. (2009) A qualitative and quantitative review of equine facilitated psychotherapy (EFP) with children and adolescents. *The Open Complementary Medicine Journal* 1, 51–57.

Ley, C. & Rato Barrio, M. (2011). Movement and sport therapy with women in Guatemalan context of violence and conflict. Body, Movement, or Dance in Psychotherapy, 6, 145-160. DOI: 10.1080/17432979.2010.546619.

Ley, C. and Rato Barrio, M. (2012) Evaluation of a psychosocial health programme in the context of violence and conflict. *Journal of Health Psychology* 18, 1371–1381. DOI: 10.1177/1359105312462435.

Macapagal, K. and House, A. (2013) *Critical synthesis package: Inventory of Beliefs About Wife Beating (IBWB)*. MedEdPORTAL Publications; 2013. Available at: https://www.mededportal.org/publication/9610

Masters, N. (2010) *Equine assisted psychotherapy for veterans with PTSD* (Unpublished Master's Thesis). Washington State University, Vancouver, Vancouver, Washington.

McClaurin (1996) *Women of Belize: Gender and change in Central America*. Rutgers University Press, Brunswick, NJ.

Menjívar, C. (2011) *Enduring violence: Ladina women's lives in Guatemala*. University of California Press, Berkeley, CA.

Nurenberg, J., Schleifer, S., Shaffer, T., Yellin, M., Desai, P. *et al.* (2015) Animal-assisted therapy with chronic psychiatric inpatients: Equine-assisted psychotherapy and aggressive behavior. *Psychiatric Services* 66, 80–86. DOI: 10.1176/appi.ps.201300524.

Office of Inspector General, U.S. Agency for International Development (2014) *Audit of USAID/Guatemala's Violence Prevention Project*. Available at: http://oig.usaid.gov/sites/default/files/audit-reports/1-520-14-009-p.pdf

Pearson, E. (1997) Effects of equine facilitated therapy on the antisocial behavior of adolescent males. In: Engel, B.T. (ed.) *Rehabilitation with the Help of a Horse: A Collection of Studies*. Barbara Engle Therapy Services, Durango, Colorado, pp. 177–206.

Pendry, P. and Roeter, S. (2013) Experimental trial demonstrates positive effects of equine facilitated learning on child social competence. *Human-Animal Interaction Bulletin* 1, 1–19.

Pendry, P., Carr, A., Smith, A. and Roeter, S. (2014) Improving adolescent social competence and behavior: A randomized trial of an 11-week equine facilitated learning prevention program. *The Journal of Primary Prevention* 35, 281–293. DOI: 10.1007/s10935-014-0350-7.

Pendry, P., Smith, A. and Roeter, S. (2014) Randomized trial examines effects of equine facilitated learning on adolescents' basal cortisol levels. *Human-Animal Interaction Bulletin* 2, 80–95.

Renner, L. and Slack, K. (2006) Intimate partner violence and child maltreatment: Understanding intra- and intergenerational connections. *Child Abuse and Neglect* 30, 599–617. DOI: 10.1016/j.chiabu.2005.12.005.

Roberts, M. (2002a) *From my hands to yours: Lessons from a lifetime of training championship horses*. Monty and Pat Roberts, Inc, Solvang, CA.

Roberts, M. (2002b) *Horse sense for people*. Penguin, New York.

Rohner, R. and Khaleque, A. (eds) (2005) *Handbook for the Study of Parental Acceptance and Rejection*, 4th ed. Rohner Research Publications, Storrs, CT.

Saunders, D., Lynch, A., Grayson, M. and Linz, D. (1987) The inventory of beliefs about wife beating: The construction and initial validation of a measure of beliefs and attitudes. *Violence and Victims* 2, 39–57.

Schultz, P., Remick-Barlow, G. and Robbins, L. (2007) Equine-assisted psychotherapy: A mental health promotion/intervention modality for children who have experienced intra-family violence. *Health and Social Care in the Community* 15, 265–271. DOI: 10.1111/j.1365-2524.2006.00684.x.

Selby, A. and Smith-Osborne, A. (2013) A systematic review of effectiveness of complementary and adjunct therapies and interventions involving equines. *Health Psychology* 32, 418–432. DOI: 10.1037/a0029188.

Trotter, K., Chandler, C., Goodwin-Bond, D. and Casey, J. (2008) A comparative study of the efficacy of group equine assisted counseling with at-risk children and adolescents. *Journal of Creativity in Mental Health* 3, 254–284. DOI: 10.1080/15401380802356880.

UNICEF (2009) *Child abuse: A painful reality behind closed doors*. Available at: http://www.unicef.org/lac/Boletin-Desafios9-CEPAL-UNICEF_eng.pdf

UNICEF (2012) *UNICEF annual report 2012 for Guatemala, TACRO*. Available at: http://www.unicef.org/about/annualreport/files/Guatemala_COAR_2012.pdf

UNICEF (2014) *Hidden in plain sight: A statistical analysis of violence against children*. Available at: http://files.unicef.org/publications/files/Hidden_in_plain_sight_statistical_analysis_EN_3_Sept_2014.pdf

United States Department of State (2013) *Guatemala 2013 human rights report*. Available at: http://www.state.gov/documents/organization/220657.pdf

United States Department of State (2014) *Guatemala 2014 crime and safety report*. Available at: https://www.osac.gov/pages/ContentReportPDF.aspx?cid=15656

Velicer, W., Huckel, L. and Hansen, C. (1989) A measurement model for measuring attitudes toward violence. *Personality and Social Psychology Bulletin* 15, 349–364.

Velzboer-Salcedo, M. and Novick, J. (2000) Violence against women in the Americas: A violation of human rights and an international public health problem. *Perspectives in Health* 5. Available at: http://www.paho.org/english/dpi/Number10_article2.htm

Vidrine, M., Owen-Smith, P. and Faulkner, P. (2002) Equine-facilitated group psychotherapy: Applications for therapeutic vaulting. *Issues in Mental Health Nursing* 23, 587–603. DOI: 10.1080/01612840290052730.

Wichstrøm, L. (1995) Harter's self-perception profile for adolescents: Reliability, validity, and evaluation of the question format. *Journal of Personality Assessment* 65, 100–116. DOI: 10.1207/s15327752jpa6501_8.

Wilson, M. (2002) Six views of embodied cognition. *Psychonomic Bulletin & Review* 9, 625–636. DOI: 10.3758/BF03196322.

Wulf, M., Aurich, J., May, A.-C. and Aurich, C. (2013) Sex differences in the response of yearling horses to handling by unfamiliar humans. *Journal of Veterinary Behavior* 8, 238–244. DOI: 10.1016/j.jveb.2012.09.002.

15 Behavioral and Physiological Effects of Equine-Assisted Early Intervention for Mother-Child Dyads with Insecure Attachment

Denise V. Hebesberger[1]*, Andrea Beetz[2] and Kurt Kotrschal[3]
[1]Department of Biology, Anglia Ruskin University, Cambridge, United Kingdom; [2]ISER, Department for Special Education, University of Rostock, Rostock, Germany; [3]Department of Behavioural Biology, University of Vienna, Vienna, Austria and Core Facility Konrad Lorenz Research Station for Behaviour and Cognition, University of Vienna, Gruenau im Almtal, Austria

Abstract
Animals, including horses may be valuable partners in many activities, pedagogy and therapy. Contact between humans and animals can facilitate an oxytocin secretion that, as a consequence, may alleviate stress-responses, increase social orientation and support attachment and caregiving. These mechanisms can be utilized in animal-assisted therapy, for example, to enhance trust between client and therapist and to help increase attachment security in clients with insecure attachment. In this study we compared the effects of an equine-assisted intervention with a conventional play-based intervention for mother-child dyads with insecure attachment, insecure caregiving and child dysregulation. Twenty mother-child dyads (with infants 12-24 months of age), with at least one part of the dyad showing an insecure attachment, were randomly assigned to eight weekly sessions of either equine-assisted or play-based intervention. Effects on mothers' caregiving behavior and physiology and on the relationship between clients and therapists were assessed via behavior coding, salivary cortisol, heart rate and heart rate variability measures. Mothers and their infants in the equine-assisted intervention had more body contact with each other ($p \leq 0.001$), a trend toward more vocal exchange ($p = 0.083$), and mothers showed a higher sympathetic activation, indicated by a higher heart rate ($p = 0.003$). In the play-based intervention, mothers showed greater parasympathetic activation than in the horse group, indicated by higher heart rate variability ($p = 0.004$) as well as enhanced rapport between mothers and the therapist ($p = 0.016$). We conclude that the main effect of the horse-assisted method was increasing positive arousal by the mother and child doing something exciting together, thus triggering attentiveness towards the child, indicated by higher rates of caregiving behavior, such as proximity and vocal contact. In contrast, the play-based intervention promoted a relaxed environment which allowed the therapists and mothers to engage more with each other. Hence, depending on intervention goals, a combination of equine-assisted and play-based interventions might be an optimal approach.

Keywords: equine-assisted intervention, early intervention, equine-assisted therapy, animal-assisted intervention, attachment, caregiving

*Corresponding author: denise.hebesberger@pgr.anglia.ac.uk

The primary caregiver functions as a secure base from which the child can explore the world (Ainsworth *et al.*, 1978; Bowlby, 1988; George and Solomon, 2008). Stress buffering in the mother-child dyad is mainly achieved by a fine-tuned mutual and complementary coordination between the attachment behaviour of the infant and the caregiving behaviour of the primary caregiver (Diamond, 2001; Beetz *et al.*, 2011). Reliable and sensitive primary caregivers are the main agents for the development of 'secure attachment', i.e., a primary social representation (the 'internal working model') on the side of the child characterized by trust in the emotional availability of the attachment figure in case of need (Bowlby, 1953; Ainsworth *et al.*, 1978; Bretherton and Munholland, 2008; Beetz *et al.*, 2011). Such relationship representations help to 'anticipate, interpret, and guide interactions with partners' (Bretherton and Munholland, 2008) later in life as they allow the person to imagine interactions with others based on previous experiences (Bretherton and Munholland, 2008).

In the case of sub-optimal caregiving, the fine-tuning of interacting with the child will be affected, resulting in 'insecure attachment', with its sub-classifications 'insecure-avoidant', 'insecure-ambivalent' or 'disorganized' (Main and Solomon, 1990; George and Solomon, 2008).

The neuropeptide oxytocin was found to play a central mediating role in the primary caregiver-offspring relationship (Heinrichs *et al.*, 2009). Oxytocin is released during labour, breastfeeding, physical contact in combination with positive emotions, sex, as well as during positive social contact (Stock and Uvnäs-Moberg, 1988; Turner *et al.*, 1999; Diamond, 2001; Heinrichs *et al.*, 2009; Handlin *et al.*, 2009; Julius *et al.*, 2013; Unväs-Moberg, 1997a, 1997b; Unväs-Moberg *et al.*, 2008). Oxytocin is a calming agent, mediates trust and social interest and supports the parent-infant bond; in fact, oxytocin has a strong antagonistic effect on stress systems (reviewed in Beetz *et al.*, 2012a).

Adolescents and adults with secure attachment are well able to cope with emotional and behavioral stress (Mickelson *et al.*, 1997; Cooper *et al.*, 1998; Diamond, 2001; Julius *et al.*, 2013) and are strengthened by the emotional social support they are able to accept from others. In contrast, individuals with insecure or disorganized attachment may exhibit dysregulated stress and calming systems (Diamond, 2001; Julius *et al.*, 2013). This could explain why individuals with insecure, and more so with disorganized attachment, may be even stressed in their company (Maunder *et al.*, 2006; Julius *et al.*, 2013). Later in life this is linked to difficulties in engaging in trusting and emotionally supporting partnerships.

Interactions throughout life can alter a person's internal working models (Bowlby, 1988; Zilcha-Mano *et al.*, 2012). Hence, attachment-based interventions may indeed be able to affect attachment representations in a positive way, towards more security (Bowlby, 1988; Julius *et al.*, 2013; Zilcha-Mano *et al.*, 2012). Animal-assisted intervention may be particularly effective, because of the close physical contact between the client and the animal. This contact may activate the oxytocin system which then induces positive physiological and psychological effects, such as reduced fear and stress, increased trust and sociability (Beetz *et al.*, 2012a; Julius *et al.*, 2013). Also, the interactions between therapist and animal may suggest to the client that the therapist is a caring and attentive person, supporting a good rapport towards the therapist (Zilcha-Mano *et al.*, 2011). Furthermore, Levinson (1969 cit. in Zilcha-Mano *et al.*, 2011) claimed that companion animals are natural objects of attachment and the internal attachment representations towards humans are usually not spontaneously transferred to animals (Beetz *et al.*, 2011; Zilcha-Mano *et al.*, 2011; Beetz *et al.*, 2012b), which is particularly important for persons with an insecure attachment (Wedl *et al.*, 2015). A companion animal might 'facilitate the formation of an attachment bond that can be relatively free of maladaptive projections' (Zilcha-Mano *et al.*, 2011). This could consequently promote a secure relationship to the therapist himself. Finally, the presence of an animal could increase the motivation of clients in therapy (Wohlfarth *et al.*, 2013).

We presently studied the effects of an equine-assisted intervention (EAI) compared to a play-based intervention (PBI) in mother-child dyads with insecure attachment, suboptimal caregiving and stress coping as well as dysregulations on side of the child. None of the participating mother-child dyads had riding experience. We therefore expected that engaging in such a mutual and probably also challenging activity within a secure environment would support caregiving behaviour like protection and mutual referencing and that the horse would specifically facilitate mutual attention towards each other. If so, attachment and caregiving behaviors like physical contact, vocal exchange, and positive reinforcement should increase over the period of the intervention. As a base for such changes in mother-child interactions, we expected a calming effect, in both mother and child, due to the contact with the animal, potentially due to the activation of the oxytocin system (Beetz *et al.*, 2012a; Beetz *et al.*, 2012b; Julius *et al.*, 2013). Hence, cortisol levels and heart rates should be lower within the EAI group than within the group receiving PBI.

Animal-assistance usually supports trust and stability in the relationship between clients and therapists (Julius *et al.*, 2013). Contact with animals may therefore affect the client's perception of the therapist, leading to an enhanced rapport (Zilcha-Mano *et al.*, 2011). We predicted that the vocal exchange between mothers and therapists would be more frequent in the horse-group as compared to the play-group.

Methods

The study was conducted at the children care centre 'Bunter Kreis' in Augsburg, Germany. This care center is certified to keep horses on its premises and to offer horse-assisted interventions.

Subjects

Twenty mother-child dyads voluntarily participated in this study. More than half of participating dyads were pointed towards our study by their social workers. The other participants found their way to our project via our newspaper adverts, where we aimed at recruiting mothers who felt challenged or had children with behavioral dysregulation, such as excessive crying, eating or sleeping problems. Dyads came from a risk-prone background with prevailing psychological disorders, a history of physical or sexual abuse, drug addiction, and teenage pregnancy. Two mothers participated with their foster children. In these two cases the children had already lived with their new family for more than ten months. At the onset of our study mothers were 19 to 46 years of age (median: 27.5), children (12 girls, 8 boys) eleven to 24 months (median 16.55 months).

One week before the therapeutic intervention started, the attachment representations of children and mothers were assessed via the Strange Situation Procedure (Ainsworth *et al.*, 1978) and the Adult Attachment Projective (George *et al.*, 1999) and coded by a reliable coder of our research team (A.B.). Dyads were randomly assigned either to the equine-assisted intervention group or to a play-based intervention group as a control (10/10). Therefore, the PBI group included one mother with a secure attachment representation (but an insecurely attached child), three with dismissing, three with preoccupied and three with disorganized attachment representation. The EAI group consisted of one mother with secure attachment (also with an insecurely attached child), four with dismissing, one with preoccupied and four with disorganized attachment representations. The PBI group comprised of two children with secure attachment, three with insecure-avoidant and five with disorganized attachment representation. In the EAI group five children showed secure attachment, one insecure-avoidant and four disorganized attachment patterns. All dyads attended one intervention session per week for the duration of 8 weeks and participated in additional data collection sessions once before the start of the intervention and directly after the end of the intervention phase.

Therapists

Two therapists, one psychologist and one occupational therapist, both with extensive work experience in equine-assisted and play-based interventions conducted the interventions in our study. To avoid possible influences of the therapists' styles, each therapist was randomly assigned to 10 mother-child dyads and worked with five dyads in the equine-assisted intervention and five dyads in the play-based intervention.

Equine-assisted intervention

Facilities

The equine-assisted intervention was conducted at a fenced area (20 x 20 meters) or in a riding pavilion (10m diameter) at the premises of the care center. Prior to our study the equine-assisted intervention had been a long-established program at this hospital.

Horses and equipment

All of the three trained therapy horses that were involved in our study belonged to one of the therapists. On days off work they spent their time in an open stable where they were kept together with other horses in a herd. On days with scheduled therapy sessions, they were brought to the premises of the hospital, where they were kept in individual box stalls. For the sessions, the horses were equipped with a well-fitted halter, a leading rope, a thick saddle pad and a vaulting girth.

Procedure

Each session of the equine-assisted intervention had three main phases.

During a welcoming phase, mother and child were given the opportunity to approach the horse and pet or talk to it in front of the stable. Thereafter, the therapist, the horse and the mother with her child went to a fenced outdoor riding area. The horse was led by a trained horse-leader. The therapist accompanied mother and child on foot. In the intervention phase, the child, the mother or both sat on horseback. Throughout the sessions, mother, child and therapist engaged in different mutual tasks, such as singing, blowing soap bubbles, playing with balloons, rings, or other toys. The main goal was to induce positive interactions between mother and child. Furthermore, therapists mirrored and verbalized the child's feelings or needs, to make them more transparent for the mothers. This aimed at facilitating caregiving behaviour in mothers. After the intervention phase, the horse was returned to the stable where mother and child could feed the horse with bread or carrots to say good-bye.

During the session two students were present. One student operated a digital video camera the other one collected the saliva samples. For videotaping the student accompanied mother, child and the horse at a short distance of about three meters. Only for saliva sampling the second student approached mother and child (for details, please see the section about saliva sampling). Otherwise, the students remained as passive as possible.

Play-based intervention

Facilities

The play-based intervention was conducted in rooms of approx. 20 m² located in one of the care center buildings close to the riding area. The rooms were especially adapted for this kind of therapy and were furnished in friendly and natural shades. The walls were painted in light yellow and a green, and a fluffy carpet covered the floor. The room contained different toys (dolls, cars, toy blocks, instruments etc..), chairs and a table.

Procedure

After arrival, mother and child were able to freely choose a play activity of their own liking. The therapist was also involved in playing, but adjusted to the interests of mother and child. As in the horse intervention group, the focus was on a positive interaction between mother and child and the mirroring of the child's emotions and needs. During the session a student equipped with a digital video camera and a saliva-collection set was present in the room for data collection (see below). During videotaping the student remained inconspicuous in a corner of the room. Only for saliva sampling the student approached mother and child. If children approached the student, she tried to stay as non-involved as possible.

Parameters investigated

Behavioral parameters

All therapeutic sessions were videotaped and later coded with the SOLOMON Coder (Péter, 2016) for the participant's behaviour: the duration of physical activity, proximity between mother and child and between therapist and child, frequency of touch between mother and child and the duration of vocal exchange between the participants.

Physiological parameters

To obtain information about the effects of the therapy situations on the modulation of the hypothalamic-pituitary-adrenal (HPA) axis and the sympathetic-adrenal-medullary (SAM) axis, we measured salivary cortisol and heart rate.

Cortisol

Saliva samples (Vining and McGinley, 1987; Woodside *et al.*, 1991) were collected via salivettes (Sarstedt, Etten-Leur, the Netherlands) at the beginning, after 20 min and after 40 min of the therapy session. As cortisol secretion follows a circadian rhythm (Mickelson *et al.*, 1997) mother-child dyads obtained their sessions at the same time of the day each week. Saliva samples were frozen at −20 degrees and later analysed by enzyme immunoassay (EIA), following the procedure described in Beetz *et al.* (2012b). Cortisol data were integrated via calculating the area under the curve with respect to the increase (AUCi) for each session (Pruessner *et al.*, 2003; Beetz *et al.*, 2011). Unfortunately, no AUCi-values were calculated for children as too many samples were of poor quality (e.g. salivettes of the children did not contain enough saliva).

Mothers were instructed not to eat prior to the intervention and to avoid physical exercise. This should ensure that the first salivary measure of the day could serve as a baseline value.

Heart rate

To measure the impact of the therapeutic sessions on the SAM-axis heart rate (HR) data were collected from which heart rate variability (HRV) was calculated (Kučera, 2006). Therefore, both, mother and child carried a POLAR ® RS800CS heart rate monitor, including a chest belt and a POLAR ® watch as a data logger. This system logged 15–240 beats per minute (bpm). Before the chest belts were put on, the electrodes were carefully moistened with a wet sponge to improve skin contact. The obtained heart rate data were entered into the Polar ProTrainer software program and automatically error corrected via the program's algorithm. Datasets with more than 5% faulty or missing data were taken out of the sample. A high error rate was found in the heart rate data lines for the toddlers, as although using the smallest version of chest belts available, they had to be additionally shortened to fit the participating toddlers in size. Therefore, the contact between electrodes and skin was not ideal. As a consequence, toddler heart rates were excluded from further analysis.

From the mothers' heart rate data, different indices for HRV were calculated, using the Polar ProTrainer program: pNN50: Percentage of successive interbeat intervals that differ in more than 50 percent. The higher the scores, the higher is the parasympathetic activity and thus relaxation (Hottenrott *et al.*, 2006; Curic *et al.*, 2007). RMSSD: This is a measure for rapid, high-frequent fluctuations of heart rate. Differences in successive interbeat intervals are calculated. The higher the scores the higher is the parasympathetic activity and thus relaxation (Hottenrott *et al.*, 2006; Curic *et al.*, 2007).

HR and HRV values were calculated for the entire therapy sessions and the phases between salivettes 1 and 2 (cortisol sampling) and between salivettes 2 and 3.

Ethical statement

The study received approval from the 'Deutsche Gesellschaft für Psychologie' (DGPs) ethics committee. This comprises risk assessments for human participants, animal wellbeing, the interventional setup, and the conducted study.

Data analysis

Data were analysed via PASW SPSS 18.00 by applying non-parametric tests: Mann-Whitney

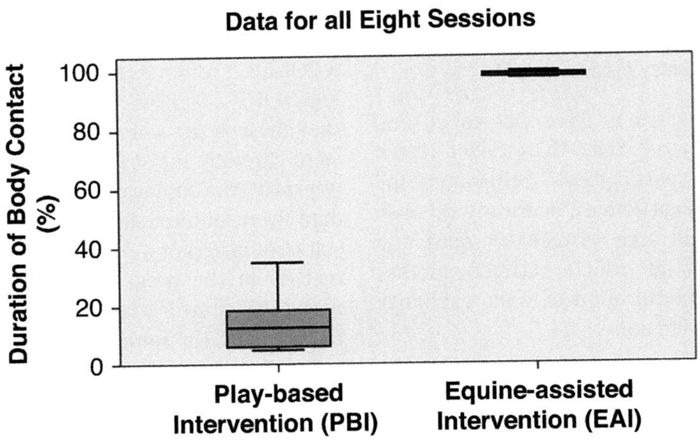

Fig. 15.1. Duration of body contact between mothers and children (in percent) across all eight sessions of the interventions grouped by intervention type.

U-Test for group comparisons and Friedman-Tests for the calculation of changes over time.

Results

Caregiving behaviour – body contact

In all sessions mother and child had significantly more body contact in the equine-assisted intervention group (EAI) (Mann-Whitney U, Z=−3.78, $p \leq 0.001$; Fig. 15.1). But we found no significant changes regarding the duration of body contact between mother and child or the frequency of touching each other over the course of the eight sessions in both intervention groups (Friedman: body contact: EAI: Chi^2 = 6.333, df = 7, p=0.501; play-based intervention (PBI): Chi^2 = 5.5, df = 7, p=0.599; touching frequency: EAI: Chi^2 = 6.415 df = 7, p=0.497; PBI: Chi^2 = 7.584 df = 7, p=0.371).

Caregiving behaviour – vocal exchange between mother and child

By tendency, more vocal exchange between mothers and children occurred in EAI as compared to PBI (Mann-Whitney U, Z=−1.436, p=0.083). Particularly in sessions 3, 5, 6 and 7 mothers of the EAI group spoke, by tendency, more to their children than mothers in PBI (sessions 3 and 6: Z=−1.814, p=0.038; session 5: Z=−1.587, p=0.062; session 7: Z=−1.587, p=0.072).

HPA-Axis modulation: salivary cortisol

We found no significant differences in salivary cortisol between the two groups at time point one, two or three, over all eight sessions. However, the AUCi values of mothers revealed a difference in session 4: values for AUCi of mothers were higher in the PBI (Mann-Whitney U, Z=−1.944, p=0.027), indicating a greater increase of cortisol levels over the course of the therapy session.

SAM-Axis modulation: heart rate data

Mothers of the EAI group showed a significantly higher heart rate than mothers of the PBI group (Mann-Whitney U, Z = −2.685, p=0.003; Fig. 15.2), but they also sat significantly less and walked more (Mann- Whitney-U, Z=−3.356, $p \leq 0.001$). Heart rate variability (pNN50, RMSSD) of mothers was significantly higher in the PBI (Mann-Whitney U, pNN50: Z=−2.570, p=0.004; RMSSD: Z=−2.797, p=0.023; Fig. 15.3). These findings indicate overall a greater parasympathetic activity and less activation in the PBI group.

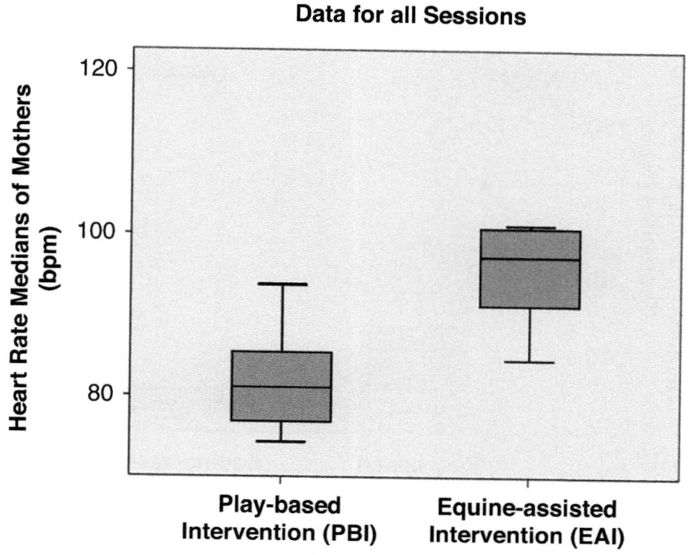

Fig. 15.2. Median heart rate values of mothers (in beats per minute, bpm) for all sessions grouped by intervention type.

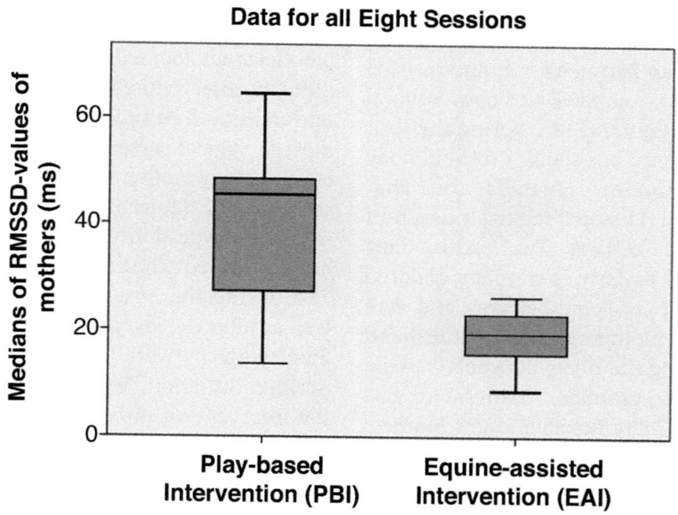

Fig. 15.3. Boxplot showing medians of RMSSD-values (ms) of mothers across all eight intervention sessions grouped by intervention type.

Interactions with the therapist

Across all eight sessions, therapists in the EAI group held more body contact with the children as compared to the PBI group (Mann-Whitney U, $Z=-3.326$, $p \leq 0.001$). Furthermore, our findings show that mothers in PBI talked significantly more to the therapist than mothers in EAI (Mann-Whitney U, $Z=-2.419$, $p \leq 0.007$; see Fig. 15.4).

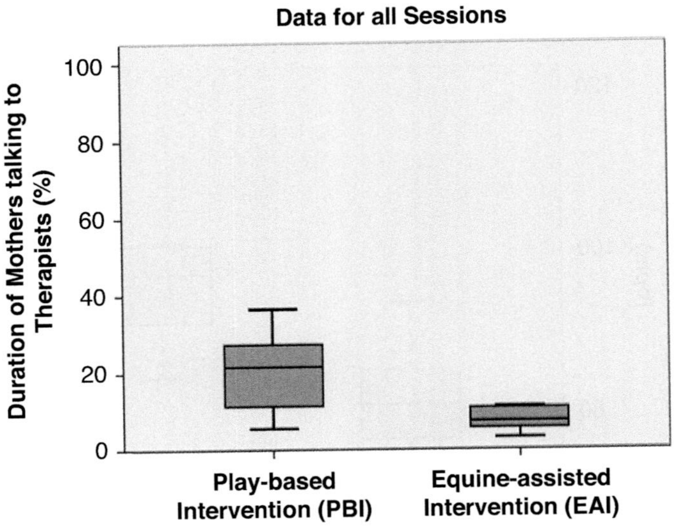

Fig. 15.4. . Boxplot showing the medians of the duration of mothers talking to therapists (in percent) across all interventional sessions arranged by intervention-group.

Discussion

In this study we found significant differences in maternal behaviour between an equine-assisted intervention (EAI), compared to a conventional play-based intervention (PBI). During the EAI, mothers found more possibilities to hold body contact, and (by tendency) to talk to their children. This pattern remained relatively constant across the eight sessions. Two factors may contribute to this pattern: first, many children were not yet able to walk on their own and were carried by their mothers or led by the hand when approaching the horse or when walking through the riding grounds. Furthermore, EAI as such provides more possibilities for mothers and children to spend time in body contact e.g. when sitting together on horseback, or when mothers safeguard their children while riding. Being carried, physical contact, rhythm and furthermore being part of motherly activity establishes an atmosphere of togetherness (Klüwer, 2008). Anisfeld *et al.* (2008) showed that carrying induces motherly responsiveness and promotes attachment security on a long-term range; this seems to be promoted to a larger extent in the equine-assisted setting. Additionally, we found a trend for more vocal exchange in the EAI setting compared to PBI. Therefore, EAI seems to provide a setting which facilitates physical and vocal contact between mothers and their children, compared to PBI. A parallel study focused on the impact of the two different interventions on maternal caregiving and maternal attitude towards the child from a psychological perspective (Beetz *et al.*, 2015). For this study mothers answered questionnaires such as the 'Maternal Attitudes towards the Own Child Questionnaire' and the 'Care-Index' and completed an 'Adult Attachment Projective Test' before the onset of the interventions as well as after the last session. Children's attachment patterns were tested via the 'Ainsworth Strange Situation Test' prior as well as after the intervention (for more details about these methods please see Beetz *et al.*, 2015). This study found an increase in sensitive caregiving and a decrease in controlling caregiving of the participating mothers as well as an increase in cooperative behaviour of the children and a reduction of child behaviour that mothers perceived as difficult. Furthermore, participating mothers improved in reading, interpreting and responding sensitively to the attachment behaviour of their infants. Overall, no group differences were found, showing that both interventions contributed towards these effects. The pre-post-comparison of the Strange

Situation Test results for the participating children indicated a shift towards higher attachment security: three children from EAI and one child from PBI, who were showing disorganised attachment representations before the onset of the intervention were classified as insecure-avoidant or secure after the intervention (Beetz *et al.*, 2015).

We initially hypothesized that the EAI would mediate calming effects in participants indicated by lower salivary cortisol levels due to oxytocin-mediated effects and decreased sympathetic activity (Julius *et al.*, 2013). However, maternal cortisol levels did not differ between the two intervention groups, nor did they change within the groups over the course of the eight intervention sessions. One possible explanation for this pattern could be that we could not exclude that some mothers arrived at the intervention site by bike or ate shortly before the start of the session. This could have impacted our baseline measurements of salivary cortisol. Furthermore, also aspects such as weather (heat, rain) or the amount of movement could have potentially increased the cortisol levels and thus, may have counterbalanced a possible activation of the oxytocin system via touch with the horse and between mother and child.

Furthermore, we found that the mothers' heart rates were higher and heart rate variability values lower during EAI as compared to the play interventions. As mentioned above, this could be explained by the greater physical activity of mother-child dyads in the EAI. They walked more and, also when sitting on horseback, muscular stimulation (balance, coordination, balance) was higher (Klüwer *et al.*, 2005; Klüwer, 2008) than when sitting e.g. on the ground, as this was the case a lot of the time in the PBI. Kaminski *et al.* (2002) found higher heart rates in a dog-assisted setting, where participants did not move much. The authors explain these findings with anticipatory arousal of subjects in expectation to spend time with the animals. This could also apply to our study as the participants were not horse-experienced prior to the trial and thus may have been more excited. Furthermore, inexperience with horse handling could also have facilitated a higher attentiveness and arousal of the mothers. Thus, in contrast to other studies (e.g. Allen *et al.*, 1991; Allen *et al.*, 2002; Barker *et al.*, 2005; Beetz *et al.*, 2011; Beetz *et al.*, 2012b) that found a significant stress-reducing effect in an animal-assisted setting, we found a stimulating effect of EAI.

Even though we did not find an alleviation in sympathetic activity, or increase in parasympathetic activity respectively in the EAI group, the questionnaire assessment of Beetz *et al.* (2015) showed that 65% of all participating mothers would have preferred the EAI to the PBI. This may reflect the greater motivational effect of animal-assisted interventions, as discussed by Wohlfarth *et al.* (2013).

The fact that mothers in PBI spent more time talking to the therapists seemingly contradicts studies which documented that the presence of an animal increased trust and bonding to the therapist and thus enhanced rapport between client and therapist (Zilcha-Mano *et al.*, 2011; Julius *et al.*, 2013). However, in our study therapists and mothers spent more time talking to the children and safeguarding them during EAI, which may have left less time for conversation between mother and therapist. Thus, the presence of the horse did not enhance rapport between mother and therapist, at least not with regard to the amount of verbal communication. However, as our findings show, EAI does provide more room for the therapist to focus on and interact with the child and for the mother to communicate with her child. Thus the therapist can serve as a model for the mother how to interact sensitively with the child.

Bearing in mind, that this was the first comparison of equine-assisted early intervention for toddlers with a standard play-based intervention, we would like to address some limitations of the current study and showing directions for future research in this field. First of all, a greater sample size would be useful to enhance generalizability of our findings. Second, the intervention was provided for only 8 weeks. Carrying out such a study over a longer period of time would help to assess if there are any long-term differences in the behavioral and physiological effects of EAI and PBI. Furthermore, a long-term study could help to reveal if potential calming effects of an animal-assisted approach were masked by novelty effects of being around horses in our study. Would participants become more familiar with being around horses, thereby improving the stress alleviating effect of such an intervention?

In line with the findings of Beetz *et al.* (2015), our results suggest that both EAI and PBI have their benefits in supporting mother-child dyads with insecure attachment. We showed that the EAI does promote different behaviors, within the therapist-mother-child triad, like physical contact between mothers or therapists and children and by tendency vocal contact between mothers and their children. Thus an equine-assisted intervention provides a setting, which could help to increase more sensitive caregiving of the mothers.

Consequently, an intervention for insecurely attached mother-child dyads could be conceptualized as a combination of equine-assisted and play-based interventions. In such a case a starting period in PBI could help to establish trust and familiarity between clients and therapists which thereafter could constitute a foundation for activities in the EAI, which then offers more opportunities for physical contact and vocal exchange between mothers and infants.

Acknowledgements

The authors want to especially thank the therapists, participants and helpers on site for taking part in our study. Furthermore, we want to thank Nora Winkler and Emily Bauer for their help with data collection and data entry.

Authors Note

This research was financially supported by a research grant received from the 'Horses and Humans Research Foundation' and by the 'Horst Rohde Stiftung fuer Mensch und Tier'.

References

Ainsworth, M.D.S., Blehar, M.C., Waters, E. and Wall, S. (1978) *Patterns of Attachment. A Psychological Study of the Strange Situation*. Lawrence Erlbaum Associates Publishers, Hillsdale, New Jersey.

Allen, K.M., Blascovich, J., Tomaka, J. and Kelsey, R.M. (1991) Presence of human friends and pet dogs as moderators of autonomic responses to stress in women. *Journal of Personality and Social Psychology* 61, 582–589. Available at: http://psycnet.apa.org/doi/10.1037/0022-3514.61.4.582

Allen, K.M., Blascovich, J. and Menes, W.B. (2002) Cardiovascular reactivity and the presence of pets, friends, and spouses: The truth about cats and dogs. *Psychosomatic Medicine* 64, 727–739. Available at: https://doi.org/10.1097/01.PSY.000 0024236.11538.41

Anisfeld, E., Casper, V., Nozyce, M. and Cunningham, N. (2008) Does infant carrying promote attachment? An experimental study of the effects of increased physical contact on the development of attachment. *Child Development* 61, 1617–1627. Available at: http://www.jstor.org/stable/113076

Barker, S.B., Knisely, J.S., McCain, N.L. and Best, A.M. (2005) Measuring stress and immune responses in health care professionals following interaction with a therapy dog: A pilot study. *Psychological Reports* 96, 713–729. Available at: https://doi.org/10.2466/pr0.96.3.713-729

Beetz, A., Kotrschal, K., Turner, D.C., Hediger, K., Uvnäs-Moberg, K. *et al.* (2011) The effect of a real dog, toy dog and friendly person on insecurely attached children during a stressful task: An exploratory study. *Anthrozoös* 24, 349–368. Available at: https://doi.org/10.2752/175303711X13159027359746

Beetz, A., Uvnäs-Moberg, K., Julius, H. and Kotrschal, K. (2012a) Psychosocial and psychophysiological effects of human-animal interactions: The possible role of oxytocin. *Frontiers in Psychology* 3, 1–15. Available at: https://dx.doi.org/10.3389%2Ffpsyg.2012.00234

Beetz, A., Julius, H., Turner, D.C. and Kotrschal, K. (2012b) Effects of social support by a dog on stress modulation in male children with insecure attachment. *Frontiers in Psychology* 3, 1–9. Available at: https://doi.org/10.3389/fpsyg.2012.00234

Beetz, A., Winkler, N., Julius, H., Unväs-Moberg, K. and Kotrschal, K. (2015) A comparison of equine-assisted intervention and conventional play-based early intervention for mother-child dyads with insecure attachment. *Journal of Occupational Therapy, Schools, & Early Intervention* 8, 17–39. Available at: https://doi.org/10.1080/19411243.2015.1026017

Bowlby, J. (1953) *Frühe Bindung und kindliche Entwicklung*, 5th ed. 2005. Ernst Reinhard, München, Basel.

Bowlby, J. (1988) *A Secure Base: Parent-Child Attachment and Healthy Human Development*. Basic Books Inc. Publishers, New York.

Bretherton, I. and Munholland, K.A. (2008) Internal working models in attachment relationships. elaborating a central construct in attachment theory. In: Cassidy, J. and Shaver, P.R. (eds) *Handbook of Attachment: Theory, Research and Clinical Applications* (pp.102–127), 2nd ed. Guildford Press, New York.

Cooper, M.L., Shaver, P.R. and Collins, N.L. (1998) Attachment styles, emotion regulation, and adjustment in adolescence. *Journal of Personality and Social Psychology* 74, 1380–1397. Available at: http://psycnet.apa.org/doi/10.1037/0022-3514.74.5.1380

Curic, A., Männer, H., Meißner, S. and Morawetz, F. (2007) Untersuchung zur herzratenvariabilität unter stress- und entspannungs- bedingung. universität regensburg, institut für experimentelle psychologie, lehrstuhl: Prof. dr. h. lukesch PWP. In: *Empirische Erhebungen Zum Bereich Der Medien- Und Gesundheitspsychologie*.

Diamond, L.M. (2001) Contribution of psychophysiology to research on adult attachment: Review and recommendations. *Personality and Social Psychology Review* 5, 276–295. Available at: http://psycnet.apa.org/doi/10.1207/S15327957PSPR0504_1

George, C., West, M. and Pettem, O. (1999) The adult attachment projective: Disorganization of adult attachment at the level of representation. In: Solomon, J. and George, C. (eds) *Attachment Disorganization*. Guilford Press, New York, NY, US, pp. 318–346.

George, C. and Solomon, J. (2008) The caregiving system. a behavioral systems approach to parenting. In: Cassidy, J. and Shaver, P.R. (eds) *Handbook of Attachment: Theory, Research and Clinical Applications*, 2nd ed. Guildford Press, New York, pp. 102–127.

Handlin, L., Jonas, W., Petersson, M., Ejdebäck, M., Ransjö-Arvidson, A.B. *et al.* (2009) Effects of sucking and skin-to-skin contact on maternal ACTH and cortisol levels during the second day postpartum-influence of epidural analgesia and oxytocin in the perinatal period. *Breastfeeding Medicine* 4, 207–220. Available at: https://doi.org/10.1089/bfm.2009.0001

Heinrichs, M., Dawans von, B. and Domes, G. (2009) Oxytocin, vasopressin and human social behavior. *Frontiers in Neuroendocrinology* 30, 548–557. Available at: https://doi.org/10.1016/j.yfrne.2009.05.005

Hottenrott, K., Hoos, O. and Esperer, H.D. (2006) Herzfrequenzvariabilitätund sport. *Aktueller Stand. Herz* 31, 544–552. DOI: 10.1007/s00059-006-2855-1.

Julius, H., Beetz, A., Kotrschal, K., Turner, D.C. and Unväs-Moberg, K. (2013) *Attachment to Pets. An Integrative View of Human-Animal Relationships with Implications for Therapeutic practice*. Hogrefe, Göttingen.

Kaminski, M., Pellino, T. and Wish, J. (2002) Play and pets: The physical and emotional impact of child-life and pet therapy on hospitalized children. *Children's Health and Care* 31–4, 321–335. Available at: https://doi.org/10.1207/S15326888CHC3104_5

Klüwer, B. (2008) Selbsterfahrung auf dem pferd. In: *In Fachgruppe Arbeit Mit Dem Pferd in Der Psychotherapie (Ed.), Psychotherapie Mit Dem Pferd. Beiträge Aus Der Praxis*. FN Verlag, Bremen-Oberneuland, pp. 10–19.

Klüwer, C. and Deutsches Kuratorium für Therapeutisches Reiten e.v. (2005) Die spezifischen wirkungen des pferdes in den bereichen des therapeutischen reitens. In: *Heilpädagogisches Voltigieren Und Reiten. Grundlagen*. Pferdesport Verlag Rolf Ehlers, Bremen-Oberneuland, pp. 5–11.

Kučera, M. (2006) Herzratenvariabilität. Einleitung. MUDr. Michael Kučera, Karla Čapka 4, 36001 Karlovy Cary, Tschechische Republik.

Main, M. and Solomon, J. (1990) Procedures for identifying infants as disorganized/disoriented attachment pattern: Procedures, findings and implications for the classification of behavior. In: Brazelton, T.B. and Yogman, M. (eds) *Affective Development in Infancy*. NJ: Ablex, Norwood, pp. 95–124.

Maunder, R.G., Lancee, W.J., Nolan, R.P., Hunter, J.J. and Tannenbaum, D.W. (2006) The relationship of attachment insecurity to subjective stress and autonomic function during standardized acute stress in healthy adults. *J. Psychosomatic Research* 60, 283–290. Available at: https://doi.org/10.1016/j.jpsychores.2005.08.013

Mickelson, K.D., Kessler, R.C. and Shaver, P.R. (1997) Adult attachment in a nationally representative sample. *Journal of Personality and Social Psychology* 73, 1092–1106. Available at: http://psycnet.apa.org/doi/10.1037/0022-3514.73.5.1092

Mommersteeg, P.M.C., Keijsers, G.P.J., Heijnen, C.J., Verbraak, M.J.P.M. and Doornen, L.J.P. (2006) Cortisol deviations in people with burnout before and after psychotherapy: A pilot study. *Health Psychology* 25, 243–248. Available at: https://doi.org/10.1037/0278-6133.25.2.243

Péter, A. (2016) Solomon Coder. A simple free solution for behavior coding. Homepage. Available at: http://solomoncoder.com/

Pruessner, J.C., Kirschbaum, C., Meinlschmidt, G. and Hellhammer, D.H. (2003) Two formulas for computation of the area under the curve represent measures of total hormone concentration versus time-dependent change. *Psychoneuroendocrinology* 28, 916–931. Available at: https://doi.org/10.1016/S0306-4530(02)00108-7

Stock, S. and Uvnäs-Moberg, K. (1988). Increased plasma levels of oxytocin in response to afferent electrical stimulation of the sciatic and vagal nerves and in response to touch and pinch in anaesthetized rats. Acta Physiologica Scandinavia, 132, 29-34. Available at: https://doi.org/10.1111/j.1748-1716.1988.tb08294.x

Turner, R.A., Altemus, M., Enos, T., Cooper, B. and McGuinnes, T. (1999) Preliminary research on plasma oxytocin in normal cycling women: Investigating emotion and interpersonal distress. *Psychiatry* 62, 97–113. Available at: https://doi.org/10.1080/00332747.1

Unväs-Moberg, K. (1997a) Oxytocin-linked antistress effects—the relaxation and growths response. *Acta Physiologica Scandinavica* 640, 38–42.

Unväs-Moberg, K. (1997b) Physiological and endocrine effects of social contact. *Annals of the New York Academy of Science* 807, 146–163. Available at: https://doi.org/10.1111/j.1749-6632.1997.tb51917.x

Unväs-Moberg, K., Bruzelius, G., Alster, P. and Lundeberg, T. (2008) The antinociceptive effect of non-noxious sensory stimulation is mediated partly through osytocinergic mechanisms. *Acta Physiologica Scandinavica* 149, 199–204. Available at: https://doi.org/10.1111/j.1748-1716.1993.tb09612.x

Vining, R.F. and McGinley, R.A. (1987) The measurement of hormones in salvia: Possibilities and pitfalls. *Journal of Steroid Biochemistry* 27, 81–94. Available at: https://doi.org/10.1016/0022-4731(87)90297-4

Wedl, M., Beetz, A., Julius, H. and Kotrschal, K. (2015) Children with avoidant or disorganized attachment relate differently to a dog and to humans during a socially stressful situation. *Anthrozoös* 28, 601–610.

Wohlfarth, R., Mutschler, B., Beetz, A., Kreuser, F. and Korsten-Reck, U. (2013) Dogs motivate obese children for physical activity: Key elements of a motivational theory of animal-assisted interventions. *Frontiers in Psychology* 4, 1–7. Available at: https://doi.org/10.3389/fpsyg.2013

Woodside, D.B., Winter, K. and Fisman, S. (1991) Salivary cortisol in children: Correlations with serum values and effect of psychotropic drug administration. *Canadian Journal of Psychiatry* 36, 746–748.

Zilcha-Mano, S., Mikulincer, M. and Shaver, P.R. (2012) Pet in the therapy room: An attachment perspective on animal-assisted therapy. *Attachment and Human Development* 13, 541–561. Available at: https://doi.org/10.1080/14616734.2

16 A Logic Model of a Dementia-Specific Program of Equine-Assisted Activities

Rebecca Lassell[1]*, Beth Fields[2], Sarah Busselman[3], Tauren Hempel[4] and Wendy Wood[5]

[1]Department of Occupational Therapy, Colorado State University; [2]Center for Caregiving Research, Education, and Policy, University of Pittsburgh and the Center for Health Equity Research and Promotion, VA Pittsburgh Healthcare System; [3]Fort Collins, Colorado; [4]Department of Equine Sciences, Colorado State University; [5]Departments of Animal Sciences and Occupational Therapy and Director of Research at the Temple Grandin Equine Center, Colorado State University

Abstract

Insufficient descriptions of *what* comprises animal-assisted interventions and *why* and *how* they are presumed to work pose significant challenges to their replication and further scientific development. The purpose of this study was to construct a comprehensive research-based logic model describing a promising program of equine-assisted activities for older adults with Alzheimer's disease and related dementias. Logic models can advance scientific development of innovative programs by elucidating their critical elements. Employing the qualitative method of directed content analysis, we conducted secondary analyses of previously collected interviews of five providers and fieldnotes of each programmatic session. Employing quantitative direct observational methods, we conducted secondary analysis of eight previously collected hour-long videotapes of four older adults with dementia during the program. The resulting logic model from our secondary analysis of mixed methods data elucidated the program's purpose, assumptions, programmatic activities, and outcomes, plus congruent linkages across these elements. These findings can guide replication of the program in multiple research and practice contexts and support its future scientific development. Next research steps include evaluating short- and long-term outcomes beyond participants' direct experiences of the program, clarifying the program's optimal dosages, and ensuring fidelity of its implementation. This logic model may also help to inform scientific development of other animal-assisted interventions.

Keywords: Equine-assisted activities, dementia, quality of life, logic model

The study reported herein sought to advance the scientific basis and replication of an innovative multifaceted program of equine-assisted activities for older adults with Alzheimer's disease and related dementias. This program was designed to enhance the quality of life (QoL) of older adults with dementia. Preliminary proof of concept that the program was associated with participants' improved QoL has been established (Fields *et al.*, 2018). The program constitutes

*Corresponding author: Rebecca.lassell@colostate.edu

a type of animal-assisted intervention (AAI). AAI is a broadly used term that describes the integration of various species of animals to help benefit human health and well-being (American Veterinary Medical Association, 2018).

AAIs incorporating birds, cats, dogs, fish, and horses have emerged for people with dementia over the past 20 years. Some of these AAIs have also shown promise for enhancing outcomes in persons with dementia related to their emotional experiences and mood, social interactions, and physical activities (Wood *et al.*, 2017; Yakimicki *et al.*, 2018). Although studies of AAIs that incorporate horses with this population are sparse, researchers have reported improved outcomes in mood and other dementia-specific indicators of QoL such as communication and participation in activities (Dabelko-Schoeny *et al.*, 2014; Fields *et al.*, 2018; Fields *et al.*, 2019). Fields *et al.* (2019) previously studied the program that is the focus of this paper, which they described as comprising a *complex intervention*.

While no exact line distinguishes simple from complex interventions, complex interventions are characterized by several distinctive elements (Craig *et al.*, 2013). For instance, some elements of complexity pertain to the existence of multiple interacting components as the intervention is being implemented. Other elements pertain to the high degrees of flexibility, tailoring, and skilled behavior that are required of providers to ensure that they deliver an intervention safely and effectively. Still other elements of complexity pertain to the range and sophistication of the measured outcomes of an intervention. In the early scientific development of a complex intervention, the Medical Research Council recommended that researchers describe the intervention's interacting components and model its processes and outcomes onto a logic model (Craig *et al.*, 2013; Moore *et al.*, 2015).

Simply described, a *logic model* visually displays the logic of an intervention by identifying its salient elements and interconnecting its key activities and desired outcomes, among other programmatic components such as guiding assumptions and resources (Mclaughlin and Jordan, 2004; WK Kellogg Foundation, 2004). Logic models are especially useful in describing the elements and processes of complex interventions in the health sciences (Moore *et al.*, 2015). Specific to older adults, for instance, health professionals and researchers have developed logic models of an elder abuse forensic center (Navarro *et al.*, 2010), of recommended practices for integrating mental health services within chronic disease prevention and health promotion programs (Lando *et al.*, 2006), and of a community-based program for recruiting volunteers among older urban African Americans (Chadiha *et al.*, 2011). Logic models such as these function as guiding scientific and practice-based frameworks that help to describe, replicate, and refine promising complex interventions (Lando *et al.*, 2006). Elucidating the logic model of a complex intervention is also foundational to subsequent assessment of the intervention's efficacy and effectiveness (Moore *et al.*, 2014).

The literature on AAIs for people with dementia shows intermittent attention to prevalent components of logic models; especially, the assumptions, inputs, activities, outputs, and outcomes of an intervention (WK Kellogg Foundation, 2004). *Assumptions* in a logic model encompass the beliefs, ideas, and principles that define how and why an intervention is thought to change or benefit people and the community to accomplish its purpose. Related to assumptions, for example, it has been presumed that horses are therapeutic because of their social, non-judgmental, and vulnerable nature, and their abilities to be in-tune with their environments (Karol, 2007; Porter-Wenzloff, 2007; Burgon, 2011). *Inputs* are the financial, human, organizational, and community resources required to implement an AAI and support the quality of its implementation (WK Kellogg Foundation, 2004). Related to inputs, Nordgren and Engström (2012) identified a flat-coated retriever, experienced nurse, and dog handler as resources needed to implement their study of a canine-assisted intervention. *Activities* encompass the processes, tools, events, technology, and actions that are integral to the implementation of an intervention (WK Kellogg Foundation, 2004). Related to activities, Dabelko-Schoeny *et al.* (2014) identified grooming activities that involved braiding, brushing, and combing the horse's hair in their study of an equine-assisted therapy for people with dementia. *Outputs* are the direct products that the program provides to the participants, such as the size and scope of services or number of participants reached (e.g. number of residents, or dosage) (Mclaughlin and Jordan, 2004; WK Kellogg Foundation, 2004). Related to outputs, Olsen *et al.* (2016) indicated that their canine-assisted intervention consisted of 30 min weekly sessions that were provided over 12 weeks to 41 participants. *Outcomes* refer to changes in the behavior, knowledge, skills, status, or levels of functioning of participants in a program (WK Kellogg Foundation, 2004). Improved engagement during activities of daily living and decreased depression and apathy have been identified as outcomes for older adults with dementia who were involved in a canine-assisted intervention (Friedmann *et al.*, 2015).

The above-described areas of attention suggest that researchers have elucidated aspects of AAIs that correspond with specific components of a logic model. At the same time, studies of AAIs often lack comprehensive descriptions of *what* comprises their respective interventions and *how* they are thought to work (Wood *et al.*, 2017); these gaps pose challenges for subsequent replication and further scientific development (see Moore *et al.*, 2014). To our knowledge, moreover, no research-based logic model has been published of an AAI for people with dementia. Hence the purpose of this study was to develop a logic model describing a promising program of equine-assisted activities for members of this population. By developing this logic model, we hope to provide an empirical basis upon which the program can be replicated across multiple research and practice contexts and also further developed scientifically. We asked these questions:

1. What was the purpose of the dementia-specific program of equine-assisted activities and on what assumptions was it based?
2. What resources facilitated implementation of the program?
3. What occupational opportunities (activities) were offered during the program?
4. What were the program's outputs?
5. What were the outcomes of the program in relation to each occupational opportunity?

Methods

Design

We conducted a descriptive, secondary analysis of data from a mixed methods study, herein referred to as the parent study; this study examined the influence of the program of equine-assisted activities on the QoL of older adults with dementia (Busselman *et al.*, 2017; Fields *et al.*, 2018; Fields *et al.*, 2019). A secondary analysis involves analyzing previously collected data for a purpose different from the original study (Cheng and Phillips, 2014). For our secondary analysis, we used previously collected qualitative and quantitative data from the parent study to richly describe the equine-assisted activities program using a logic model framework. A mixed methods approach is often situated in the philosophy of pragmatism (Creswell and Plano, 2017). Pragmatists find truth in solving practical problems to best answer their research questions, substantiating the use of a wide range of research methods (Huber, 1973; Cherryholmes, 1992; Shaw *et al.*, 2010). Guided by this philosophy, we sought to bridge the divide between research and community-based AAIs by using a logic model to describe the program in order to guide future refinement and replication in other contexts.

Setting and sample

The parent study was conducted at two sites that included a PATH Intl. certified therapeutic riding center and a long-term care (LTC) facility. Research participants in the parent study were purposively recruited and included direct service providers and older adults with dementia. Providers had to meet two inclusion criteria: (a) a frontline provider or facility administrator and (b) history of in-depth involvement with the program. Two providers were from the LTC facility who regularly attended program sessions. The other three providers delivered the program of equine-assisted activities and held certifications in therapeutic riding from PATH Intl. Older adults with dementia had to meet the following inclusion criteria: (a) 45 years or older, (b) able to ambulate with minimal assistance from a caregiver (by walking or wheelchair), (c) stable regimen of medications (including psychotropic medications), (d) diagnosed with mild-moderate stage dementia (e) resident at a participating LTC facility for four weeks or longer prior to the study, (f) not allergic to horses, (g) English-speaking, and (h) interest in horses. Approval from the Institutional Review Board of Colorado State University was obtained. All participants and their legally authorized representatives provided informed consent.

Data collection

Five direct service providers took part in individual interviews in the parent study. These interviews were semi-structured, ranged from 30 to 60 min, and were audio-recorded and transcribed. Interview data were collected to learn about providers' perceptions of positive and negative influences of the program on the QoL of older adults with dementia. Research team members recorded field notes after each session to describe what occurred. In addition to the interview and field note data, four older adults with dementia were videotaped during the program. These older adults had been selected for videotaping because they evidenced a range of dementia severity from mild to moderate. Videotaped data were collected for the purpose of documenting QoL outcomes for the older adults with dementia. Videotaping occurred across four sessions, resulting in 16 hours of video data.

Data analysis and trustworthiness

To answer research questions one through four, we uploaded the previously collected interviews and field notes into NVivo (www.qsrinternational.com/nvivo/home), a qualitative data analysis software. For secondary analysis of these data, we then used the method of directed content analysis. Content analysis is considered a basic form of qualitative description, providing the facts of an event and the meaning attributed to the event by participants in 'everyday language' (Sandelowski, 2000). Content analysis differs from other qualitative methods in that it is pragmatically informed by an existing body of knowledge on a topic or event and the experiences of the research team, versus phenomenological and hermeneutical traditions (Neergaard *et al.*, 2009). Further, content analysis allows for the subjective interpretation of text using a 'systematic classification process of coding' to help identify themes (Hsieh and Shannon, 2005, p. 1278). When using *directed content analysis*, researchers apply a deductive approach to incorporate qualitative codes established in previous research in order to more thoroughly describe an event. We created a priori parent codes that corresponded to each component of a logic model defined in this paper's introduction (i.e. purpose and assumptions, inputs, activities, and outputs). We then mapped data from the parent study onto these codes, while also developing and applying relevant sub-codes. For example, the code, *Assumptions*, contained the sub codes horse and nature connection, social participation, and holistic experience. (The Supplementary materials Table S16.3 contains names and definitions of all codes and sub codes).

We used the strategies of triangulation, peer-debriefing, and negative case analysis to ensure trustworthiness of findings resulting from this secondary analysis (Savin-Baden and Major, 2013). Triangulation consisted of revisiting interviews and field notes to ensure consistency with emerging logic model findings. Weekly peer-debriefings were held among researchers to reach consensus and ensure consistent interpretation of the data. Finally, we conducted negative case analysis to search for any conflicting accounts of the program evident in the interview and field note data.

To answer research question five, we uploaded videotape data into Noldus Observer XT 13 (www.noldus.com), a behavioral analysis software. We selected eight hour-long videotapes for secondary analysis because they had been previously coded for dementia-specific indicators of QoL using a direct observational measure, the Activity in Context and Time (ACT) (WK Kellogg Foundation, 2004). Table 16.1 presents these codes and definitions. These indicators of QoL reflected the actual activities of participants during the program and their concomitant expressed affects.

Our secondary analysis of these videotapes involved coding them for occupational

Table 16.1. Parent Study Codes and Definitions using the Modified Activity in Context in Time

Domain: Sub domain	Code	Definition	Indicator Category
Time Use: Gaze	Engaged gaze	Sustained intentional scanning, watching, or visually orientating to some person, event, object or physical environmental attribute in the absence of agitation or distress.	Positive
	Unengaged gaze	No evidence of attention to something in environment; eyes can be open.	Negative
	Eyes closed	Eyes are closed.	Negative
	Sitting	In a seated position.	Neutral
Time Use: Position and Movement	Stairs	Ambulating up or down stairs.	Positive
	Standing	Standing in an upright position.	Positive
	Walking	Ambulating across an area (regardless of assistance or ambulation device). Participant must be continuously walking with less than a five second pause to be coded as walking.	Positive
	Sitting on horse	Seated on a saddled horse, regardless of whether the horse is moving or not.	Neutral
	Lying down	Reclined or lying down.	Negative
Time Use: Communication	Yes communication	Any active exchange of information that is either verbal or nonverbal, that can include physical contact, physical gestures, spoken language, or turn taking.	Positive
	No communication	There is no evidence of an exchange of information.	Neutral
Time Use: Participation	Yes participation	Active sustained engagement in activities that are appealing or acceptable to an individual and not externally coerced or driven.	Positive
	Modifier: Put on or take off helmet	Putting on or taking off riding helmet or hat; participation in putting helmet on is complete when the chin strap is closed and the helmet is adjusted	Positive
	Modifier: Ride the horse	Riding the horse inside or outside, can be sitting on the horse, but does not include mounting. Includes games, the sensory trail, and obstacle courses such as weaving in and out of cones or walking over ground poles.	Positive
	Modifier: Mount or dismount	Process of getting off the horse that begins when the participant is physically touching the horse or tack with intent to get on the horse. Mounting ends when the horse begins to move forward out of the mounting block. Dismounting begins when the participant begins to swing their leg or move off of the horse. Dismounting ends when the participant has two feet on the ground.	Positive

Continued

Table 16.1. Continued

Domain: Sub domain	Code	Definition	Indicator Category
	Modifier: Pet	Using hands or arms to stroke the horse (not using any type of equipment such as a brush). Activity beings when the participant touches the horse and ends when the series of stroking the horse has ended.	Positive
	Modifier: Groom	Grooming the horse with some type of equipment (i.e. brush); activity begins when participant touches horse with a brush and ends when the series of brush strokes has ended.	Positive
	No participation	Absence of participation in activity.	Neutral
Apparent Affect	Anger	Clenching teeth, grimacing, shouting, yelling, cursing, berating, pushing, physical aggression or implied aggression such as fist shaking, pursed lips, narrowed eyes, or knit brow.	Negative
	Anxiety/Fear	Furrowed brow, motoric restlessness, repeated or agitated movement, facial expression of fear or worry, withdrawal from other, tremor, tight facial muscles, calls repetitively, hand wringing, leg jiggling, or eyes wide.	Negative
	Interest	Eyes following object, intent fixation on object or person, visual scanning, facial, motoric, or verbal feedback to other, eye contact maintained, body or vocal response to music, turn body or move toward person or object.	Positive
	Pleasure	Smiling, laughing, stroking, touching with 'approach' manner, nodding, singing, arm or hand outreaching, open-arm gesture clapping, any signs of warmth or affection towards a person or horse.	Positive
	Sadness/Depression	Cry, tears, sigh, mouth turned down at corners, eyes/head turned down and face expressionless, wiping eyes.	Negative
	No apparent affect	No movement of eyes, face, or body.	Negative
Agitation	Yes agitation	Inappropriate verbal, vocal, or motor activity that is not judged by the outside observer to result directly from needs or confusion of the individual; agitated behaviors are always socially inappropriate and can manifest as 1) abuse or aggression towards others, 2) appropriate behavior performed with inappropriate frequency, or 3) inappropriate according to social standards for the specific situation (most frequently manifestations are restlessness, pacing, complaining, repetitive sentences or questions, negativism, constant requests for attention, and cursing or verbal expression).	Negative

Continued

Table 16.1. Continued

Domain: Sub domain	Code	Definition	Indicator Category
	No agitation	Absence of agitation behaviors.	Positive
Missed Observations	Missed observation	Occurred anytime a participant was blocked from view for more than ten seconds and could apply to any code.	Neutral

Gaze and participation definitions were based on Wood (2005) definitions. Apparent Affect definitions were based on Lawton *et al.* (1996) and Lawton *et al.* (2000). Definitions of agitation were based on Cohen-Mansfield *et al.* (1989).

opportunities. As defined for the purpose of this coding, *occupational opportunities* encompassed the options that were offered to participants to engage in specific activities throughout each session of the program (Wood *et al.*, 2017). To honor participants' preferences for activities, staff and volunteers always presented these opportunities as options rather than as requirements or expectations. Codes for occupational opportunities were created through an iterative process of watching videotape data, generating initial codes, modifying these codes, creating new ones, and reaching consensus on all existing codes until we met saturation. Table 16.2 illustrates these codes and definitions. We used a continuous sampling approach in Noldus, where one occupational opportunity code was assigned for every second of time (see Noldus Information Technology, 2016). An occupational opportunity began when it was offered to a participant with dementia and ended when it was no longer available. For example, we started coding the opportunity of petting when a handler led a pony within reach of a participant with dementia and ended when the pony was no longer within reach. Researchers coded missed observation if a participant with dementia was blocked from view.

The nesting function in Noldus allows researchers to merge data from two separate coding strands of the same video with the option of inserting one coding strand within another (Noldus Information Technology, 2016). Hence, after completing coding for occupational opportunities, we used the nesting function in Noldus to merge previously coded indicators of QoL within the coded occupational opportunities. We used this nesting function to document how proffered occupational opportunities (e.g. the opportunity to ride the horse) did or did not overlap with participants' actual activities in response to the opportunities (e.g. actually mounting and riding the horse versus walking away from the horse). We then used the analysis function in Noldus to generate frequencies and durations of each occupational opportunity and its corresponding indicators of QoL across all program sessions. Percentages of each code for the total program duration were calculated using a formula in Excel and were then mapped onto the logic model. For example, the most prevalent identified occupational opportunity, observing, offered older adults' opportunities to talk with people and also to watch horses and people interact before or after their own opportunities to interact with a horse. This opportunity accounted for 54% of the total duration of the program. During observing, participants demonstrated an engaged gaze (i.e. a positive indicator of QoL) 93% of the time; we then mapped these findings onto the logic model.

To ensure trustworthiness of this coding process, two researchers underwent extensive training in Noldus. Inter-coder reliability was calculated based on hours of all videotaped data, or 20%. A kappa coefficient of 0.82 was achieved indicating nearly perfect agreement (Viera and Garrett, 2005). Two subsequent checks for drift reliability were also completed among the secondary-analysis study team.

Results

Findings from our secondary data analysis are mapped onto the logic model framework (Fig. 16.1) and described in the subsequent sections.

Table 16.2. Secondary Data Analysis: Occupational Opportunity Codes and Definitions

Code	Definition
Grooming	The opportunity to groom a horse began when a participant (who was not mounted on a horse) was within reaching distance of a horse and grooming tools were available for use. Grooming the horse ended when the participant no longer had a grooming tool in his or her hand and began walking away from the horse.
Observing	Observing was defined solely by the fact that participants were sitting on the bench or in a chair, facing the activities that were ongoing. Hence observing offered opportunities to watch horses and people involved in the program, and to survey many and oftentimes dynamically interacting elements of its equine environment.
Petting	The opportunity to pet a horse began when a horse was within reaching distance of a participant (who was not mounted on a horse) and there were no physical barriers hindering the participant from petting the horse. The opportunity ended when the horse was no longer within reaching distance of the participant.
Riding	The opportunity to ride a horse began when a participant was standing on the mounting platform and the horse began walking between the blocks towards the rider. The rider would then be assisted by program volunteers and staff to mount the horse. Riding included games such as cones, obstacle courses, walking over poles, reaching for rings or a toy, etc. This opportunity ended when the participant had dismounted, defined as having both feet on the ground, and began walking away from the horse.
Transitions	The opportunity of transitions began when a participant ended an occupational opportunity by walking to the next one. This opportunity included chances to put on or take off a helmet, which were previously coded as participation. Transitions ended when the participant entered a different occupational opportunity.

Each occupational opportunity was unique and mutually exclusive, meaning that one occupational opportunity could be assigned at a time. Transitions were not a program activity, however, in order to capture every moment of time, transitions were coded to cover the amount of time traveling between activities

Purpose and assumptions

The stated purpose of the dementia-specific program of equine-assisted activities was to improve the QoL of people with Alzheimer's disease and related dementias. Consistent with this purpose, providers hypothesized that specific qualities of the program favorably influenced older adults' QoL. Namely, providers posed that shared activities among older adults and horses fostered social connections and participation. Grooming, for instance, created opportunities for older adults to get to know a volunteer on a one-on-one basis and also to become acquainted with a horse. Additionally, providers reported that horses evoked rich sensory experiences and the natural setting of the therapeutic riding center rekindled prior positive memories of horses and nature, helping to uplift older adults' emotions. Furthermore, providers believed that experiences of the program as a whole enhanced older adults' QoL. For instance, many components of the program synergistically gave rise to simultaneous positive experiences for participants, such as connecting with a horse, re-living previous memories in nature, socializing, and being physically active.

Inputs

Personnel and training

The Director of Special Activities at the therapeutic riding center was responsible for managing the program, overseeing paperwork, and establishing and maintaining collaborative partnerships between the therapeutic riding center and participating LTC facilities. This role involved visiting each LTC facility in which older adults in the program lived, learning about

Table 16.3. Definitions of qualitative codes and sub codes.

Codes and sub codes	Definitions
Purpose	Text clearly stating the mission or purpose of the program.
Assumptions	Text describing the beliefs, ideas, and principles that define how and why the program is thought to change or benefit older adults with dementia (see WK Kellogg Foundation, 2004).
Horse and nature connection	Comments describing older adults' positive behaviors by being around horses and the natural environment during the program. Includes phrases involving the sensory aspects of interacting with horses and nature (e.g. sights, smells, touch) and tapping into previous memories of being with horses and in nature.
Social participation	Phrases related to attributing the benefits of the program to social interactions and connections.
Holistic experience	Text explaining older adults' positive behaviors by detailing the interdependence of the different experiences available in the program.
Inputs	Text describing the financial, human, organizational, and community resources required to implement the program (see WK Kellogg Foundation, 2004).
Personnel and training	Phrases describing the roles of the therapeutic riding center staff and volunteers and the training they received.
Horses	Phrases describing the horses including their breed, age, personality, and their roles in the program.
Facility and equipment	Phrases describing the therapeutic riding facility and the equipment used during the sessions.
Cost and transportation	Text describing the cost of program sessions and how the LTC facilities brought older adults with dementia to program sessions.
Participant screening and measures	Descriptions of the screening process for older adults with dementia to be enrolled in the program as well as text describing the measures used to document older adults' outcomes.
Occupational Opportunities	Text describing the options that were offered to participants to engage in specific activities throughout each session of the program (see Wood *et al.*, 2017).
Ground experience	Phrases referring to activities offered off of the horse.
Riding experience	Text describing activities that occur when mounted on horseback.
Outputs	Text describing the direct products that the program provided to the participants, such as the size and scope of services or number of participants reached (e.g. number of residents, or dosage) (Mclaughlin and Jordan, 2004; WK Kellogg Foundation, 2004).

the facility's care approach, and addressing any miscommunication or conflict that arose. Two therapeutic riding instructors certified by the Professional Association of Therapeutic Horsemanship, International (PATH Intl.) led the program and were assisted by volunteers. Volunteers generally held a 3:1 ratio with older adults for riding activities and a 1:1 ratio for ground activities including grooming and observing. All activities occurred concurrently where volunteers and therapeutic riding instructors presented opportunities to participate in activities to each participant throughout the duration of the program. Riding was selectively presented to older adults who had been cleared by their physician. A therapeutic riding

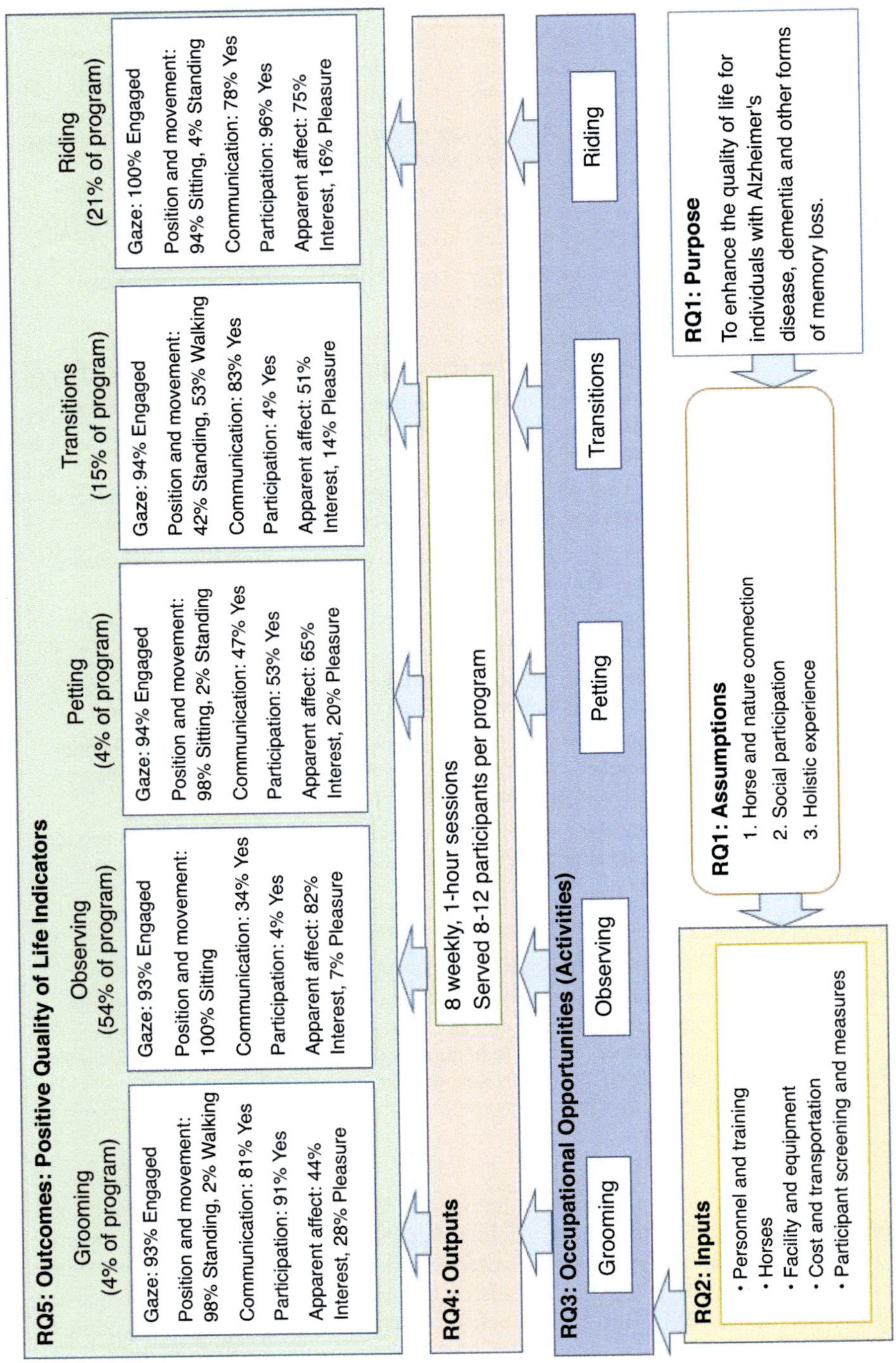

No negative quality of life indicators were observed. Percentages in each category do not add up to 100% due to missed observations.

Fig. 16.1. Logic Model of a Dementia-specific Program of Equine-assisted Activities

instructor directed riding activities while two side-walkers and a horse leader accompanied the older adults. Groundwork activities, such as petting and grooming, were facilitated by a volunteer. Specific training was provided to volunteers for the roles of barn-worker, horse-leader, and side-walker. Volunteer responsibilities included program set-up and implementation. For example, volunteers arranged seating in the arena and gathered equipment prior to each session. Volunteers implemented the program by interacting with the older adults and adhering to safety procedures, such as helping the older adults to put on riding helmets.

Horses

In accordance with the standards of PATH Intl. (www.pathintl.org/quick-links/standards-manual), all horses utilized in the program underwent a rigorous screening and training process. Therapeutic riding instructors selected horses to deliver the program based on (a) narrow to normal width, (b) slow and steady movement, (c) tolerance for imbalanced riders, and (d) comfort being surrounded by four to five people. Horses with a narrow to normal width were chosen as a precaution for older adults' hips. Three horses and one pony participated in the program. All were mares with the following ages and breeds: a 14-year-old Spanish Barb/Quarter Horse, 18-year-old Halfinger, 21-year-old Appaloosa, and a 15-year-old Shetland/Welsh pony.

Facility and equipment

The program occurred at a premier accredited therapeutic riding facility, the highest level of accreditation possible through PATH Intl. A PATH Intl. certification requires the facility to uphold industry standards for safety and administration of equine-assisted activities. The therapeutic riding center is located on the outskirts of the Rocky Mountains in Loveland, Colorado with scenic views of the foothills. Resources utilized during the program included an indoor arena, handicap accessible mounting ramp, and an outdoor sensory trail. Materials for groundwork included benches and chairs facing the ongoing activities. Grooming equipment involved a hitching post for the horses, standard halters, lead ropes, and grooming buckets with two to four brushes each. Equipment and tack utilized during riding included a standard riding helmet, western saddles, saddle pads, bridles, and, in some cases, colored reins clipped to a halter. Materials for games during riding included colored rings, cones, ground poles, basketball hoop, and a ring tree. The outdoor sensory trail, located at the base of the foothills of the Colorado Rockies, included wooden logs, ground poles, jump standards, barrels, and a bridge.

Cost and transportation

Cost of the program included administration fees for participant screenings, meetings, volunteer and staff trainings, as well as the cost of each session. Costs were based on program frequency and duration in accordance with each LTC facility's needs. Most commonly, a LTC facility would enroll in a four-week program with hour-long sessions. However, some LTC facilities attended weekly sessions, spanning eight weeks. LTC facilities were responsible for transporting older adults via bus.

Participant screening and measures

To begin the program's screening process, a participant's healthcare proxy was required to complete an enrollment packet that included a physician's consent (see Appendix A). The enrollment packet contained background information on the participant's history with horses as well as a list of contraindications involving medical conditions that would deem them unsafe to participate in the program, such as a spinal cord injury. Once older adults passed the initial screening process, riding instructors traveled to the LTC facility to complete an in-person evaluation of each potential participant. The screening involved wearing a riding helmet and sitting on a barrel to assess the older adults' hip flexibility. Riding instructors determined whether the older adult was able to ride comfortably, and assessed their ability to follow directions, respond to cues, and safely mount and dismount. During this screening, staff looked to see if the older adult was interested in participating in the program and if they demonstrated signs of distress, pain, agitation, or potentially dangerous behaviors to help determine if the older adult was appropriate for the program. Furthermore, riding instructors began to formulate which program activities would best suit each participant. Staff from the LTC facility such as nurses and therapists were encouraged to attend the screening to answer questions related to older adults' health and functional abilities. Upon passing both screening phases, older adults were officially enrolled as participants in the program. Currently, therapeutic riding instructors do not utilize a standardized outcome measure to document outcomes of the program.

Occupational opportunities

Ground experiences

Four distinct occupational opportunities to interact with horses on the ground were offered to participants: observing, grooming, petting, and transitioning from one activity to another. During the opportunity for observing, participants sat in a chair with a back or on a bench, facing the ongoing activities in the indoor arena. Observing involved opportunities to socialize with volunteers or other participants, watch the horses and ongoing activities of the program, and survey the equine environment in which the program occurred. Volunteers and staff in the program also presented participants with opportunities to groom, pet, and ride a horse. If a participant expressed interest in any of these opportunities, a volunteer would assist them in putting on a helmet prior to participating. Grooming involved opportunities to brush a horse, comb a horse's mane, and become acquainted with the horse and volunteers. During this opportunity, volunteers and program staff helped participants successfully sequence and complete grooming tasks by offering simple steps and tactile cues. For example, a staff member would guide a participant's hand to brush the horse's neck. The opportunity to pet a horse or pony was presented while participants were seated on the bench. Staff led a miniature pony to participants, providing the opportunity to engage in petting. Petting created opportunities for conversations with staff, volunteers, the horse or pony, and other older adults. Transitions occurred when volunteers escorted participants from one activity to the next.

Riding experiences

Two providers described riding as the most challenging activity, posing the highest risk. Offering the opportunity to ride involved having a therapeutic riding instructor accompany participants onto the mounting platform and the horse into the mounting area. If participants chose to ride, then two therapeutic riding instructors on either side of the ramp would assist them to mount the horse. Once the participant mounted the horse, the horse was led out of the mounting platform. The horse leader checked the participant's weight distribution and guided the instructors to adjust the stirrups if necessary. Next, two side-walkers accompanied the participant on either side of the horse, to ensure safety during riding activities. The horse leader and the side-walkers instructed the participant on riding techniques such as steering with reins and giving the horse verbal commands such as 'walk.' Riding included games involving upper body exercises, steering around cones and barrels, throwing balls into a basketball hoop, placing rings on a ring tree, and riding over ground poles. Typically, five of 12 participants, who had been cleared by their physician, chose to ride each session. Riding lasted roughly 15 min to avoid fatigue and began in an indoor arena, and weather permitting, included riding outside over obstacles on the sensory trail. Therapeutic riding instructors assisted participants in dismounting using either a crest or croup technique as described in the PATH Intl. riding instructor manual (www.pathintl.org/images/pdf/resources/Evaluator%20Update%20pdfs/Riding-on-site-Workshop-Manual.pdf).

Outputs

Outputs of the program included eight-weekly sessions, lasting approximately 1 hour. All four older adults who participated in the video observations attended the entirety of each of the four video-taped sessions. Furthermore, videotape data aligned with providers' reports of serving 8–12 participants per program. The program began by serving one local LTC facility. Currently, the program collaborates with four local care agencies, ranging from LTC to adult day programs for people with dementia and is offered in the spring, summer, and fall seasons.

Outcomes

All occupational opportunities were associated with high levels of positive and neutral QoL indicators pertaining to time use (gaze, position and movement, communication, participation) and apparent affect and agitation (Fig. 16.1). No negative QoL indicators were observed at any time throughout the program. Therefore, each occupational opportunity supported older adults' QoL, demonstrating congruence with the purpose of the program.

With respect to opportunities provided on the ground, observing was most pervasively offered in the program and also associated with the highest frequencies of an observed positive affect of interest. Thus, while not actually interacting with horses during the opportunity called observing, older adults watched and showed interest in various activities going on around them while sitting on the bench. Opportunities to groom a horse, pet a horse, or transition from one activity to another were offered for much shorter durations than opportunities for observing. Yet these opportunities were also distinctively linked with particular positive QoL indicators. Namely, the opportunity to groom a horse was related to the greatest proportions of time that participants spent standing and communicating with the horse or volunteers, the second greatest proportion of time spent participating in the offered opportunity (grooming), and the greatest frequencies of an observed positive affect of pleasure. Therefore, grooming supported indicators of QoL related to an uplifted emotional state, communication, and the use of participants' physical capacities. Grooming seemed to offer opportunities for more complex engagements than petting and observing, in that participants often simultaneously groomed and talked to their horses while also conversing with other nearby people. The opportunity to pet a horse was linked to the second greatest levels of the observed affect of pleasure. Petting was accessible to all older adults with varying physical abilities and occurred while older adults were seated. Like grooming, petting demonstrated variances in positive and neutral QoL indicators, such as an uplifted emotional state, communication, and participation in a short period of time. Lastly, the opportunity to transition from one activity to another was linked to the greatest proportions of time that older adults spent walking and communicating with people and horses.

The opportunity to ride a horse was the second most pervasively offered opportunity. Three of the four older adults who were videotaped during the program had previously been cleared for riding, all of whom chose to ride once given the opportunity. This opportunity was associated with the greatest proportions of time that older adults spent with an engaged gaze and actively participated in all the opportunities offered; namely, mounting a horse, riding the horse while oftentimes doing various other activities such as games or obstacle courses, and dismounting a horse. Riding was also linked to large proportions of time spent communicating with the horse, therapeutic riding instructor, and horse handlers. This opportunity offered the longest bouts of time where older adults were observed to engage in more complexly layered activities simultaneously. For example, older adults were observed to communicate with staff and the horse, while demonstrating an engaged gaze, and following directions to participate in riding, such as steering the horse around cones.

Discussion

Findings from this study align with the Medical Research Council's recommendation that researchers describe an intervention's interacting components and model its processes and outcomes onto a logic model in the early phases of scientific development (Craig *et al.*, 2013; Moore *et al.*, 2015). As next developed, the logic model resulting from this study depicts *what* comprised the purpose, assumptions, programmatic activities, and outcomes of the program of equine-assisted activities for older adults with dementia; it illustrates congruent linkages across these components; and it suggests a foundational understanding of *why* and *how* the program worked. For these reasons, the logic model can serve as a guiding framework to replicate and refine the program across multiple sites (see Moore *et al.*, 2014). This research-based logic model can also help to justify further scientific investment in developing the program and clarify next research steps.

Creating the logic model allowed us to make providers' assumptions describing *why* the program worked explicit. Providers, who assumed that the many elements of the program synergistically contributed to the positive QoL experiences of participants, emphasized in particular connections with horses and nature and social participation as mechanisms of change. Much literature pertaining to AAIs support these assumptions. For example, investigators of AAIs have drawn from Wilson's biophillia hypothesis (Kellert and Wilson, 1993), which assumes that humans are hardwired to connect with animals and nature (Beck and Katcher, 2002; Beetz, 2017). Furthermore, being in nature has been connected with positive emotional states in older adults with dementia (Duggan *et al.*, 2008) and linked to health (Kaplan, 1995; Kuo, 2015).

Providers in this study also credited the experience of getting to know a horse as enhancing the QoL of older adults who participated in the program, a presumption that aligns with the fairly widely expressed view that the horse-human bond can elicit positive outcomes (Yorke *et al.*, 2008; Dell *et al.*, 2011; Burgon, 2011). Other researchers of AAIs have similarly assumed that interactions with animals can help to facilitate social connections, serve as a non-threatening topic of conversation (Hunt *et al.*, 1992; Wells, 2009), and provide social support (Gee *et al.*, 2017). Providers in the current study also assumed that the sensory experience of engaging with a horse and nature favorably influenced positive outcomes of QoL. In like fashion, Dabelko-Schoeny *et al.* (2014) posed that the multi-sensory experience of being around a horse and a farm can contribute to positive outcomes in older adults with dementia.

The assumption of providers that many programmatic components as a whole helped to enhance participants' QoL was reflected in how they delivered programmatic activities. For example, the screening process allowed therapeutic riding instructors to collaborate with LTC staff to select older adults who were interested and able to safely participate in the program. The different roles of personnel (e.g. therapeutic riding instructor, side-walker), their respective training processes (e.g. staff training on dementia care) and availability of different types of equipment (e.g. mounting platform) were all needed to deliver the program safely. These inputs, which made it possible to deliver the occupational opportunities during the program and implement the entire program safely, collectively contributed to realizing the program's purpose of enhancing older adults' QoL.

Other examples of the holistic nature of the program pertained to the delivery of specific programmatic elements. To honor each participant's preferences and needs, providers offered a dynamic array of possible interactions with horses rather than prescribing a set of group activities. This method of offering opportunities is congruent with Kitwood's (1997) person-centered approach to dementia care, one in which caregivers honor each person's preferences by providing choices. Accordingly, because providers in the present study intentionally offered different ways to watch or engage with horses throughout the session, the resulting logic model emphasized *occupational opportunities* rather than *activities*, a common element in logic models (see WK Kellogg Foundation, 2004). Use of the term, occupational opportunity (Wood *et al.*, 2017), in the model underscores the importance of providing possibilities for action. Detailed descriptions of *how* each occupational opportunity was implemented may also help to guide future replication and evaluation of the program. Wood *et al.* (2017) noted that AAI studies targeting older adults with dementia provide little description of the intervention and instead focused on outcomes, such as improved mood and decreased negative behaviors. The providers' description of each occupational opportunity in the present study may possibly serve as a template for elucidating the elements and implementation processes of other dementia-specific animal-assisted interventions, thereby addressing a crucial step in scientific development.

Lastly, the logic model that resulted from this study linked providers' assumptions about why the program worked with their programmatic activities and also associated indicators of QoL exhibited by participants in the program. Nested within each occupational opportunity, for instance, were the specific ways in which participants responded to the opportunities, showing that all five occupational opportunities had distinct value as gauged by participants' positive expressions of engagement, interest and pleasure. Hence, each occupational opportunity was shown to be distinct by variations in positive and neutral indicators of QoL among participants with varying levels of cognitive and physical abilities. For example, participants with more severe cognitive deficits were provided with simple directions and tactile cues during grooming, such as guiding their hands to brush the horse's neck. Participants with limited mobility were offered the occupational opportunity of petting a horse while seated on the bench.

Directions for future research

Findings from our logic model revealed several areas for future research. Namely, there are needs to explicate a more in-depth theory of change; investigate factors that pose risk to or protect program implementation and outcomes; examine fidelity of implementation and optimal dosages of the program; document short and long-term outcomes beyond the immediate experiences of participants during the program; and document the impact of the program on the local community.

Researchers should investigate the missing linkages of *how* providers' proposed mechanisms were active in producing outcomes of QoL during the program. Simplistic depictions of causal relationships are weaknesses associated with logic models (Rogers, 2008)—especially those created during the early phases of development. However, the Medical Research Council underscores the importance of first describing an intervention to determine *what* ingredients are present, identifying which ones are presumed to bring about change, prior to testing causal explanations (Craig *et al.*, 2013; Moore *et al.*, 2015). Therefore, the developed logic model can be refined to illustrate complex pathways to QoL outcomes through future causal testing to build a more in-depth theory of change (see Mclaughlin and Jordan, 2004; Moore *et al.*, 2014).

We suggest a process evaluation as a next step for the future research of the program. A process evaluation guided by a logic model can help illuminate which ingredients are active and how they produce change within a complex intervention, building a more in-depth theory to drive future research (Moore *et al.*, 2014). The Medical Research Council provides guidance on key components of a process evaluation framework (Moore *et al.*, 2015). Relying on this guidance, we suggest these next research steps: (a) external factors that may protect or pose risk to participants' outcomes or the success or failure of the program in multiple contexts could be assessed through interviews of volunteers, therapeutic riding staff, family members and other care partners, among other stakeholders, and (b) the program's fidelity, or its quality and consistency in implementation, could be assessed using checklists of observed core components of the program. Assessment of fidelity would help to ensure that future replications of the program will be of a high quality. During replication of the program, researchers should also examine the program's fidelity, or its quality and consistency in which each occupational opportunity is implemented. Fidelity could be assessed using checklists of observed core components of the program (Steckler and Linnan, 2002). Assessment of fidelity would help to ensure that future replications of the program will be of a high quality and is crucial during a full-scale evaluation, such as a randomized controlled trial (see Moore *et al.*, 2014).

Future research should examine dosage as a next step. The outputs of the program illustrated by the logic model, provide an understanding of the dosage, format of delivery, and the number of participants reached. The delivered product of eight weekly sessions to 8–12 participants delivered immediate outcomes of positive and neutral indicators of QoL. However, it is unknown what the most effective dosage and format of delivery (individual, small group, large group, etc.) of the program is, based upon a person's stage of dementia. This unknown is also reflected in the AAI literature with variances in dosage and delivery formats across participants with varying stages of dementia (see Yakimicki *et al.*, 2018). For example, Thodberg *et al.* (2016) found that individual visits with a dog twice a week over six weeks increased prosocial behaviors for older adults with mild to severe dementia. In contrast, Olsen *et al.* (2016) canine-assisted intervention delivered twice a week for 12 weeks in small group format (3–7 people) found significant improvements in QoL in older adults with severe dementia, but not for those in the mild to moderate stages. Hence, more research of the program is needed to examine the most effective dosage based on the older adults' stage of dementia.

Other next steps concern short-term outcomes of one to 3 years and long-term outcomes of four to 6 years, plus community impact. Both levels of these outcomes could be explored qualitatively through interviews of with family members or other care partners, or quantitatively using dementia-specific measures such as Dementia Quality of Life Instrument (Mulhern *et al.*, 2013). Lastly, the program's impact at a level of the local community could be captured by conducting interviews with still other stakeholders such as community leaders or individual and corporate donors to the program. Addressing each of these areas for future research, will advance the scientific merit of the program and inform the development of other dementia-specific AAIs.

Limitations

A limitation of our secondary analysis was that the qualitative interviews were originally conducted for a purpose other than developing a logic model. Therefore, interview questions were not designed to elicit responses for the desired logic model components, meaning the data were fitted to the logic model retrospectively. However, Mclaughlin and Jordan (2004) highlight that logic models can be created retrospectively for existing programs to guide future evaluation. Furthermore, our sample was not inclusive of the full range of participants served by the program. Our study captured findings for older adults in the mild to moderate stages of dementia and did not include those with mild cognitive impairment or those in the late stages of the disease. Lastly, we did not include the perspectives of informal care partners who could have illustrated additional program characteristics due to the limited scope of the parent study.

Conclusion

While AAIs are growing in popularity for older adults with dementia, their scientific development remains in early phases. Specifically, AAIs for this population lack a thorough description of what occurs during these types of interventions, posing challenges for their replication and subsequent evaluation. We therefore developed a logic model of a promising program of equine-assisted activities to describe *what* the program is and *why* and *how* it works. The developed logic model elucidated the program's essential components and processes, and its most immediate outcomes as measured by directly observed indicators of older adults' QoL, while they were engaging in the program. This logic model provides a basis for replication and further refinement of the program across multiple sites. It is our hope that the logic model may serve as a guide for preliminary scientific development of other AAIs whose purpose is to also enrich the QoL of people with dementia.

Acknowledgements

With permission, the authors would like to gratefully acknowledge Hearts and Horses Therapeutic Riding Center, Seven Lakes Memory Care, and the Temple Grandin Equine Center for their ongoing support of this research. Furthermore, the authors would like to thank the Carl and Caroline Swanson Foundation for generously funding this study.

Author's Notes

Beth Fields is now at the Department of Kinesiology, Occupational Therapy Program, University of Wisconsin, Madison. Tauren Hempel is now working in long-term care in Chetek, Wisconsin.

Supplementary Content

Refer to supplementary material on the CABI Digital Library: https://www.cabidigitallibrary.org/doi/book/10.1079/9781836991601.0000

References

American Veterinary Medical Association (2018) Animal-assisted Interventions: Definitions. Available at: https://www.avma.org/KB/Policies/Pages/Animal-Assisted-Interventions-Definitions.aspx?PF=1

Beck, A. and Katcher, A. (2002) Future directions in human-animal bond research. *American Behavioral Scientist* 41(1), 79–93. Available at: https://doi.org/10.1177/0002764203255214

Beetz, A.M. (2017) Theories and possible processes of action in animal assisted interventions. *Applied Developmental Science*. Available at: https://doi.org/10.1080/10888691.2016.1262263

Burgon, H.L. (2011) "Queen of the world": Experiences of "at-risk" young people participating in equine-assisted learning/therapy. *Journal of Social Work Practice* 11. Available at: https://doi.org/10.1080/02650533.2011.561304

Busselman, S.L., Wood, W., Hooper, B. and Bruemmer, J. (2017) *An exploratory study of an equine-assisted intervention for people with dementia living in a residential care facility*.

Chadiha, L.A., Washington, O.G.M., Lichtenberg, P.A., Green, C.R., Daniels, K.L. *et al*. (2011) Building a registry of research volunteers among older urban African Americans: Recruitment processes and outcomes from a community-based partnership. *Gerontologist* 51(Suppl.1). Available at: https://doi.org/10.1093/geront/gnr03

Cheng, H.G. and Phillips, M.R. (2014) Secondary analysis of existing data: Opportunities and implementation. *Shanghai Archives of Psychiatry* 26(6), 371–375. Available at: https://doi.org/10.11919/j.issn.1002-0829.214171

Cherryholmes, C.H. (1992) Notes on pragmatism and scientific realism. *Educational Researcher* 1006013(6), 13–17. Available at: https://doi.org/10.3102/0013189X02

Cohen-Mansfield, J., Marx, M.S. and Rosenthal, A.S. (1989) A description of agitation in a nursing home. *Journal of Gerontology* 44(3), M77–M84.

Craig, P., Dieppe, P., Macintyre, S., Michie, S., Nazareth, I. *et al.* (2013) Developing and evaluating complex interventions: The new medical research council guidance. *International Journal of Nursing Studies* 50(5), 587–592. Available at: https://doi.org/10.1016/j.ijnurstu.201

Creswell, J.W. and Plano, C.V.L. (2017) *Designing and conducting mixed methods research*, 3rd ed. SAGE Publications, Thousand Oaks, California.

Dabelko-Schoeny, H., Phillips, G., Darrough, E., DeAnna, S., Jarden, M. *et al.* (2014) Equine-assisted intervention for people with dementia. *Anthrozoos* 27(1), 141–155. Available at: https://doi.org/10.2752/175303714X

Dell, C.A., Chalmers, D., Bresette, N., Swain, S., Rankin, D. *et al.* (2011) A healing space: The experiences of First Nations and Inuit youth with Equine-Assisted Learning (EAL). *Child and Youth Care Forum* 40(4), 319–336. Available at: https://doi.org/10.1007/s10566-011-9140-z

Duggan, S., Blackman, T., Martyr, A. and Schaik, P. (2008) The impact of early dementia on outdoor life. *Dementia* 7(2), 191–204. Available at: https://doi.org/10.1177/1471301208091158

Fields, B., Bruemmer, J., Gloeckner, G. and Wood, W. (2018) Influence of an equine-assisted activities program on dementia-specific quality of life. *American Journal of Alzheimer's Disease and Other Dementias* 72052(5), 309–317. Available at: https://doi.org/10.1177/1533317518772052

Fields, B., Wood, W. and Lassell, R. (2019) Impact of a dementia-specific program of equine-assisted activities: Providers' perspectives. *Quality in Ageing and Older Adults* QAOA-10-2018-0047. Available at: https://doi.org/10.1108/QAOA-10-2018-0047

Friedmann, E., Galik, E., Thomas, S.A., Hall, P.S., Chung, S.Y. *et al.* (2015) Evaluation of a pet-assisted living intervention for improving functional status in assisted living residents with mild to moderate cognitive impairment: A pilot study. *American Journal of Alzheimer's Disease and Other Dementias* 45477(3), 276–289. Available at: https://doi.org/10.1177/1533317514545477

Gee, N.R., Mueller, M.K. and Curl, A.L. (2017) Human–animal interaction and older adults: An overview. *Frontiers in Psychology*. Available at: https://doi.org/10.3389/fpsyg.2017.01416

Hsieh, H.-F. and Shannon, S.E. (2005) Three approaches to qualitative content analysis. *Qualitative Health Research* 76687(9), 1277–1288. Available at: https://doi.org/10.1177/1049732305276687

Huber, J. (1973) Symbolic interaction as a pragmatic perspective: The bias of emergent theory. *American Sociological Review* 38(2), 274–284. Available at: https://www.jstor.org/stable/2094400

Hunt, S., Hart, L. and Gomulkiewicz, R. (1992) Role of small animals in social interactions between strangers. *The Journal of Social Psychology* 132(2), 245–256.

Kaplan, S. (1995) The restorative benefits of nature. *Journal of Environmental Psychology* 71057(1), 169–182. Available at: https://doi.org/10.1177/1359104507071057

Karol, J. (2007) Applying a traditional individual psychotherapy model to equine-facilitated psychotherapy (EFP): theory and method. *Clinical Child Psychology and Psychiatry* 12(1), 77–90. Available at: https://doi.org/10.1177/1359104507071057

Kellert, S.R. and Wilson, E. (1993) *Biophillia Hypothesis*. Island Press, Wasington D.C.

Kitwood, T. (1997) *Dementia reconsidered: The person comes first*. Open University Press, Buckingham, UK.

Kuo, M. (2015) How might contact with nature promote human health? Promising mechanisms and a possible central pathway. *Frontiers in Psychology* 6(August), 1–8. Available at: https://doi.org/10.3389/fpsyg.2015.0

Lando, J., Williams, S., Williams, B. and Sturgis, S. (2006) A logic model for the integration of mental health into chronic disease prevention and health promotion. *Prev Chronic Dis (Serial Online)* 3(2), 1–4. Available at: http://www.cdc.gov/pcd/issues/2006/apr/05_0215.htm

Lawton, M., Haitsma, K. and Perkinson, M. (2000) Emotion in people with dementia: A way of comprehending their preferences and aversions. In: Lawton, P.M. and Rubinstein, R.L. (eds) *Interventions in Dementia Care: Toward Improving Quality of Life*. Springer Publishing Company, New York, NY, pp. 95–119.

Lawton, M.P., Haitsma, K. and Klapper, J. (1996) Observed affect in nursing home residents with alzheimer's disease. *The Journals of Gerontology Series B: Psychological Sciences and Social Sciences*. Available at: https://doi.org/10.1093/geronb/51B.1.P3

Mclaughlin, J.A. and Jordan, G.B. (2004) Using logic models. In: *Handbook of Practical Program Evaluation*, Vol. 1968. pp. 7–32.

Moore, G., Audrey, S., Barker, M., Bond, L., Bonell, C. *et al.* (2014) *Process evaluation of complex interventions: UK Medical Research Council (MRC) guidance*. Available at: https://doi.org/http://www.population healthsciences.org/MRC-PHSRN-Process-evaluation-guidance-final-2-.pdf

Moore, G.F., Audrey, S., Barker, M., Bond, L., Bonell, C. *et al.* (2015) Process evaluation of complex interventions: Medical research council guidance. *Bmj 350*(mar19 6), h1258–h1258. Available at: https://doi.org/10.1136/bmj.h1258

Mulhern, B., Rowen, D., Brazier, J., Smith, S., Romeo, R. *et al.* (2013) Development of DEMQOL-u and DEMQOL-PROXY-u: Generation of preference-based indices from DEMQOL and DEMQOL-PROXY for use in economic evaluation. *Health Technology Assessment* 17(5), 1–160. Available at: https://doi.org/10.3310/hta17050

Navarro, A.E., Wilber, K.H., Yonashiro, J. and Homeier, D.C. (2010) Do we really need another meeting? Lessons from the Los Angeles county elder abuse forensic center. *Gerontologist* 50(5), 702–711. Available at: https://doi.org/10.1093/geront/gnq018

Neergaard, M.A., Olesen, F., Andersen, R.S. and Sondergaard, J. (2009) Qualitative description-the poor cousin of health research. *BMC Medical Research Methodology* 9(1), 1–5. Available at: https://doi.org/10.1186/1471-2288-9-52

Noldus Information Technology (2016) *The observer XT the next generation of observation software: Reference manual version 13*. Wageningen, The Netherlands.

Nordgren, L. and Engström, G. (2012) Effects of animal-assisted therapy on behavioral and/or psychological symptoms in dementia: A case report. *American Journal of Alzheimer's Disease and Other Dementias*. Available at: https://doi.org/10.1177/1533317512464117

Olsen, C., Pedersen, I., Bergland, A., Enders-Slegers, M.J. and Ihlebæk, C. (2016) Effect of animal-assisted activity on balance and quality of life in home-dwelling persons with dementia. *Geriatric Nursing*. Available at: https://doi.org/10.1016/j.gerinurse.2016.04.002

Porter-Wenzloff, L. (2007) Finding their voice: Developing emotional, cogntive, and behavioral congruence in female abuse survivors through equine faciliated therapy. *Nursing* 3(4), 529–534.

Rogers, P.J. (2008) Using programme theory to evaluate complicated and complex aspects of interventions. *Evaluation*. Available at: https://doi.org/10.1177/1356389007084674

Sandelowski, M. (2000) Focus on research method: Whatever happend to qualitative description. *Research in Nursing & Health* 23, 334–340. Available at: https://doi.org/10.1002/1098-240X(200008)23:4<334::AID-NUR9>3.0.CO;2-G

Savin-Baden, M. and Major, C. (2013) Quality. In: *Qualitative Research: The Essential Guide To Theory and Practice*. Routledge, New York, NY, pp. 471–484.

Shaw, J.A., Connelly, D.M. and Zecevic, A.A. (2010) Pragmatism in practice: Mixed methods research for physiotherapy. *Physiotherapy Theory and Practice* 26(8), 510–518. Available at: https://doi.org/10.3109/09593981003

Steckler, A.B. and Linnan, L. (2002) *Process evaluation for public health interventions and research*. Available at: https://doi.org/10.1016/j.evalprogplan.2003.09.006

Thodberg, K., Sørensen, L.U., Videbech, P.B., Poulsen, P.H., Houbak, B. *et al.* (2016) Behavioral responses of nursing home residents to visits from a person with a dog, a robot seal or a toy cat. *Anthrozoos* 29(1), 107–121. Available at: https://doi.org/10.1080/08927936.20

Viera, A.J. and Garrett, J.M. (2005) Understanding interobserver agreement: The kappa statistic. *Family Medicine* 37(5), 360–363.

Wells, D.L. (2009) The effects of animals on human health and well-being. *Journal of Social Issues* 65(3), 523–543. Available at: https://doi.org/10.1111/j.1540-4560.2009.01612.x

WK Kellogg Foundation (2004) WK Kellogg foundation logic model development guide. Available at: www.wkkf.org (accessed 29 April 2025).

Wood, W. (2005) Toward developing new occupational science measures: An example from dementia care research. *Journal of Occupational Science* 12(3), 121–129. Available at: https://doi.org/10.1080/14427591.2005.9686555

Wood, W., Fields, B., Rose, M. and McLure, M. (2017) Animal-assisted therapies and dementia: A systematic mapping review using the lived environment life quality (LELQ) model. *American Journal of Occupational Therapy*. Available at: https://doi.org/10.5014/ajot.2017.027219

Wood, W., Lampe, J.L., Logan, C.A., Metcalfe, A.R. and Hoesly, B.E. (2017) The lived environment life quality model for institutionalized people with dementia. *Canadian Journal of Occupational Therapy* 84(1), 22–33. Available at: https://doi.org/10.1177/00084174166

Yakimicki, M.L., Edwards, N.E., Richards, E. and Beck, A.M. (2018) Animal-assisted intervention and dementia: A systematic review. *Clinical Nursing Research* 1–21. Available at: https://doi.org/10.1177/10547738187

Yorke, J., Adams, C. and Coady, N. (2008) Therapeutic value of equine – human bonding in recovery from trauma. *Anthrozoos* 274038. Available at: https://doi.org/10.2752/089279308X

Appendix A

HEARTS & HORSES

Therapeutic Riding Center

est. 1997

Welcome!

Thank you for your interest in Hearts & Horses, a non-profit therapeutic horsemanship program, which creates a supportive and dynamic environment for the development of children and adults with disabilities living in our area. Through the teaching of horsemanship skills, physical, psychological, cognitive, behavioral, and communication goals are achieved and personal strengths are emphasized.

Children and adults with a wide array of disabilities including cerebral palsy, autism, and other spectrum disorders, sight and hearing deficits, multiple sclerosis, Down syndrome, emotional, developmental and learning disabilities find success in meeting their goals thanks to the powerful interaction with a horse. At present, we have an average of 150 riders per week participating in equine assisted activities at Hearts and Horses.

Physicians, therapists, teachers and friends may refer participants to Hearts & Horses. Our highly trained instructors and therapists design individual lesson plans for each participant tailored to their ability level and goals. Through a variety of equine-related activities, our participants realize many benefits that include greater confidence, self-awareness, increased balance, muscle strength and self-esteem.

Hearts & Horses offers a year-round program that includes group riding lessons and one-on-one intensive therapy with a licensed physical or occupational therapist.

To find more information about enrolling in an upcoming session at Hearts & Horses, please call 970-663-4200 or via email: info@heartsandhorses.org

Sincerely,

Jan Pollema
Executive Director

2013 Non Profit of the Year • A PATH Intl. Premier Accredited Facility

163 North County Road 29 • Loveland, Colorado 80537 • Phone: 970.663.4200 • Fax: 970.663.3891 • www.heartsandhorses.org

EXPLANATION OF PROGRAM SERVICES, PARTICIPANT ELIGIBILITY AND POLICIES

****KEEP FOR YOUR REFERENCE****

Page 1 of 3

PROGRAM SERVICES

Therapeutic Riding (Ages 4 and up) –Therapeutic riding lessons are equestrian skill based lessons for individuals with disabilities. The focus of the lessons is skill development and progression while improving the rider's physical, cognitive, emotional and/or social skills. Lessons are taught by a Professional Association of Horsemanship International (PATH Intl.) Certified Therapeutic Riding Instructor and assisted by volunteers. Helping the participant reach their full potential is of paramount importance. The length of the lessons includes mounting and dismounting in addition to any pre or post mounted horse care planned by the instructor. Partial scholarships may be available based on family's financial need.

- **Group Classes -** 3-6 participants; generally one hour in length. Participants will engage in pre-mounted and post-mounted horse care (if appropriate). Participants are scheduled by age, skill level and availability.
- **Private Lessons -** Private lessons are typically 30-45 minutes in length depending on the needs and abilities of the participant. Due to arena space, private lessons are rare and only available with prior approval only.
- **Un-mounted Activities -** For participants who may be unable to ride a horse, un-mounted activities may be available.

Hippotherapy (Ages 2 and up) - Hippotherapy is a clinical, physical or occupational therapy using the horse as the modality and is administered by a licensed occupational, physical, or speech therapist. The horse's movement promotes active responses in the client and facilitates activation of postural control, balance, motor and sensory systems. Participants who have movement, speech and/or motor deficits can benefit from Hippotherapy.

A physician's prescription for physical therapy and/or occupational therapy with Hippotherapy is required. Hippotherapy fees are $75 per 30 minute session and are to be paid to the therapist at each treatment session. This service may qualify for insurance reimbursement. At this time Hearts & Horses does not bill through insurance, so any reimbursement must be processed through the individual's insurance, by the individual and is solely their responsibility.

TO GET STARTED

1. Contact Hearts & Horses for a Participant Intake Interview and schedule a short visit to Hearts & Horses.
2. Read carefully the Participant Enrollment Packet, which also includes a Physician Assessment.
3. Please complete the Participant Enrollment Packet and have your physician complete the Physician's Assessment.
4. Set up evaluation once forms are returned to Hearts & Horses.
5. After the evaluation we will discuss our services and your goals to determine the right placement in a class (if appropriate) and new participants can register for the next upcoming session during the open registration period.

New Participants: Once all completed forms are returned to Hearts & Horses along with a $35 evaluation fee, you will be contacted to schedule an evaluation. If it is determined that a physical or occupational therapist is needed, there will be a $45 evaluation fee. If the evaluation fee is a hardship, scholarship funds may be used to help offset the cost.

Prior to and during the evaluation, we will ensure that our program is appropriate for you and that there are no contraindications to your participation in mounted equestrian activities.

Un-mounted equestrian activities are also available for those unable to ride.

***ALL enrollment forms must be updated annually in January.**

HEARTS & HORSES Therapeutic Riding Center • est 1997 • A PATH Intl Premier Accredited Facility • *2013 Non Profit of the Year*

163 North County Road 29 • Loveland, Colorado 80537 • Phone: 970.663.4200 • Fax: 970.663.3891 • www.heartsandhorses.org

EXPLANATION OF PROGRAM SERVICES, PARTICIPANT ELIGIBILITY AND POLICIES

****KEEP FOR YOUR REFERENCE****

Page 2 of 3

ELIGIBILITY GUIDELINES AND DISCHARGE POLICY

Minimum Age: Therapeutic Riding – 4 years of age; Hippotherapy – 2 years of age. There is no maximum age for services at Hearts & Horses, as long as the Participant has no physical or medical contraindications.

Weight Maximums: Due to safety considerations for staff, participants and volunteers, we may not be able to accept a participant into the program who weighs over 170 pounds.

Postural Control: Riders over 60 pounds must be able to maintain a sitting position; at least by holding on with one hand.

Discharge Policy: Hearts & Horses strives to provide the safest possible conditions for participants, volunteers and employees. The acceptance, and continued participation, of a participant in our program depends on the availability of instructors, volunteers and suitable horses and is based on our determination that we can safely accommodate the participant. Hearts & Horses adheres to precautions and contraindications for participants established by the Professional Association of Therapeutic Horsemanship International (PATH Intl.). Hearts & Horses retains the right to refuse any participant that we cannot safely accommodate. Participants must inform us of changes in health status.

SAFETY RULES & GUIDELINES

Our program has an excellent safety record. Please observe these safety rules.

- If weight of participant is over 170 pounds, please let us know. For safety considerations, we may not be able to accept participants over 170 pounds.
- Children must be supervised at all times; please do not leave children unattended or allow then to run and play loudly.
- Parents must stay on the property during their child's lesson.
- Dogs and smoking are not allowed on Hearts & Horses property.
- Do not visit the horses without permission and supervision, including entering the horses stalls.
- Please observe quietly; your family members and friends are always encouraged to visit. When lessons are in progress or horses are out, guests are required to remain in the appropriate spectator area.
- Our horses and volunteers should be treated kindly - they work very hard for us all.
- Please drive slowly and park in designated areas.

ATTIRE AND EQUIPMENT

- Appropriate clothes are long pants and appropriate shoes for being around horses, preferably hard soled boots with a low heel. No sandals, flip flops, or Crocs please! Dress for comfort and according to the weather. Wear close-fitting clothing for safety as well as comfort. Loose or baggy clothing can get caught and tangled in equipment. No dangling jewelry is permitted.

- All participants are required to wear an ASTM/SEI approved Equestrian helmet when near/on horses. If you don't have your own helmet, Hearts & Horses will provide you with one. Should you choose to purchase your own, we can recommend several tack shops in our area.

SCHOLARSHIP INFORMATION

If you need financial assistance, we have a limited number of partial scholarships available for those who qualify. It is our policy to keep our services accessible, so please contact the Program Coordinator if you need assistance to help cover the cost of your lessons. ***A scholarship application which MUST INCLUDE verification of your income (ex: tax return) must be submitted** **<u>by the registration deadline of the session you wish to register for.</u>**

HEARTS & HORSES Therapeutic Riding Center • est 1997 • A PATH Intl Premier Accredited Facility • *2013 Non Profit of the Year*
163 North County Road 29 • Loveland, Colorado 80537 • Phone: 970.663.4200 • Fax: 970.663.3891 • www.heartsandhorses.org

EXPLANATION OF PROGRAM SERVICES, PARTICIPANT ELIGIBILITY AND POLICIES
****KEEP FOR YOUR REFERENCE****
Page 3 of 3

VOLUNTEERS

Hearts & Horses instructors are assisted by a team of incredible volunteers. They groom and tack the horses and help the participant during class. Many volunteers help in other aspects of the organization, from facility maintenance, to office work, to fundraising. If you or someone you know is interested in volunteering, please refer them to the volunteer page on our website for further information. Please remember to thank your volunteers.

CANCELLATION POLICY

Hearts & Horses will hold classes inside to allow riding during adverse weather conditions. However, if any conditions arise that may create unsafe circumstances in which to hold class, classes may be cancelled. The safety of our participants, volunteers and equines is of upmost importance! ***Refunds or credits are not issued for classes that are cancelled for circumstances beyond Hearts & Horses control**. Classes may be cancelled due to the following:

- Larimer County is on accident alert status or weather conditions have the potential to deteriorate significantly by the end of class time.
- Extreme winds, heat (above 95) or cold (below 20)

ATTENDANCE AND PROMPTNESS

- Regular attendance is important! Please call if you cannot make your scheduled time as soon as possible (24 hours is ideal) so that we do not have horses, volunteers and instructors waiting for you. **Two "no call/no shows" in a session may result in you being asked to forfeit your spot to someone on the waiting list.** Hearts & Horses does not refund money for classes that you must cancel.
- Please arrive on time or a few minutes early for your class. We will make every effort to accommodate each rider; however, late arrivals may result in not being able to ride that day. A rider that arrives 15 minutes of more after the start of their class will not be allowed to ride.

THANK YOU!

Thank you for taking the time to review our policies and procedures, which are designed to provide our participants with a safe, effective and enjoyable environment. We encourage your input and suggestions. Please feel free to forward comments to Jan Pollema, Executive Director, at 970-663-4200 or email: jan@heartsandhorses.org

***Please keep this information (pages 1-3) for your reference.**

HEARTS & HORSES Therapeutic Riding Center • est 1997 • A PATH Intl Premier Accredited Facility • *2013 Non Profit of the Year*
163 North County Road 29 • Loveland, Colorado 80537 • Phone: 970.663.4200 • Fax: 970.663.3891 • www.heartsandhorses.org

2016 Hearts & Horses Registration Form

This writable PDF form can be saved and emailed
Register by Mail, email, fax or in Person

Session Dates • Registration Deadlines • Pricing

Session		Session Dates	Registration Begins	Registration Deadline	Price: 1.5 HR Group	Price: 1 HR Group	Price: 30 min Private	Price: 45 min Private
Winter	8 Weeks	Jan 11-Mar 5	11/20/15	12/11/15	$360	$240	$320	$400
Spring	8 Weeks	Mar 21-May 14	2/5/16	2/19/16	$360	$240	$320	$400
Summer	8 Weeks	Jun 6-Aug 6	4/15/16	4/29/16	$360	$240	$320	$400
Fall	8 Weeks	Aug 20-Oct 17	7/1/16	7/15/16	$360	$240	$320	$400
Holiday	6 Weeks	Nov 2-Dec 17	9/23/16	10/7/16	$270	$180	$240	$300

Participant's name ____________________ Weight (*Please call if over 170#*)______ Height ______ Age ______

Responsible Party ____________________ Female Head of Household? ☐ Yes ☐ No For grant reporting purposes only

Phone(s)____________________ Email Address____________________

Mailing Address ____________________ City ________________ Zip ________

Emergency Contact Name & Phone ____________________

Has participant's medical status changed? (seizures, etc?) ☐ Yes ☐ No If yes, please describe below:

____

Program (Check One) ☐ Therapeutic Riding ☐ Hippotherapy ☐ Changing Leads ☐ Hearts & Horses for Heroes (Veterans)

Type of Lesson (Check One) ☐ Group ☐ 30 min. Private ☐ 45 min. Private

* (Private lessons by prior arrangement only)

Returning Participants

☐ I would prefer to keep my current day/time.

☐ I need to change my class; List your availability in the box →

☐ **I am a NEW participant;** List your availability in the box →

Availability:

Method of Payment (Please check one) *Payment or arrangement for payment must be made at the time of registration.

Please specify amount paid: $____________ Checks can be made payable to Hearts & Horses

☐ **Check#** ________ OR ☐ **Cash** *Please hand deliver to Hearts & Horses Office (M-F 9:00 am-5:00 pm)

☐ **Credit Card** *Call office or pay online at www.heartsandhorses.org If paying online, please attach your receipt.

☐ **Paid for by other agency or benefactor** *Please note you are still responsible for securing payment

Name of agency/contact person/address/phone ____________________

☐ I have **2016 scholarship** forms on file. I understand that funds are limited.

Office Use Only: Date Received __________ Initials ________

HEARTS & HORSES Therapeutic Riding Center • est 1997 • A PATH Intl Premier Accredited Facility • *2013 Non Profit of the Year*

163 North County Road 29 • Loveland, Colorado 80537 • Phone: 970.663.4200 • Fax: 970.663.3891 • www.heartsandhorses.org

2016 Enrollment Packet – Hippotherapy & Therapeutic Riding
(Page 1 of 5)

In order to ensure coordinated care, Hearts & Horses staff and volunteers may be provided with information about participant's abilities/disabilities.

Today's Date ______________

Participant's Name ______________________________ Birthdate ______________ Age ________

Weight ____________ *(Please call if weight over 170#)* Height __________ Sex ________ Grade ______

Parent / Caregiver / Self (*please circle*) Name ______________________________________

Email ______________________________ Phone ______________________________

Address ______________________________ City______________ Zip Code ____________

Income Range (*for grant purposes – kept private and confidential*): ☐ Less than $30,000 ☐ $30,000 to $47,000 ☐ Above $47,000 ☐ Decline to Answer

Ethnicity (*for grant purposes – kept private and confidential*): ________________ ☐ Decline to answer

Client Goals: Personal ______________________________________

School ______________________________________

Family ______________________________________

Horsemanship ______________________________________

Strengths and Abilities ______________________________________

____

____

Presenting Problems/Concerns ______________________________________

____

____

Primary Diagnosis ______________________________________

Secondary/Tertiary Diagnoses ______________________________________

____

Physical Disabilities/Limitations ______________________________________

____

Allergies to medications or foods ______________________________________

____

- -

HEARTS & HORSES Therapeutic Riding Center • est 1997 • A PATH Intl Premier Accredited Facility • *2013 Non Profit of the Year*

163 North County Road 29 • Loveland, Colorado 80537 • Phone: 970.663.4200 • Fax: 970.663.3891 • www.heartsandhorses.org

2016 Enrollment Packet – Hippotherapy & Therapeutic Riding
(Page 2 of 5)

EMERGENCY CONTACTS – IN THE EVENT OF AN EMERGENCY, HEARTS & HORSES SHOULD CONTACT:

Name ____________ Relation ____________ Phone(s) ____________

Name ____________ Relation ____________ Phone(s) ____________

Name ____________ Relation ____________ Phone(s) ____________

POLICY OF CONFIDENTIALITY

I agree to respect and observe privacy and confidentiality of the participants, volunteers, and donors of Hearts & Horses, Inc. and will not discuss or disclose any sensitive information about any person or their family.

____________ Minor Participant Signature ____________ Parent or Legal Guardian Signature

CANCELLATION POLICY

Hearts & Horses will hold classes inside to allow riding during adverse weather conditions. However, if any conditions arise that may create unsafe circumstances in which to hold class, classes may be cancelled. The safety of our participants, volunteers and equines is of upmost importance. ***Refunds or credits are not issued for classes that are cancelled for circumstances beyond Hearts & Horses control**. Classes may be cancelled due to extreme heat, wind or cold.

I have read and understand Hearts & Horses cancellation policy: ____________
Signature of Responsible Party

PHOTO & VIDEO RELEASE

Name of Participant ____________

For valuable consideration given and which is hereby acknowledged, the undersigned hereby grants to Hearts & Horses, Inc. permission to take, or have taken, still and moving photographs and films including television pictures of the above named Participant, and consents and authorizes Hearts & Horses, Inc., its advertising agencies, news media, and any other persons interested in Hearts & Horses, Inc. and its programs, to use and reproduce the photographs, films, videos and pictures, and to circulate and publicize the same by any means deemed appropriate by Hearts & Horses, Inc., including, without limitation the generality of the foregoing: newspapers, web sites, social media, television media, brochures, pamphlets, instructional materials, books, and clinical materials.

With respect to the foregoing matters, no inducements or promises have been made to secure this signature to this release other than the intention of Hearts & Horses, Inc. to use, or cause to be used, such photographs, films, videos and pictures for the primary purpose of promoting and aiding Hearts & Horses, Inc. and the field of equine assisted activities and therapies.

I give consent ____________ Date ____________
Signature of adult participant or parent/guardian/caregiver of minor participant

I **do not** give consent ____________ Date ____________
Signature of adult participant or parent/guardian/caregiver of minor participant

HEARTS & HORSES Therapeutic Riding Center • est 1997 • A PATH Intl Premier Accredited Facility • *2013 Non Profit of the Year*
163 North County Road 29 • Loveland, Colorado 80537 • Phone: 970.663.4200 • Fax: 970.663.3891 • www.heartsandhorses.org

2016 Enrollment Packet – Hippotherapy & Therapeutic Riding
(Page 3 of 5)

RELEASE OF LIABILITY

Name of Participant ______________________________

WARNING

Under Colorado Law, an equine professional is not liable for an injury to or the death of a participant in equine activities resulting from the inherent risks of equine activities, pursuant to section 13-21-119, Colorado Revised Statutes.

RELEASE AND INDEMNIFICATION

I am aware that any activities involving horses are hazardous and I am voluntarily participating in these activities with knowledge of the danger involved, and hereby agree to accept any and all risks of injury, including death, and damage to property arising from participation. I **hereby promise not to sue, and hereby release**, to the fullest extent permitted by law, Hearts & Horses, Inc. and its agents, officers, directors, members, representatives, instructors, volunteers, coordinators, insurers, independent contractors, therapists and employees (collectively the "Released Parties"), from, **and hereby waive**, all claims of whatsoever kind that may be asserted against the Released Parties for personal injury and property damage arising from or in connection with participation in equine activities, and from the condition of the real property and personal property used in connection with such equine activities. By way of example, and not in limitation, this Waiver and Release includes releasing and waiving claims based upon: any negligent acts or omissions of the Released Parties and any other person; contract; warranty; premises liability; products liability; subrogation; contribution; and loss of consortium or loss of society.

I also hereby agree to indemnify, defend, and hold and save harmless the Released Parties from any claims, damages, expenses and costs incurred of whatsoever nature (including by way of example, and not in limitation, attorney fees and expenses), which may be made against or incurred by the Released Parties, arising from or in connection with my participation, including without limitation, any claims made by me or any other person.

It is intended that this Release and Indemnification shall release the Released Parties from, and waive, any and all claims, and indemnify the Released Parties, to the greatest extent allowed by law. In the event for any reason a Court determines that any portion of this Release and Indemnification is not enforceable, that provision shall be modified so as to give it the greatest effect allowed by law, or if it cannot be so modified shall be severed and the balance of the Release and Indemnification shall be given the greatest force and effect available under law. Furthermore, in the event that notwithstanding this Release and Indemnification, it is determined that any Released Party has any liability for any claim, in no event shall the liability exceed the amount of $500 in total aggregate for all claims arising from or in connection with my participation.

I acknowledge that by signing this document I am waiving important legal rights. I also acknowledge that the Released Parties would not allow me to participate in equine activities unless I have agreed to the waivers, releases, indemnifications and limitations contained in this Release and Indemnification. I acknowledge that the Released Parties are relying upon these provisions as a primary material consideration for allowing my participation in equine activities. I acknowledge and agree that the terms hereof are binding upon me, and my heirs, successors, representatives, insurers, and assigns.

If signing on behalf of another person, I represent and warrant to the Released Parties that I am the parent or legal guardian with the capacity to execute and make the foregoing waivers and indemnifications on behalf of such person; and I further acknowledge and agree that I am also personally bound by and make the releases and waivers as above set forth, and that I am jointly and severally liable for the indemnifications to the Released Parties.

Signature ______________________________ Date ______________

Signature of adult participant or parent/guardian/caregiver of minor participant

2016 Hearts & Horses – Physician Assessment and Health History (Page 4 of 5)

Name of Participant ____________ Date of Birth ________ Age ______ Gender ☐ M ☐ F

To Be Completed by Physician

Height __________ Weight ______ Date of Last Tetanus Shot __________

Mobility: ☐ Independent ☐ Assistive Device __________

Primary Diagnosis __________ Date of Onset ______

Secondary Diagnosis __________ Date of Onset ______

Seizures ☐ No ☐ Yes Type __________ Date of last seizure ______

Shunts/Impants: __________

Past/Prospective Surgeries (include dates and reasons) __________

PLEASE LIST ALL CURRENT MEDICATIONS

1. __________ Taken for: __________
2. __________ Taken for: __________
3. __________ Taken for: __________

In order to safely provide this service, Hearts & Horses requests that you please note that the following conditions may suggest precautions and contraindications to equestrian activities. Therefore, when completing this form, please indicate whether these conditions are present and to what degree.

Orthopedic
Atlantoaxial Instability (include neurologic symptoms)
Coxarthrosis
Cranial defects
Heterotopic Ossification/ Myositis Ossificans
Joint subluxation/dislocation
Osteoporosis
Pathologic fractures
Spinal Joint Fusion/Fixation
Spinal Joint instability/Abnormalities

Medical/Psychological
Allergies
Animal Abuse
Cardiac Conditions
Physical/Sexual/Emotional Abuse
Blood Pressure Control
Dangerous to Self or Others
Exacerbations of Medical Conditions (e.g. RA,MS)
Fire Setting
Hemophilia
Medical Instability
Migraines
PVD
Respiratory Compromise
Recent Surgeries
Substance abuse
Thought Control Disorders
Weight Control Disorder

Neurologic
Hydrocephalus/shunt
Paralysis due to spinal cord injury
Seizure
Spina Bifida/Chiari II Malformation
Tethered cord/Hdromyelia

Other
Age - Under 4 years
Indwelling Catheters/Medical Equipment
Medications – e.g., Photosensitivity
Poor Endurance
Skin Breakdown

*Next Page →

HEARTS & HORSES Therapeutic Riding Center • est 1997 • A PATH Intl Premier Accredited Facility • *2013 Non Profit of the Year*
163 North County Road 29 • Loveland, Colorado 80537 • Phone: 970.663.4200 • Fax: 970.663.3891 • www.heartsandhorses.org

2016 Hearts & Horses – Physician Assessment and Health History (Page 5 of 5)

When complete with ALL signatures, please return BOTH pages of this form

_______________________ Date of Birth __________ Age ________ Gender ☐ M ☐ F

Name of Participant

As thoroughly as possible, please indicate current or past difficulties/symptoms in the following systems/areas that apply (include surgeries).

Area	No	Yes	Comments
Auditory			
Visual			
Tactile/Sensory			
Speech			
Cardiac			
Circulatory			
Integumentary/Skin			
Immunity			
Pulmonary			
Neurologic			
Muscular			
Balance			
Orthopedic			
Allergies			
Learning Disability			
Cognitive			
Emotional/Psychological			
Pain			
Other			

*****FOR PARTICIPANTS WITH DOWN SYNDROME*****

An annual neurological exam to exclude Atlantoaxial instability is required for all participants with Down Syndrome over the age of three. Please provide the following information:

Date of Exam ____________ Results of Exam ______________________________

Given the above diagnosis and medical information, this person is not medically precluded from participation in equine-assisted activities and/or therapies. I understand that Hearts & Horses, Inc. will weigh the medical information indicated above against any existing precautions and contraindications before accepting this person for mounted equestrian activities. Therefore, I refer this person to Hearts & Horses for ongoing evaluation to determine eligibility for participation.

Physician Name/Title______________________ MD DO PA NP Other ________

Signature ______________________________ **Date** ____________

Address____

Phone ______________________ License/UPIN Number ____________

HEARTS & HORSES Therapeutic Riding Center • est 1997 • A PATH Intl Premier Accredited Facility • *2013 Non Profit of the Year*

163 North County Road 29 • Loveland, Colorado 80537 • Phone: 970.663.4200 • Fax: 970.663.3891 • www.heartsandhorses.org

Conclusion

Lori R. Kogan
Colorado State University, USA

The collection of research presented in this book highlights the growth and potential of EAS as an intervention to support human health and wellbeing across the lifespan. From children with autism spectrum disorder to adolescents developing essential life skills to adults addressing complex trauma, the horse-human bond offers unique therapeutic possibilities that can complement traditional interventions. As the field continues to develop, it is essential to critically examine where we are, where we need to go, and how we can ensure that our practices honor both the human participants and the equine partners.

Ethical Foundation

The ethical foundation of EAS rests on the recognition that horses are sentient beings with their own needs, preferences, and feelings - not merely tools for human benefit. It is important to recognize that in animal assisted services, not all physical interactions are enjoyed by the animal and some types may actually increase their mental and physical stress (Ferlazzo *et al.*, 2023). As opposed to purpose-bred therapy dogs, horses in EAS are often retired from previous careers before working as EAS animals (Rankins *et al.*, 2023). Evidence of high turnover of horses working in EAS has been reported (Rankins *et al.*, 2021; Watson *et al.*, 2020), with a median length of time of seven years. This short period of time suggests additional attention to welfare is needed (Baxley *et al.*, 2024).

Throughout this book, we see examples of how interactions between horses and humans—characterized by mutual respect, attunement, and positive engagement—form the foundation for therapeutic benefits. As noted in Hauge *et al.*'s study of adolescents' persistence during horse-related tasks, for example, the horses' responses to human contact were predominantly positive or neutral, suggesting that well-designed interventions can create positive experiences for both human and equine participants.

This commitment to equine welfare must be evident in all aspects of EAS, including, but not limited to, the training and selection of the horses. Positive training methods and the development of trust-based relationships between horses and handlers have been shown to increase a horse's contact-seeking behaviors towards unfamiliar people and a more positive attitude towards their handlers (Briefer Freymond *et al.*, 2014; Lundberg *et al.*, 2020). These approaches not only benefit the horses but also model respectful, attuned interactions that many human participants need to experience and learn, laying the foundation for both equine and human wellbeing in EAS programs.

Not all horses are suited for EAS work, and matching horse temperament and experience to the specific demands of different types of services is essential. The desired traits of EAS horses depend on the type of program being delivered (Brubaker *et al.*, 2021). For example, horses participating in therapeutic riding for children with cerebral palsy, as examined in Kaiser *et al.*'s study, or autism spectrum disorder in Peters, Bentele, & Wood's study, require different physical and temperamental characteristics than those participating in equine-assisted psychotherapy for trauma survivors, such as Hershberger *et al.*'s work with sexually exploited children or very young children like those included in Hebesberger, Beetz and Kotrsachal's study. In addition, positive welfare mandates careful attention to the behaviors of each horse every day to assess for any avoidant behaviors, injury, or illness that might preclude their participation in EAS (French and Fournier, 2023).

Ongoing monitoring of equine stress indicators and workload are crucial aspects of welfare-centered EAS. It is important, however, that we continue to strive towards good welfare and positive affect rather than just the absence of stress. Some signs of a positive affective state include relaxed eyes, extended upper lip, and backward pointing, unpinned ears in the nose line, and allogrooming (Feh and de Mazières, 1993; Lansade *et al.*, 2018). To increase positive affect, human-horse interactions such as grooming should be based on the length and frequency the horse prefers (Fry, 2021).

Future research should continue to investigate reliable indicators of positive and negative equine experiences during EAS activities, incorporating both physiological measures such as heart rate variability and cortisol levels with behavioral observations. Reciprocal benefits in EAS require an understanding of equine cognition, emotion, and social behavior. While considerable research has examined horse behavior and welfare in traditional equestrian contexts, less attention has been given to understanding horses' experiences specifically within therapeutic environments (Scopa *et al.*, 2019). Pioneering work has begun to explore how horses perceive and respond to humans exhibiting different emotional states (Nakamura *et al.*, 2018), findings that have profound implications for EAS, where participants often present with anxiety, depression, or trauma-related symptoms. The development of standardized protocols for monitoring equine welfare in EAS settings should be a top priority. By explicitly incorporating equine welfare considerations into program design, implementation, and evaluation, EAS practitioners can strengthen the ethical foundation of their work while potentially enhancing its effectiveness.

In addition, as the field of EAS continues to develop, greater attention should be paid to developing comprehensive plans for equine retirement and end-of-life care. Programs demonstrating ethical excellence should have protocols for regular assessment of each horse's suitability to continue in therapeutic work, recognizing that this may change over time due to age, health conditions, or shifting preferences. One recent study found that behavior during lessons, interactions with participants, soundness, and overall health are the most common factors used when making retirement decisions (Rankins *et al.*, 2021). These factors should also help inform initial selection decisions, carefully evaluating such factors as age at program entry and existing health conditions when determining a horse's suitability for EAS work.

Standardization of Terminology and Practice Guidelines

The standardization of terminology and practice guidelines represents a critical step forward for the field of EAS. The 2020 consensus document on EAS terminology, as described in the introduction to this book (Wood *et al.*, 2021), has provided a common language that facilitates clearer communication across disciplines, stronger research methodologies, and more consistent implementation of services. This framework, which categorizes twelve types of equine-assisted services under the broad categories of therapy, learning, and horsemanship, offers a useful structure needed to advance both research and practice. An additional change in terminology to note is how to refer to EAS. It has been suggested that the intervention should be defined by the profession held by the provider, not the fact that an animal is included (e.g., occupational therapy incorporating equines) (Stewart *et al.*, 2016). A therapy-first approach can help minimize confusion and clearly indicate that the professional is working within their field (Wood *et al.*, 2021).

Building on this terminology foundation, the next phase of development should focus on establishing evidence-based practice guidelines for each type of EAS. These guidelines would ideally include required qualifications and training for human service providers (Fry, 2021; Stewart *et al.*, 2016); appropriate selection, training, and care of equine partners; and safety protocols and risk management procedures. Several chapters in this book offer promising starting points for such guidelines. For instance, Peters, Bentele, and Wood's transactional analysis of occupational therapy in an equine environment highlights the importance of creating intervention manuals that standardize core components while allowing for individualization.

The future of EAS research will also depend on successful interdisciplinary collaborations that bring together experts from diverse fields including psychology, neuroscience, veterinary medicine, occupational therapy, education, and public health. Each of these disciplines offer unique methodological approaches, theoretical frameworks, and measurement techniques that can enrich our understanding of the processes associated with EAS. It is encouraging to note that many universities and research institutions are beginning to establish human-animal interaction research centers that can facilitate such interdisciplinary work, enable multi-site studies, share data repositories, and allow for collaborative grant proposals.

Methodological diversity

The studies presented in this book demonstrate considerable methodological diversity, ranging from randomized controlled trials (e.g., Pendry, Smith and Roeter; Pendry and Roeter), qualitative analysis (e.g., chapter by Carlsson *et al*), phenomenological analysis (e.g., Hershberger *et al.*) to case studies (e.g., Adams *et al.*). This array of methodological techniques is welcome as we continue striving to understand both outcomes and processes. To continue advancing the field, it is suggested that several methodological challenges should be addressed including sample size and power, control conditions, longitudinal follow-up, and mediators and mechanisms that account for observed outcomes.

In addition, advancing EAS will require continued methodological innovation to address the complexities of interventions involving horses, humans, and natural environments. One promising approach involves adaptive trial designs that allow for intervention modifications based on interim findings while still maintaining scientific rigor (Pallmann *et al.*, 2018). Such designs might be particularly suited to EAS research where personalization based on participant and equine characteristics is essential. Similarly, factorial designs could help us identify the components, or combinations of components, within EAS that contribute to observed outcomes. Given the multi-component nature of EAS programs, understanding which elements are essential versus optional could help optimize both effectiveness and efficiency. Mixed-methods approaches that combine quantitative measures with qualitative explorations of participant experiences, as exemplified by Fournier *et al.*'s chapter can also be a valuable tool for capturing the multidimensional nature of EAS outcomes. Additionally, systematic reviews and meta-analyses can help identify gaps in the research and areas needing further study.

The development and validation of standardized outcome measures that are sensitive to the unique effects of EAS represents another important direction for future research. While existing psychometric instruments can capture many relevant outcomes (e.g., social competence, anxiety, depression), they may not fully capture the distinctive experiences and changes that occur through horse-human interactions (Hanson *et al.*, 2023). The need for specialized measures is evident in several chapters of this book. For instance, Fournier *et al.*'s study of metaphor in equine-assisted psychotherapy identified unique metaphorical processes that cannot be captured by standard psychological outcome measures. Similarly, Carlsson, Ranta, and Traeen's qualitative findings regarding emotional mirroring in equine-assisted social work point to relational processes that might benefit from specialized assessment approaches.

The growing body of EAS research provides the foundation needed to integrate it into the mainstream healthcare system and be eligible for insurance coverage. Several chapters in this book point to the potential complementarity between EAS and traditional healthcare approaches. For example, Berg, Bach-Gorman, and Hammer's study of equine-assisted psychotherapy for anxiety in college students found higher attendance rates compared to office-based therapy groups, suggesting that EAS might be a useful tool for clients who are less receptive to conventional approaches. In addition, the results from studies involving participants with various conditions, such as autism spectrum disorder (e.g., Erdman, Miller, & Jacobson; Peters, Bentele, & Wood), cerebral palsy (Kaiser *et al.*), and mental health challenges (Fournier *et al.*; Pasiuk *et al.*), highlight the potential for EAS to address diverse healthcare needs.

Educational applications

The educational applications of EAS, particularly in the domain of social-emotional learning, represent another promising area for expansion. Pendry and Roeter's study demonstrating the positive effects of equine-facilitated learning on child social competence highlights the potential for school-EAS partnerships to support developmental outcomes. Zaker *et al.*'s study on EAS and social-emotional learning for at-risk youth provides a model for how such educational applications might be structured and evaluated. Their findings of improved emotional regulation and decreased problem behaviors suggest that EAS could be a valuable addition to school-based intervention programs, particularly for students at risk of academic disengagement or behavioral difficulties. Further incorporation of EAS in educational settings could benefit from aligning EAS programming with educational standards and curricula, developing school-based EAS interventions for students with diverse needs, establishing research partnerships between schools and EAS providers, and creating professional development opportunities for educators to incorporate EAS.

Community-Based Prevention and Intervention

Several chapters in this book illustrate the potential for EAS to address community-level challenges such as violence prevention (Gibbons *et al.*'s study in Guatemala) and substance misuse (Adams *et al.*'s study with First Nations youth). Adams *et al.*'s work with First Nations youth in treatment for volatile substance misuse provides a compelling example of how EAS can be culturally aligned and integrated with traditional healing approaches. These examples suggest that EAS may have particular value in contexts where conventional approaches are often not effective. With care to ensure that these programs are culturally appropriate and thoughtfully integrated into existing community-based organizations, they offer much promise.

Barriers and Challenges

Despite its promising applications, EAS faces significant accessibility barriers including geographic limitations, financial constraints of participants, physical accessibility for some populations, and cultural barriers in communities with limited exposure to horses. Future development of the field can benefit from exploring options to address these barriers including mobile programs to bring equine experiences to communities, sliding-scale or free fee structures, and cultural humility approaches that honor diverse perspectives on human-animal relationships. Chapters by Kaiser *et al.* and Lassell *et al.* demonstrate innovative approaches to addressing accessibility challenges through adaptive equipment, tailored programming, and community engagement. Kaiser *et al.*'s examination of adaptive devices for riders with physical disabilities, for example, highlights the ongoing effort to make EAS physically accessible to diverse participants, while Lassell *et al.*'s exploration of implementation factors for residents of long-term care facilities identifies both barriers and facilitators to program accessibility.

A Vision for the Future

A bright future of EAS is one characterized by ethical excellence, scientific rigor, professional integration, and inclusive accessibility. The diverse studies presented in this book represent important steps toward this vision. By continuing to refine our understanding of the complex and powerful dynamics of horse-human relationships, we can cultivate a field that honors both scientific advancement and compassionate care.

References

Baxley, B.H., Rodriguez-Mori, H. and Anderson, N.C. (2024) Welfare and stress of horses used for Equine-assisted services: A systematic review. *Applied Animal Behaviour Science*, *280*, 106424. DOI: 10.1016/j.applanim.2024.106424.

Briefer Freymond, S., Briefer, E.F., Zollinger, A., Gindrat-von Allmen, Y., Wyss, C. and Bachmann, I. (2014) Behaviour of horses in a judgment bias test associated with positive or negative reinforcement. *Applied Animal Behaviour Science*, *158*, 34–45. Available at: DOI: 10.1016/j.applanim.2014.06.006.

Brubaker, L., Schroeder, K., Sherwood, D., Stroud, D. and Udell, M.A.R. (2021) Horse Behavior towards Familiar and Unfamiliar Humans: Implications for Equine-Assisted Services. *Animals*, *11*(8), Article 8. DOI: 10.3390/ani11082369.

Feh, C. and de Mazières, J. (1993) Grooming at a preferred site reduces heart rate in horses. *Animal Behaviour*, *46*(6), 1191–1194. DOI: 10.1006/anbe.1993.1309.

Ferlazzo, A., Fazio, E., Cravana, C. and Medica, P. (2023) Equine-assisted services: An overview of current scientific contributions on efficacy and outcomes on humans and horses. *Journal of Veterinary Behavior*, *59*, 15–24. DOI: 10.1016/j.jveb.2022.11.010.

French, M. and Fournier, A. (2023) Striking a Balance Between Welfare and Therapeutic Progress in Equine-assisted Services. In *Animal-assisted Interventions* (pp. 42–44). DOI: 10.1079/9781800622616.0009.

Fry, N.E. (2021) Conceptualization of Psychotherapy Incorporating Equine Interactions in the United States. *Human-Animal Interaction Bulletin*, *2021*. DOI: 10.1079/hai.2021.0036.

Hanson, M.R., Alm, K., Fields, B., Gabriels, R., Schmid, A.A., Stallones, L. and Peters, B.C. (2023) The potential usefulness of standardized assessments to measure participant outcomes of adaptive/therapeutic horseback riding: A survey study. *Frontiers in Veterinary Science*, *10*. DOI: 10.3389/fvets.2023.1303991.

Lansade, L., Nowak, R., Lainé, A.-L., Leterrier, C., Bonneau, C., Parias, C. and Bertin, A. (2018) Facial expression and oxytocin as possible markers of positive emotions in horses. *Scientific Reports*, *8*(1), 14680. DOI: 10.1038/s41598-018-32993-z.

Lundberg, P., Hartmann, E. and Roth, L.S.V. (2020) Does training style affect the human-horse relationship? Asking the horse in a separation–reunion experiment with the owner and a stranger. *Applied Animal Behaviour Science*, *233*, 105144. DOI: 10.1016/j.applanim.2020.105144.

Nakamura, K., Takimoto-Inose, A. and Hasegawa, T. (2018) Cross-modal perception of human emotion in domestic horses (*Equus caballus*). *Scientific Reports*, *8*(1), 8660. DOI: 10.1038/s41598-018-26892-6.

Pallmann, P., Bedding, A.W., Choodari-Oskooei, B., Dimairo, M., Flight, L., Hampson, L.V., Holmes, J., Mander, A.P., Odondi, L., Sydes, M.R., Villar, S.S., Wason, J.M.S., Weir, C.J., Wheeler, G.M., Yap, C. and Jaki, T. (2018) Adaptive designs in clinical trials: Why use them, and how to run and report them. *BMC Medicine*, *16*(1), 29. DOI: 10.1186/s12916-018-1017-7.

Provan, M., Ahmed, Z., Stevens, A.R. and Sardeli, A.V. (2024) Are equine-assisted services beneficial for military veterans with post-traumatic stress disorder? A systematic review and meta-analysis. *BMC Psychiatry*, *24*(1), 544. DOI: 10.1186/s12888-024-05984-w.

Qin, Z., Guo, Zizhao and Wang, J. (2024) Multiple Equine Therapies for the Treatment of Gross Motor Function in Children with Cerebral Palsy: A Systematic Review. *Journal of Community Health Nursing*, *41*(2), 82–95. DOI: 10.1080/07370016.2024.2304825.

Rankins, E.M., McKeever, K.H. and Malinowski, K. (2023) Equids in Equine Assisted Services: A Scoping Review. *Journal of Equine Veterinary Science*, *127*, 104825. DOI: 10.1016/j.jevs.2023.104825.

Rankins, E.M., Wickens, C.L., McKeever, K.H. and Malinowski, K. (2021) A Survey of Horse Selection, Longevity, and Retirement in Equine-Assisted Services in the United States. *Animals*, *11*(8), Article 8. DOI: 10.3390/ani11082333.

Scopa, C., Contalbrigo, L., Greco, A., Lanatà, A., Scilingo, E.P. and Baragli, P. (2019) Emotional Transfer in Human–Horse Interaction: New Perspectives on Equine Assisted Interventions. *Animals*, *9*(12), Article 12. DOI: 10.3390/ani9121030.

Stewart, L., Chang, C., Parker, L. and Grubbs, N. (2016) *Animal-Assisted Therapy in Counseling Competencies*. Animal-Assisted Therapy in Mental Health Interest Network. Available at: https://www.counseling.org/docs/default-source/competencies/animal-assisted-therapy-competencies-june-2016.pdf

Watson, E., Davis, A., Splan, R. and Porr, C.A.S. (2020) Characterization of Horse Use in Therapeutic Horseback Riding Programs in the United States: A Pilot Survey. *Journal of Equine Veterinary Science*, *92*, 103157. DOI: 10.1016/j.jevs.2020.103157.

Wood, W., Alm, K., Benjamin, J., Thomas, L., Anderson, D., Pohl, L. and Kane, M. (2021) Optimal Terminology for Services in the United States That Incorporate Horses to Benefit People: A Consensus Document. *The Journal of Alternative and Complementary Medicine*, *27*(1), 88–95. DOI: 10.1089/acm.2020.0415.

Index

Printed and bound by CPI Group (UK) Ltd, Croydon, CR0 4YY
13/09/2026
14961085-0003